The Cognitive Basis of Institutions

Perspectives in Behavioral Economics
and the Economics of Behavior

The Cognitive Basis of Institutions

A Synthesis of Behavioral and
Institutional Economics

Shinji Teraji
Department of Economics, Yamaguchi University,
Yamaguchi, Japan

Series Editor
Morris Altman

ACADEMIC PRESS

An imprint of Elsevier

Academic Press is an imprint of Elsevier
125 London Wall, London EC2Y 5AS, United Kingdom
525 B Street, Suite 1800, San Diego, CA 92101-4495, United States
50 Hampshire Street, 5th Floor, Cambridge, MA 02139, United States
The Boulevard, Langford Lane, Kidlington, Oxford OX5 1GB, United Kingdom

Notices
Knowledge and best practice in this field are constantly changing. As new research and experience broaden
our understanding, changes in research methods, professional practices, or medical treatment may become
necessary.

Practitioners and researchers must always rely on their own experience and knowledge in evaluating and
using any information, methods, compounds, or experiments described herein. In using such information or
methods they should be mindful of their own safety and the safety of others, including parties for whom they
have a professional responsibility.

To the fullest extent of the law, neither the Publisher nor the authors, contributors, or editors, assume any
liability for any injury and/or damage to persons or property as a matter of products liability, negligence or
otherwise, or from any use or operation of any methods, products, instructions, or ideas contained in the
material herein.

British Library Cataloguing-in-Publication Data
A catalogue record for this book is available from the British Library

Library of Congress Cataloging-in-Publication Data
A catalog record for this book is available from the Library of Congress

ISBN: 978-0-12-812023-1

For Information on all Academic Press publications
visit our website at https://www.elsevier.com/books-and-journals

Working together
to grow libraries in
developing countries

www.elsevier.com • www.bookaid.org

Publisher: Candice Janco
Acquisition Editor: Graham Nisbet
Editorial Project Manager: Barbara Makinster
Production Project Manager: Swapna Srinivasan
Cover Designer: Matthew Limbert

Typeset by MPS Limited, Chennai, India

Contents

Preface

Societies display large differences in economic performance. The institutional characteristics of societies underlie these differences. From a cognitive viewpoint, this book asks: what are institutions and how do they evolve? The book highlights some basic mechanisms of interaction between mental phenomena and institutions. Understanding institutions is ultimately about understanding how we think.

Institutions are a matter of the mind. The mind is not a completely external object of analysis. Behavioral decision theory can enhance the analysis of institutional features. Since institutions are composed of individuals who are subject to heuristics and biases, institutions and their decision-making rules can either magnify or mitigate the impact of heuristics and biases. Furthermore, in an institutional position, its decision-making rules and procedures may themselves lead to the emergence and reinforcement of heuristics and biases. Institutional structures are complex: individual elements are interdependent and self-organizing. Past experience is encapsulated in institutional rules. By following rules, individuals reduce their uncertainty about the possible outcomes that their social interactions can bring about. Institutions then become uncertainty-reducing devices.

Cognitive psychology has shown that people utilize heuristics both in judgment and choice, under conditions of certainty, as well as conditions involving uncertainty. Experimental studies of human judgment under uncertainty show that decision-making is guided by particular heuristic rules that may give rise to errors. Human beings rely on a set of heuristics for their decision-making, and the use of these heuristics sometimes leads to systematic deviations from the normatively correct decision. Decision frames are specific mental models that result partly from the formulation of the problem at hand, and partly from the particular norms, habits, and other individual characteristics of the decision-maker.

Human behavior is largely governed by cognitive rules. Friedrich Hayek's theory of mind sheds light on the process of choice; it describes the human mind as an adaptive classification system by which individual behavior is shaped. The essence of Hayek's cognitive theory is the proposition that all of an organism's experience is stored in network-like systems of connections between the neurons of its cerebral cortex. Individual reasoning differs from person to person, as it is based on subjective perceptions. The subjectivity of individual knowledge finds its foundation in Hayek's explanation of the construction of the mind. The perception of the world around us is conjectural, in the sense that it is informed by a set of classificatory dispositions which is itself the product of a kind of accumulated experience. The human mind is the product of the environment in which it has grown up. The mind emerges from the complexity of the connections between the physical world and the sensory world.

Herbert Simon's statement on the notion of bounded rationality emphasizes the limits in the information and computational capacities of an economic agent. Individuals are presumed to attempt to act rationally, but to be bounded in their ability to achieve rationality. An agent uses only the information that is explicitly displayed in making a decision, without considering other things. As a consequence, individuals exhibit a very large measure of docility. Society is sustained by processes favorable to individuals endowed with some docility in following rules.

Spontaneous orders are formed when individuals follow shared rules of behavior. However, the individuals themselves may not be able to articulate the rules they follow. For Hayek, one of the main characteristics of human behavior consists of following rules of conduct. The rules one individual is expected to follow influence the choices made by other individuals. If people have widely divergent expectations, some of their actions will invariably fail and need to be revised. Culture limits the range of actions that people are likely to take in a particular situation, making their conduct more predictable, and thereby facilitating the formation of reliable expectations. Shared mental models can give rise to behavioral regularities to the extent that they can be observed in the population. As a consequence, following rules of conduct mutually reinforces sets of expectations to maintain a degree of social order.

Individuals are institutionally embedded. Institutions are means by which agents are able to gather sufficient information in order to coordinate. On the other hand, institutions are roughly regularities of behavior. Individuals, interacting with others, are assumed to continue to change their responses to the actions of others until no improvement can be obtained in their expected outcomes from independent actions. Repeated patterns of behavior create expectations about future behavior, and ensure a degree of predictability in social interaction. Institutions affect individual choices, individuals choose their actions, and institutions emerge through individual actions. We must reconcile institutions as structures, and individual actions as rule-following behavior. Institutions influence individuals' behavior and, in turn, their behavior shapes these institutions. While institutions represent phenomena that coordinate human activities at the macro-level, they are reproduced by individuals in activities on the micro-level.

At the level of the individual, the cognitive processes enable the individual to adjust his or her actions to external reality. Individuals adjust their actions to achieve a better fit with reality. In this sense, the mind is endogenous to the individual's external environment. Shifts in mental models change individuals' plans and actions, which in turn leads to institutional evolution. Thus, a key to understanding institutional evolution is an understanding of how individuals modify their mental models. Institutional evolution is an endogenous phenomenon with a cognitive dimension.

Chapter 1, Introduction: What Are Institutions?, explains what institutions are. Institutions are defined as a system of rules, beliefs, norms, and organizations that conjointly generate a regularity of behavior. Institutions are considered as durable rules which govern human interactions. They form the incentive structure, and define the choice set for individuals. Following rules helps people make decisions with some degree of certainty about which behavior is acceptable, and which is not. It reduces the amount of information individuals must collect, and enhances their ability to make plans and to coordinate with each other. Institutions coordinate the individual's actions at a lower cost, because they reduce the volatility in the plans of others. Norms describe the uniformity of behavior that characterizes a group. Norms specify a limited range of behavior that is acceptable in a situation, and facilitate confidence in the course of action.

Chapter 2, Institutions and the Economics of Behavior I, focuses on the micro-foundations of the link between institutions and economic outcomes. The new institutional economics seeks to account for the emergence and persistence of institutions on the basis of their efficiency. One of key concepts of the new institutional economics is transaction costs. Institutions emerge and persist when the benefits they confer are greater than the transaction costs involved in creating and sustaining them. Without the concept of transaction costs, it might be impossible to understand how an economic system works. Poorly constructed institutions are identified as a source of higher transaction costs. The assignment of property rights matters, because of positive transaction costs. Transaction costs are more generally considered to be the resources spent on delineating, protecting, and capturing property rights.

Chapter 3, Institutions and the Economics of Behavior II, studies how individuals behave, and how thinking and emotions affect individual decision-making. By adding insights from psychology, behavioral economics tries to modify the conventional economic approach, and to analyze how "flesh-and-blood" people act in social contexts. Anomalies are often labeled as decision-making failures or mistakes. Anomalies arise from the way humans process information to form beliefs. Heuristics describe how people make judgments and decisions based on approximate rules of thumb. Heuristics require less effort compared to a rational, calculated choice. Human decision-making is prone to phenomena like anchoring, framing, status quo bias, and inertia. Systematic biases and temporally inconsistent motivations lead to poor choices. By changing the choice architecture as the context in which people make decisions, outcomes can be improved in a way that makes choosers better off.

Chapter 4, Why Bounded Rationality?, considers four principles about decision-making by following Herbert Simon's arguments:

1. **The principle of bounded rationality.** Bounded rationality is an alternative conception of rationality that models the cognitive processes of decision-makers more realistically. The capacity of the human mind for formulating and solving complex problems rationally is bounded.
2. **The principle of satisficing.** Optimizing is replaced by satisficing—the requirement that satisfactory levels of the criterion variables be attained. An individual establishes his or her goal as an aspiration level.
3. **The principle of search.** Alternatives of action and the consequences of action are discovered sequentially through search processes. An individual sequentially searches for alternatives, and selects one that meets the aspiration level.
4. **The principle of adaptive behavior.** An individual continually adjusts his or her behavior to changing environments. Human rationality cannot be understood merely by considering the mental mechanisms that underlie human behavior. Instead, we should elucidate the relationship between the mental mechanisms and the environments in which they work.

In Chapter 5, Emergence of Prosocial Behavior, self-interest is not the only human motivation. Throughout their lifetime, people depend on frequent and varied cooperation with others. In the short-run, it would be advantageous to cheat. In the long-run, people are likely to cooperate in a mutually beneficial manner. This would make them better able to resist the temptation to cheat in the first place, and would enable them to generate a reputation for being cooperative. People are, in real life, affected by psychological and emotional factors. People exhibit prosocial behavior when they do not always make choices that maximize their own pecuniary payoffs. Formal models of social preferences assume that people are self-interested, but are also concerned about the payoffs of others. That is, a player's utility function not only depends on his or her material payoff, but may also be a function of the allocation of resources within his or her reference group. People have social preferences if, and only if, they exhibit prosocial behavior, and have relatively stable social preferences.

Chapter 6, Cognition and Order, covers Friedrich Hayek's theory of mind. Hayek provides a theory of the process by which the mind perceives the world around it. Hayek's theory of mind shows how a structure discriminates between different physical stimuli, and generates the sensory order that we actually experience. What we know at any moment about the world is determined by the order of the apparatus of classification which has been built up by previous sensory linkages. Hayek's theory of mind explains how different pieces of cognitive information cause different perceptions and, therefore, different actions. The subjectivity of individual knowledge finds its foundation in the construction of the mind. The mind is an adaptive system interacting with and adapting to its external environment by performing a multi-level classification on the stimuli it receives from the environment.

Chapter 7, Society and Knowledge, deals with social order. Hayek's thought largely rests on the concept of spontaneous order. Spontaneous orders in human affairs are patterns that arise as the unintended consequences of individual actions. The dissemination of knowledge is crucial in society. People live in a world of expectations about interactions with others' actions. Rules are behavioral patterns that individuals expect each other to follow. Relying on rules is a device we have learned to use, because our reason is insufficient to master the detail of complex reality. If rules are recognized as recurrent patterns of behavior, individuals act according to rules of conduct. The diffusion of shared behavioral patterns is necessary to obtain social order. Shared rules facilitate decision-making in complex situations by limiting the range of circumstances which individuals have to pay attention to.

In Chapter 8, Understanding Institutional Evolution, theories of institutions can be classified into two broad approaches: institutions-as-rules, and institutions-as-equilibria. According to the first approach, institutions are conceived as rules that guide the actions of individuals engaged in social interactions (institutions-as-rules). On the other hand, the second approach views institutions as behavioral patterns (institutions-as-equilibria). In order to have a complete picture of institutions, we need to take both approaches into consideration. The main purpose is to develop a general framework, within which it is possible to analyze the coevolution of individuals' mental models and institutions. In Hayek's theory of cultural evolution, societies have developed through a process in which individuals choose the rules that form the social order. New rules undergo some kind of decentralized selection process, as a consequence of which some spread through the population.

Some chapters in this book were presented as articles at international conferences sponsored by the Society for the Advancement of Behavioral Economics, the International Association for Research in Economic Psychology, and the Herbert Simon Society. I have benefited from the ideas and comments of colleagues and experts in the fields. I am especially grateful to Morris Altman, who has helped me understand behavioral and institutional economics.

I would like to acknowledge Graham Nisbet, Senior Acquisitions Editor at Elsevier, who is critical to the success of this project. Many thanks to Barbara Makinster, Senior Editorial Project Manager at Elsevier, for her patience and support.

Shinji Teraji
Yamaguchi, Japan
May, 2017

Chapter 1

Introduction: What Are Institutions?

1.1 INSTITUTIONS AND RULES

The idea that institutions matter is commonplace. The underlying institutional framework is possibly the key to understanding why some societies have remained mired in poverty, and others have attained high welfare levels. It is difficult to understand economic history without paying attention to institutions. For example, Acemoglu and Robinson (2012) explain divergent economic histories with qualitative measures of institutional quality. A history of poor political institutions frequently leads to weak economic performance.[1] North (1990, p. 54) states that "the inability of societies to develop effective, low-cost enforcement of contracts is the most important source of both historical stagnation and contemporary underdevelopment in the Third World." According to Roland (2004), the reason why transplanted institutional blueprints fail to achieve their objectives in many countries of the Global South is that they clash with the host country's institutions such as entrenched power structures and social norms. Power structures change slowly, because ruling elites maintaining their power in societies with inefficient institutions prefer not to give up that power. Social norms, such as attitudes towards the death penalty or acceptance of corruption, are rooted in values that tend to change slowly. The sticky nature of beliefs helps explain why imported laws and constitutions have been so unsuccessful.

What are institutions actually? Various explanations of institutions abound. However, there is no consensus on how to conceptualize institutions themselves. Opinions range from the view that institutions determine individual behavior, to the idea that institutions are the unplanned outcome of the interplay of individual behavior.

Institutions can influence and frame the decisions individuals make when interacting with others. Individuals behave differently than they would in the absence of institutions. Institutions exist when something exerts an external

---

1. According to North and Weingast (1989), the institutions maintained in England after its Glorious Revolution in 1688 gave its Government the "credible commitment" to property rights that later facilitated its access to crucial funds.

The Cognitive Basis of Institutions. DOI: https://doi.org/10.1016/B978-0-12-812023-1.00001-6

influence over the behavior of social actors. Rules indicate what to do or what not to do in a given environment. Rules can be considered as prescribed patterns that guide actual behavior and thought. Greif (2006) defines institutions as a system of social factors that conjointly generate a regularity of behavior:

> *An institution is a system of rules, beliefs, norms and organizations that together generate a regularity of (social) behavior.*

(Greif, 2006, p. 30)

A regularity of behavior means behavior that is followed and is expected to be followed in a given social situation by most individuals. The expected aggregate behavior in society constitutes a structure that influences each individual's behavior. Greif's (2006) definition of institutions suggests a causal link between the regularity of behavior and the sets of rules, beliefs, norms, and organizations that give rise to it (it should be noted that organizations are included in his definition of institutions).[2] People are coordinated through the regularity of behavior in a recurrent situation. Such regularities can be described as noncooperative equilibria of strategic games: out-of-equilibrium actions are unstable and are unlikely to be repeated in the course of human interactions. An equilibrium in game theory is a profile of strategies (or actions), one for each player participating in a strategic interaction. The defining characteristic of equilibrium is that each strategy must be a best response to the actions of the other players or, in other words, no player has an incentive to change his or her strategy unilaterally. Institutionalized rules aggregate private information and provide shared information. These rules provide the necessary clues for each individual to form his or her beliefs. According to Greif and Mokyr (2017), institutions are based on shared cognitive rules, that is, social constructs that convey information which distills and summarizes society's beliefs and experience. It is crucial that individuals have a common understanding of the same situation or action, and attribute to it the same meaning. Thus, these rules provide the cognitive and informational resources necessary for the common understanding.

The most popular and widely cited characterization of institutions can be found in Douglass North's *Institutions, Institutional Change and Economic Performance* (1990):

> *Institutions are the rules of the game of society or, more formally, the humanly devised constraints that shape human interactions.*

(North, 1990, p. 3)

---

2. Dealing with medieval European trade, Greif (2006) establishes a causal connection between institutions and long-distance seaborne trade. Medieval Europe's initial institutions facilitated the expansion of long-distance trade, and the resulting expansion of trade created growth-enhancing institutions.

According to this view, institutions are durable rules which govern human interactions. Institutions are considered as society's rules of the game, that is, self-imposed restrictions governing social relationships. Institutions form the incentive structure and define the choice set for purposeful individuals.[3] Institutions reduce uncertainty by imposing constraints on human actions. A constraint implies that there are situations in which an individual who is subject to an institution prefers not to abide by the rule. Institutions include both formal rules (such as laws and constitutions) and informal ones (such as conventions and norms). Formal rules are explicit or written down, and they are enforced by the state or actors with specialized roles. Informal rules are, on the other hand, implicit; they help agents to coordinate their behavior and expectations, and are enforced endogenously by the members of the relevant group. The enforcement of the rules is then considered as a distinct issue from the formation and content of the rules themselves. The formal and informal rules constitute the institutional structure within which human interactions occur.[4] Here, a "rule" is fundamentally the others' expected behavior which motivates our behavior, rather than the rule itself (Greif and Kingston, 2011). Agents structure their activities and operate transactions within rules defined at the institutional level.

North (1990) focuses on individual values and incentives interacting with the social context. The same logic of individual choice does not prevail in every society. Individuals are constrained by the existing institutional structure. As North (1990) suggests:

> *Economic (and political) models are specific to particular constellations of institutional constraints that vary radically both through time and cross sectionally in different economies. The models are institution specific and in many cases high sensitive to altered institutional constraints.*

> (North, 1990, p. 110)

Consequently, individual action is socially and historically determined. Ideas and ideologies affect individuals' behavioral patterns. Incentives, perceptions, and ways of thinking are "socially transmitted ... and are a part of the heritage we call culture" (North, 1990, p. 37). Then, culture, which is considered as the set of past experiences, ideas, and ideologies, defines the ways individuals process and utilize information. Culture shapes the subjective mental constructs that individuals use to interpret the real world around them and make choices.

---

3. For North, institutions are differentiated from organizations; organizations may act as "players" of the political game, attempting to alter broader institutional rules for the benefit of their members.
4. Building on his analysis of informal contract enforcement among Maghribi traders, Greif (1994) emphasizes the role of formal institutions which support the larger-scale economic exchange. That is, "a formal legal code is likely to be required to facilitate exchange by coordinating expectations and enhancing the deterrence effect of formal organizations" (Greif, 1994, p. 936).

For Friedrich Hayek (1973), one of the main characteristics of human behavior consists of following rules of conduct. Humans have a tendency to fit themselves into prescribed behavior in specific circumstances: *Homo sapiens* are "a rule-following animal" (Hayek, 1973, p. 11). The word "rules" is sometimes used broadly to refer to any kind of directive for decision-making. Like prices, rules coordinate and motivate interdependent behavior. By emphasizing the importance of rules as well as prices, Hayek's thought develops an interdisciplinary approach, as opposed to a narrow, economic one, to the explanation of coordination.[5] Because of our ignorance, we ought necessarily to rely largely on traditional rules, instead of attempting to design the system of rules. Following rules helps people make decisions with some degree of certainty about which behavior is acceptable and which is not. The rules, guiding the individual actions as a whole, are abstract and unconscious. Individuals are, therefore, unable to explain the actions of others. The knowledge need not be articulable. We can know more than we can tell.[6] However, they can understand the actions of others, because others act according to a similar mode of categorization of the real world. Hayek asks whether inarticulate rules always guide our mental activity. Hayek (1973) relates tacit knowledge and practical learning of rules as follows:

> So long as individuals act in accordance with rules it is not necessary that they be consciously aware of the rules. It is enough they know how to act in accordance with the rules without knowing that the rules are such and such in articulated terms.
>
> (Hayek, 1973, p. 99)

Rules of conduct are largely tacit. People are not capable of explaining these rules. However, respecting these rules, to a great extent, conditions the coordination of individual actions. Hayek (1967) argues that there tacit rules must exist for regularity in the conduct of individuals to exist:

> [T]he term rule is used for a statement by which regularity of the conduct of individuals can be described, irrespective of whether such a rule is known to the individuals in any other sense than that they normally act in accordance with it.
>
> (Hayek, 1967, p. 67)

---

5. According to Caldwell (2014, p. 33), "[s]cientism was the source of economists' errors by causing them to think that a theory must make reference to quantitative data to be truly scientific. A second problem was their assumption that simple relationships between aggregate statistical concepts exist … But with markets economists are dealing with essentially complex phenomena."

6. Michael Polanyi (1969) uses the concept of "tacit knowledge" to refer to all those kinds of scientific knowledge that cannot be expressed in explicit form (spoken words, formulae, maps, graphs, mathematical theory, and so on). The concept of tacit knowledge has been widely used in social sciences in order to incorporate it within a more comprehensive theory of practices and their role in social reality.

Rules of conduct govern our perceptions and, in particular, our perceptions of others' actions. In his *The Counter-Revolution of Science*, Hayek states that "[t]here is a great deal of knowledge which we never consciously know implicit in the knowledge of which we are aware, knowledge which yet constantly serves us in our actions, though we can hardly be said to 'possess' it" (Hayek, [1952] 1979, p. 217, original emphasis). We cannot articulate much of what we know. It is important to realize that tacit knowledge serves in the internal structure of action. "Knowing how" corresponds to tacit knowledge, while "knowing that" relates to conscious knowledge. Knowing how consists in using habits and following rules whose nature and definitions do not need to be explained in the individual's mind. In fact, Hayek contends that "many of the greatest things man has achieved are the result not of consciously directed thought, and still less the product of a deliberately coordinated effort of many individuals, but of a process in which the individual plays a part which he can never fully understand" (Hayek, [1952] 1979, p. 150).

Rules of conduct determine what range of actions is permissible in a specific situation. Rules are a device for coping with our ignorance of the effects of particular actions. Hayek (1967) points out that rules often limit the range of possibilities within which the choice is made consciously:

> By eliminating certain kinds of action altogether and providing certain routine ways of achieving the object, they merely restrict the alternatives on which a conscious choice is required. ... [T]he rules which guide an individual's action are better seen as determining what he will not do rather than what he will do.
>
> (Hayek, 1967, p. 56)

Interactions of people in society make adaptation possible. At an individual level, actions are based on subjective perceptions of what exists. However, a correspondence between individual actions and an overall order is inherently problematic. People live in a world of expectations about interactions with others' actions. The rules of conduct governing our actions are adaptations to our ignorance of the external environment in which we have to act. Institutions facilitate social interaction since they restrict individual agents concerning their dispositions to behave. In order to act in society, individuals must accept certain rules without consciously thinking about them. Such rules are themselves part of a spontaneous order that is not the product of conscious reason but, nonetheless, facilitates reasoned actions. Institutions lead to regularities in human behavior and serve to coordinate the interaction between individuals. It is meaningful to discuss the social order only when all agents share the same perception of existing reality which includes others' actions.

1.2   BELIEF SYSTEMS

Since Carl Menger's analysis of the emergence of money, institutions have been a central theme of investigation for Austrian economic theory.[7] The Austrian analysis of institutions resides in the subjective point of view.[8] The subjectivity of individual knowledge finds its foundation in Hayek's (1952) explanation of the construction of the mind. Individuals are fundamentally heterogeneous. The analysis of coordination of individual actions is the main problem for economists supporting the Austrian tradition. According to Ludwig Lachmann (1976), the market process is characterized by inconsistency of individual plans. Inconsistency of individual plans is the direct consequence of the introduction of subjectivism to expectations. The particular problem Lachmann, heavily influenced by George Shackle (1972), emphasizes is that of divergent expectations. Lachmann (1970) is interested in the influence of institutions on the formation and revision of individual plans. The main task of social theory is to explain observable phenomena by reducing them to the individual plans that typically give rise to them. Lachmann (1970) explains institutions as follows: "[t]hey [institutions] are ... co-ordinating the actions of millions whom they relieve of the need to acquire and digest detailed knowledge about others and form detailed expectations about their future action" (Lachmann, 1970, p. 50).

There always exists a large variety of beliefs, and which of them spreads is a key problem. For Lachmann, a social structure is just "recurrent patterns of events" (Lachmann, 1970, p. 23). Lachmann (1970) regards people as social beings whose beliefs, expectations, and plans are profoundly shaped by the institutional environment in which they live. Individual actions are guided by intersubjectively shared social structures. This implies that people are able to understand and anticipate how others will act in particular situations, thereby facilitating the formation of reliable expectations and mutually compatible plans. If people have no grounds for believing that a particular set of

---

7. Menger's story about the emergence of money is widely known. Consider three traders, Adam, Bill, and Charles, in a barter economy with three goods, X, Y, and Z. As their current endowments and preferred outcomes, Adam has X and wants Y, Bill has Y and wants Z, and Charles has Z and wants X. A direct barter among any of the traders is impossible (none of the traders has a double coincidence of wants). Suppose that Charles attempts to engage in indirect barter. He trades his own Z for Bill's Y. Bill gains more of what he prefers. Now with Y, Charles trades that directly with Adam. Adam gains more of what he prefers and so, too, does Charles. Over time, as the practice of indirect exchange continues, a most-marketable good will eventually serve as a general medium of exchange. Thus, some type of money emerges as a social institution.

8. Rizzo (2013, pp. 50−51) lists eight highly interrelated themes about Austrian economics: (1) the subjective (yet socially embedded) quality of human decision-making; (2) the individual's perception of the passage of time; (3) the radical uncertainty of expectations; (4) the decentralization of explicit and tacit knowledge in society; (5) the dynamic market processes generated by individual action; (6) the function of the price system in transmitting knowledge; (7) the supplementary role of cultural norms and other cultural products (institutions) in conveying knowledge; and (8) the spontaneous evolution of social institutions.

institutions will endure into the future, they will have no reason to orient their plans toward those institutions. In order to serve their "rule-influenced" purpose, institutions must display stability. Institutions play a crucial role in enabling people to formulate expectations that are sufficiently accurate for the successful coordination of plans. Lachmann (1970) describes as follows:

An institution provides means of orientation to a large number of actors. It enables them to co-ordinate their actions by means of orientation to a common signpost.

(Lachmann, 1970, p. 49)

People face great uncertainty when they assess the acceptability of their actions, make predictions about how others will act, and project likely results. Expectations are, in the context of ignorance about reality, formed as a result of rule-governed creativity. That is, by following rules, people could make decisions with some degree of certainty about which behavior is acceptable and which is not. Following rules reduces the amount of information individuals must collect, and enhances their ability to make plans and to coordinate with each other. Institutions coordinate individual actions at a lower cost, because they reduce the volatility in the plans of others. In order for knowledge to be disseminated among individuals, there must be some underlying society that is knowledge-based. It is considered as a social network that shares beliefs, providing the information channels that circulate ideas. People learn not only explicitly, but also implicitly, from languages, customs, traditions, and so on.

According to Ostrom (2005), whenever one addresses questions about institutional change in contrast to ongoing actions within institutional constraints, it is necessary to recognize the following two points. First, changes in the rules used to order actions at one level occur within a currently "fixed" set of rules at a deeper level. Second, changes in deeper-level rules are usually more difficult and more costly to accomplish, which increases the stability of mutual expectations among individuals interacting according to the deeper set of rules. Ostrom (2005) distinguishes three levels of rules: "operational rules" (rules that govern day-to-day interactions); "collective-choice rules" (rules for choosing operational rules); and "constitutional rules" (rules for choosing collective-choice rules). That is, when operational rules are being chosen, collective-choice and constitutional rules are treated as exogenous. Whether institutional change occurs ultimately depends on the deeper-level rules and on how individuals perceive the likely effects of a change in rules.

Williamson (2000) distinguishes four levels of institutions, according to how quickly they change (Fig. 1.1). The arrows connect a higher level with a lower level; the higher level imposes constraints on the level immediately below. The top level (L1) is the social embeddedness level. This is where informal institutions, customs, traditions, norms, religion, etc., are included. Institutions at L1 change very slowly (on the order of centuries or millennia), and this level is taken as given by most economists. Level 2 (L2), which is construed by the institutions of embeddedness, is referred to as the institutional

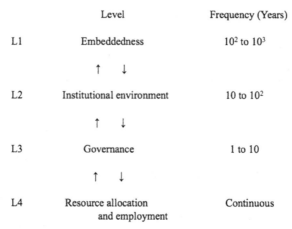

	Level	Frequency (Years)
L1	Embeddedness	10^2 to 10^3
	↑ ↓	
L2	Institutional environment	10 to 10^2
	↑ ↓	
L3	Governance	1 to 10
	↑ ↓	
L4	Resource allocation and employment	Continuous

FIGURE 1.1   Economics of institutions.

environment: the formal rules of the game. Design opportunities (the executive, legislative, judicial, and bureaucratic functions of government, as well as the distribution of powers) are posed at this level. The definition and enforcement of property rights and contract laws are important features. Thus, the informal constraints are located at L1, while the formal rules are located at L2. Level 3 (L3) is governance, at which the sets of rules govern day-to-day interactions. These rules are assumed to adjust so as to minimize transaction costs. The new institutional economics has been covered principally with Levels 2 and 3. The economics of property rights operates at L2, while transaction cost economics, at L3, operates by taking the rules of the game at L2 as shift parameters. Finally, Level 4 (L4) is resource allocation and employment. Neoclassical economics and agency theory have been concerned with this level. Adjustments to prices and quantities, specified in individual contracts, occur more or less continuously in response to changing market conditions.

Human actions are imprinted by their history in a way. Institutional change can be seen as path dependent. Path-dependent processes occur because of the "network effect," that is, when the benefit of consuming a good or adopting a technology varies directly with the number of others who consume the good or adopt the technology (Katz and Shapiro, 1985). A key finding of path dependence is a property of "lock-in" by historical events. Small chance events have durable consequences in the long run. Paul David and Brian Arthur expose increasing returns to scale in their papers that are regarded as the foundation of path dependency (David, 1985; Arthur, 1989). Neoclassical economics assumes the consumption and production sets to be convex; increasing returns to scale are excluded by convexity. Decreasing returns to scale exclude the possibility of self-reinforcing mechanisms in economic processes. In a world of increasing returns to scale, on the other hand, initial and trivial circumstances can have important and irreversible influences on the ultimate market allocation of resources. A new technology, subject to increasing returns, generates higher payoffs for users as it becomes more prevalent in the market. With increasing returns to scale, economic agents have strong incentives to continue down a specific path once initial steps are taken in that direction. Current choices of techniques become the link through which prevailing economic conditions may influence the future dimensions of technology and knowledge. The probability of further movement in the same direction increases with each move down that direction. The form of path dependence conflicts with conventional economics where efficient outcomes are attained.[9] Suboptimal or inefficient technologies can be locked-in as industry standards. Hence, path-dependent externalities prevent unregulated market-based economies from achieving optimal efficiency.

Social interactions often lead to patterns of what are called "positive feedbacks." Once an initial advantage is gained, positive feedback effects lock in a particular technology and other competitors are excluded. The term "positive feedbacks" refers more broadly to any situation in which the payoff to taking an action is increasing in the number of people taking the same action. For example, the payoff to learning a particular language depends on the number of speakers. To distinguish the large class of positive feedback cases from the subset based on increasing returns to scale in production, Bowles (2004) uses the term "generalized increasing returns," rather than increasing returns to scale. Generalized increasing returns appear to be a source of multiple equilibria. Changes in norms, habits, laws, technologies, or organizational forms can be analyzed as path-dependent processes.

Institutions do not evolve just by themselves; the direction in which they are changed depends on the beliefs of those who change them. Beliefs are at the heart of understanding the process of institutional change. Radical reforms are constrained by societies' inherited belief systems. North (2005) focuses on the evolution of belief systems that individual agents hold. Individual agents perceive "human landscape" and interpret it. There is a link between institutions and individual agents. North (2005) summarizes this relationship as follows:

> There is an intimate relationship between belief systems and the institutional framework. Belief systems embody the internal representation of the human landscape. Institutions are structure that humans impose on that landscape in order to produce the desired outcome. Belief systems therefore are the internal representation and institutions the external manifestation of that representation.
>
> (North, 2005, p. 49)

---

9. Liebowitz and Margolis (1995) claim that, where there is a feasible improvement to be gained from moving on to a better path, economic agents are willing to pay to bring the improvement about. According to their argument, if one of two options is superior in the long run, but not in the short run, market arrangements will generally ensure the adoption of the superior path. However, many of the benefits of increasing returns are external to individual agents, and the best outcome will not be selected in the long run.

Institutions structure individual agents. Especially, ideas and ideologies shape the mental models that individuals use to interpret the world around them and make choices. As Denzau and North (1994) point out, individuals with common cultural backgrounds and experiences will share reasonably convergent mental models. Thus, agents who belong to the same cultural group are exposed to the same external representation of knowledge, which produces shared mental models. Culture, as a system of beliefs, provides shared collective understandings in shaping individual strategies. Culturally shared mental models expedite the process by which people learn directly from experiences and facilitate communication between people. Each agent has little motivation to deviate from such similarities, as long as the consequences of similar behavior do not systematically diverge. The resulting belief system constrains the repertoire of possible reactions to changes in the external environment. Cognition may have more subjective aspects, while culture enables individuals to develop intersubjectively shared mental models.

Established institutions generate powerful inducements that reinforce their own stability. Generally, the idea of institutions is understood as systems of established and prevalent rules that structure social interactions. Path dependence includes features such as sustained persistency and lock-in. Path dependence in the evolution of belief systems results from a "common cultural heritage" which "provides a means of reducing the divergent mental models that people in a society possess and constitutes the means for the intergenerational transfer for unifying perceptions" (North, 2005, p. 27). Beliefs are mental constructs derived from learning through time. Institutional change is a slow, path-dependent process, with current institutions having long-term, dynamic effects on the set of institutions, economic behavior, and outcome in a given society. As North (2005, p. 77) posits, "the belief system underlying the institutional matrix will deter radical change." Here, an institutional matrix is considered as a framework of interconnected institutions that together make up the rules of the economy. Institutional influences on an economy will occur through a number of formal and informal institutional components. The structure formed by several institutional arrangements will define a set of interrelated incentives for individual agents. The presence of a particular institution may or may not be compatible with the presence of another one. The conditions for the existence of an institution must be determined by taking into account a large set of institutional arrangements. The aggregate coherence given by a set of institutional arrangements is defined by their complementary character. The influence of one institution is reinforced when the other complementary institution is present. Complementary institutions reinforce each other. According to Aoki (2001), institutional complementarities are situations of "synchronic interdependence" across distinct institutional domains. Aoki (2001, p. 225) argues "one type of institution rather than another becomes viable in one domain, when a fitting institution is present in another domain and vice-versa." The institutional choices in one domain act as exogenous parameters in another domain and constitute the "institutional environment" under which choices are made.

Consider two institutional domains, σ and φ, with two different sets of agents who do not directly interact (Pagano and Rossi, 2004; Pagano and Vatiero, 2015). Suppose that the agents in domain σ (e.g., the domain of property rights) face the choice of a rule from σ^X or σ^Y, while the agents in domain φ (e.g., the domain of technology) choose a rule from φ^X or φ^Y. Using a payoff function u, consider the following conditions:

$$u(\sigma^X, \varphi^X) - u(\sigma^Y, \varphi^X) \geq u(\sigma^X, \varphi^Y) - u(\sigma^Y, \varphi^Y) \qquad (1.1)$$

$$u(\varphi^X, \sigma^X) - u(\varphi^Y, \sigma^X) \geq u(\varphi^X, \sigma^Y) - u(\varphi^Y, \sigma^Y) \qquad (1.2)$$

These two conditions express the idea of complementarity between two different domains.[10] The condition (1.1) implies that the additional benefit for the agents in domain σ from choosing σ^X rather than σ^Y increases, as their institutional environment in φ is φ^X rather than φ^Y, while the condition (1.2) implies that the additional benefit for the agents in domain φ from choosing φ^X rather than φ^Y increases, as their institutional environment in σ is σ^X rather than σ^Y. It can be proved that there are two pure Nash equilibria, or institutional arrangements, for the system which comprises σ and φ.[11] That is, two strategy profiles (σ^X, φ^X) and (σ^Y, φ^Y) can be Nash equilibrium profiles. Then, σ^X and φ^X are institutional complements, and so are σ^Y and φ^Y.

The complementary character is fundamental for defining the pattern of evolution of an economic system. Understanding institutional change necessitates an understanding of path dependence, in order to appreciate the nature of the limits it imposes on change. Path dependence includes features such as persistence and lock-in. Thus, "at any moment of time the players are constrained by path dependence—the limits to choices arising from the combination of beliefs, institutions, and artifactual structure that have been inherited from the past" (North, 2005, p. 80).

---

10. These are, in terms of institutional choices, the super-modularity conditions among strategies analyzed by Milgrom and Roberts (1990).

11. A Nash equilibrium is, by definition, a prediction of a feasible strategy for each player, such that each player's strategy maximizes his or her expected payoff, given what the other players are predicted to do. Given any game, a prediction of the players' behavior that is not a Nash equilibrium could not be commonly believed by all the players. If everyone believed such a non-equilibrium prediction, then at least one player would rationally prefer to choose some other strategy which is different from his or her predicted one.

1.3 NORMS AND COORDINATION

There is now growing recognition that social norms play a significant role in behavioral prediction and explanation. Social norms are rules of behavior shared by the members of a social group that "specify what actions are regarded by a set of persons as proper and correct, or improper and incorrect" (Coleman, 1990, p. 243). Social norms specify a limited range of behavior that is acceptable in a situation, and facilitate confidence in the course of action. Social norms can impact our judgments. For example, according to Cialdini et al. (1990), people are more likely not to litter when the floor is nearly spotless in public places. When the apparent social norm is not to litter, people are more likely to conform to that norm by refraining from littering. Norms describe the uniformity of behavior that characterizes a group. Norms make people stick to prescribed behavior, even if new and apparently better options become available (Elster, 1989). Norms come into existence as a product of group interactions. A norm is the behavior that is performed by the majority of the relevant group. A norm is constructed as the tendency of the distribution of a certain attribute within a group. People learn information about a social category that represents the collective, and this information is used to infer the distribution of the members' attitudes within the collective. How these interactions take shape depends on the specific institutions of the society. Habit and tradition become "institutionalized" behavior more or less in accordance with the norms of prior generations, which is again passed over to the next generations. Different norms provide the contours of different groups.

Norms serve to guide an individual's behavior, and are thus an attractive method of social control. Examples range from table manners to business practices. A rule against poor table manners is not suitable for embodiment in law. A norm specifies a behavior that is seen as desirable in the shared view of group members, and whose violation elicits at least informal disapproval. It is regularity such that people conform to it, and they generally approve of conformity to it and disapprove of deviance from it. For example, according to Ostrom (2000), social norms are defined as "shared understandings about actions that are obligatory, permitted, or forbidden" (Ostrom, 2000, pp. 143−144).[12] It is necessary that people believe a social norm exists and know the class of situations to which the norm pertains. This allows that conflicting norms may exist simultaneously. People must be aware that they are in a situation where a particular norm applies.

---

12. In the common-pool resource extraction problem, the self-interested action is to engage in resource over-extraction. Common-pool resource exploitation is popularly known as the Hardin's (1968) "tragedy of the commons." Ostrom (1990) shows how local communities may succeed in overcoming the tragedy of the commons. Cooperation requires action by the community to define rules, and to monitor and enforce them.

Individuals are able to make action plans when they are able to predict what their environment will be like over time. They have to update expectations about what the future holds, in order to make more accurate predictions.

Norms broadly encompass conventions. According to McAdams and Rasmusen (2007), behavioral regularities that lack normative attitudes are referred to as conventions. In social psychology, norms consist of two major categories: they are both descriptive ("is" statements) and injunctive ("should" statements) (Cialdini and Trost, 1998). Descriptive norms describe what people generally do in a situation, while injunctive norms describe what people ought to do there. Conventions are thought of as descriptive norms that are simply what people do. A specific convention is an actual behavioral regularity in a given group. It implies the behavioral pattern, actually followed by the group members in a recurrent situation of social interaction. In the theory of conventions, norms are considered as coordinating devices, and one of several alternatives that enable people to achieve their goals (Young, 1998). A convention is a stable solution to a recurrent coordination problem; it is arbitrary in the sense of being the realization of only one of multiple equilibria that a coordination game displays. An injunctive norm is, on the other hand, referred to as any norm that one is obliged to follow due to the threat of sanctioning its violation. It is considered a belief system about what constitutes morally approved and disapproved conduct.

In game theory, a social norm is seen essentially as a convention.[13] Without the normative obligation, the behavioral regularity would be simply an equilibrium that results when each person takes his or her best step considering the actions of others. Norms emerge not from a collective need, but from the decentralized interaction of agents according to their own interests. There must be at least one alternative norm which could have served more or less equally well as a solution to the coordination problem. There are multiple equilibria—two or more outcomes that satisfy the Nash criterion that no player would benefit by unilaterally switching strategies. The notion of convention is thus characterized in terms of coordination equilibrium, that is, a strategy combination in which no one would have been better off had any one player alone acted otherwise. Every player strictly prefers that all conform to the equilibrium. Pure coordination games are games with a perfect coincidence of interests, having at least two Nash equilibria with equal payoffs. A pure coordination game is commonly illustrated by the choice

---

13. A game consists of the following three components: a set of players who take part in the game; a set of strategies for each player; and a set of payoff functions for each player that give a payoff value to each combination of the players' chosen strategies. Standard game theory assumes that (1) all players form beliefs based on an analysis of what others might do (strategic thinking); (2) all players choose the best response given those beliefs (optimization); and (3) all players adjust strategies and beliefs until they are mutually consistent (equilibrium). It is widely accepted that not every player behaves rationally in complex situations. Therefore, assumptions (1) and (2) are sometimes violated.

	A	*B*
A	*UA, UA*	0, 0
B	0, 0	*UB, UB*

FIGURE 1.2 Payoff matrix (where $0 < UA < UB$).

between driving on the left and right side of the road. There are two pure strategy Nash equilibria with identical payoffs—where everyone drives on the left and where everyone drives on the right. All players share some interests in avoiding the non-coordinated outcome—where some drive on the left and others drive on the right. Once the determination is made, either by explicit agreement or spontaneous convergence on a salient rule, the convention is stable because no one has an interest in deviating from it all by themselves. It is costly to try to change the convention once it is established. Nash equilibria of a strategic-form game then correspond to population equilibria of individuals who interact in accordance with the rules of the game. The evolution of conventions is an instance of players learning to play an equilibrium. In an evolutionary explanation of a trait, its frequency in a population is used in terms of selection pressures that favor (or disfavor) its replication relative to its alternatives in a particular environment. The primary determinant of the payoff of any particular trait is the frequency distribution of itself and its alternatives in the population. In successive rounds of the game, people are assumed to adjust their social traits in response to the payoffs from expressing those traits in the previous round. When an evolutionary model endogenies strategies, it explains the frequency distribution of these strategies.

Following Boyer and Orléan (1992), we consider how conformity effects can lead to the emergence of a group consensus around one convention. Let us consider the symmetrical game defined by the payoff matrix as given in Fig. 1.2. It is assumed that there is a large population from which pairs of individuals are repeatedly drawn at random to play this symmetrical game. Players do not understand that their own behavior potentially affects the future play of their opponents. Let p_t be the frequency of individuals who play strategy A at time t in the population. Then, an individual playing strategy A will obtain the utility $U(A, p_t)$, given by the following formula:

$$U(A, p_t) = p_t UA$$

Likewise, for the frequency of individuals who play strategy B at time t in the population $1 - p_t$, an individual playing strategy B will obtain the following utility $U(B, p_t)$:

$$U(B, p_t) = (1 - p_t)UB$$

Two utilities, $U(A, p_t)$ and $U(B, p_t)$, are described in Fig. 1.3.

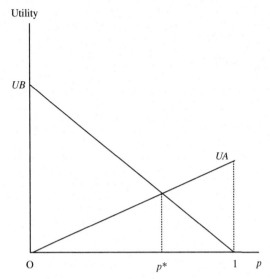

FIGURE 1.3 Two utilities.

There exists a value p^*, such that $U(A, p^*)$ is equal to $U(B, p^*)$:

$$p^* = \frac{UA}{UA + UB}$$

For $p > p^*$, $U(A, p)$ is greater than $U(B, p)$, even if UA is smaller than UB. Successful behavior becomes more prevalent just because players imitate successful behavior.

The frequency p_t increases if $U(A, p_t)$ is greater than $U(B, p_t)$, which can be formulated in the following way:

$$\frac{dp_t}{dt} = G[U(A, p_t) - U(B, p_t)]$$

where G is a nondecreasing, sign-preserving function. This is a differential equation that, together with an initial condition, determines a path for the population that describes, for any time t, the state of the population.

Thus, for the sufficient number of agents having chosen strategy A, its representation in the population grows ($dp_t/dt > 0$), and the system eventually converges to the convention A that is Pareto-inefficient. That is, for $p_t > p^*$, the frequency p_t converges to 1. If no fraction of the population plays strategy B at any point of time, then it is never played. Equilibrium can be viewed as the steady state of a community whose members are myopically groping toward maximizing behavior.

People are, in real life, skilled at coordinating their actions. An equilibrium in game theory is a profile of strategies, one for each player participating in a strategic interaction. What distinguishes an equilibrium from other profiles is that no player can do better by changing his or her strategy unilaterally. If others do their part in the equilibrium, no player has an incentive to deviate. When a game has only one equilibrium, Nash equilibrium must be the only rational prediction of players' behavior. However, a game can have multiple Nash equilibria. Then, players need to coordinate between multiple equilibria. When more than one equilibria are possible, a player's choice of strategy is not fully determined by the payoffs.

Schelling (1960) has developed the study of mechanisms allowing people to coordinate. What is necessary is to coordinate predictions:

> *People can often concert their intentions and expectations with others if each knows that the other is trying to do the same. Most situations—perhaps every situation for people who are practiced at this kind of game—provide some clue for coordinating behavior, some focal point for each person's expectation of what the other expects him to expect to be expected to do.*
>
> (Schelling, 1960, p. 57, original emphasis)

Schelling (1960) proposes that, when the problem is selecting one means of coordinating among many, certain solutions stand out from the others as the sort that will attract the attention. "Some kind of prominence or conspicuousness" (Schelling, 1960, p. 57) or "uniqueness" (p. 58) of a coordination equilibrium is "focal," in the sense that it creates convergent expectations.[14] This is a property that captures the attention of players as a result of common perceptions, impressions, and associations. Players tend to select an equilibrium that is unique in some non-payoff dimension, merely because that uniqueness causes each player to expect every player to focus on it. In a coordination game, all players are better off as common effort increases.

Schelling's (1960) "focal-point" idea is important as an explanation of how players coordinate. Each member of the population expects every other member to behave in accordance with the relevant regularity. Individuals often coordinate at a point that, in some sense, "stands out" from the others. A shared vision of what should be obvious to each player leads to the emergence of a focal point. Schelling (1960) contends that coordination is inherently dependent on empirical evidence. There is sometimes a logic to decide what solution has the unique property. However, the uniqueness often depends on the experience and contingent associations of the individuals involved.

Lewis (1969) gives a more formal structure to characterize coordination problems. Individuals' interests roughly coincide in pure coordination games. In order to have a sufficient reason for choosing a particular action, an agent needs to have a sufficient degree of beliefs that the other agent will choose a

---

14. See Sugden (1995) for a formal theory of focal points. In his model, there is a one-to-one relationship between labels and strategies, so that players are always able to distinguish between their strategies.

certain action. When people interact with each other, their actions and the deliberate reasoning by which they choose their actions are interdependent. In Lewis' (1969) argument, coordination may be achieved with the aid of mutual expectations about action. Past experience of a convention, or "precedent," is the source of such expectations. A specific convention is assumed to be arbitrary, in the sense of being the realization of only one of the multiple potential regularities that could emerge in a recurrent situation of social interaction.[15] Lewis (1969) justifies the stability of conventions through the reconstruction of the actual processes of reasoning of those involved in the interaction situation where a particular convention emerges. Lewis (1969) adds this common knowledge requirement to the definition of convention to rule out cases in which agents coordinate as the result of false beliefs regarding their opponents. Salience plays a crucial role in the emergence and reproduction of conventions. Conventions are understood as salient regularities of behavior, common knowledge of which enables people to successfully negotiate the complex landscape of ordinary social interactions. Lewis (1969) argues that the precedent would make one of the coordination equilibria salient. When players have no reason to choose one strategy rather than another, their default choices tend to favor strategies with certain properties of salience. Knowing this, rational players of coordination games choose salient strategies deliberately.

1.4   ORGANIZATIONS

As Hayek (1976) suggests, the working of the market can be understood best by looking at it as an "exchange game." It is a game that "proceeds, like all games, according to rules guiding the actions of individual participants" (Hayek, 1976, p. 71). The market can be defined as an institutionalized arena for exchange. The reason for the participants to play the exchange game is that it is wealth-creating or positive-sum. Participants can expect to realize better outcomes than they could expect from feasible alternative games. Profit-seeking is not unconditional or unconstrained; the exchange game is played under certain rules. Institutions that are based on transparent rules, such as property rights and the rule of law, are universally good for wealth enhancement, because they promote individual choice and factor mobility.

Uncertainty arising from others' actions generates economic problems. The individual ability to coordinate their plans deteriorates if the external

---

15. As an example, in Lewis' home town of Oberlin, Ohio, there was a time when local calls were cut off without warning after three minutes. Then, in the population of phone users, an interaction occurs between two phone users when a call between them is cut off. One convention is that the original caller calls back. Another one is that the person called calls back. In fact, the first of these was the convention.

environment becomes volatile. Therefore, individuals are always eager to reduce uneasiness arising from uncertainty. Institutions coordinate individual actions at a lower cost, because they reduce the volatility in the plans of others. Wealth can expand through cooperative interaction. A respect for private property, a well-functioning rule of law, and a stable monetary order are crucial for individual experimentation and widespread coordination. Institutions must provide a predictable environment within which people can orient their behavior.

Hayek (1976, p. 109) defines the notion of "catallaxy" as "the special kind of spontaneous order produced by the market through people acting within the rules of the law of property, tort and contract." Wealth grows because people can innovate with their own property, and the returns to their efforts accrue to them. As Hayek (1976, p. 115) suggests, "[t]he best way to understand how the operation of the market system leads not only to the creation of an order, but also to a great increase of the return which men receive from their efforts ... as a game which we may now call the game of catallaxy. It is a wealth-creating game (and not what game theory calls a zero-sum game), that is, one that leads to an increase of the stream of goods and of the prospects of all participants to satisfy their needs." Wealth enhancement must be anticipated in order to participate in a positive-sum game. Therefore, "[w]hile within an organization the several members will assist each other to the extent that they are made to aim at the same purpose, in a catallaxy they are induced to contribute to the needs of others without caring or even knowing about them" (Hayek, 1976, p. 109).

Herbert Simon ([1969] 1996) argues about Hayek's perspective on the role of the market:

No one has characterized market mechanisms better than Friedrich von Hayek ... His defense did not rest primarily upon the supposed optimum attained by them but rather upon the limits of the inner environment—the computational limits of human beings.

(Simon, [1969] 1996, p. 34)

Markets are important mechanisms of coordination. However, for Simon, markets cannot fulfill their function if they do not possess "a high degree of economic stability and a low level of externalities" (Simon, 2000, p. 751).

In modern economies, organizations perform a more relevant role of coordination than the price system does. As March and Simon (1958, p. 154) argue, "[m]any of the central problems for the analysis of human behavior in large scale organizations stem from the operation of subsystems within the total organizational structure." An organization is a set of regularly interacting individuals who are bound by a set of internal rules. These rules determine the organization's structure. The structure of an organization consists of the following key elements: the assignment of members to tasks and jobs; the division of the organization into departments and other groups of members; the ordering of members based on superior—subordinate relations for integrating their activities; the incentives to which members are subject in fulfilling their tasks; the monitoring of members; and the arrangements for coordinating the organization's activities.

As Simon (1991) points out:

Prices perform their informational function when they are known or reasonably predictable. Uncertain prices produced by unpredictable shifts in a system reduce the ability of actors to respond rationally.

(Simon, 1991, p. 40)

Simon (1991) recognizes that prices reduce, under specific conditions, the need of individuals for information. However, if uncertainty on prices prevails, the market reduces the capacity of individuals to make correct decisions. Then organizations have a key role to reduce ever-prevailing uncertainty.

Simon was, first and foremost, concerned with modeling how people think. Following Simon (1987, p. 222), bounded rationality denotes "the whole range of limitations on human knowledge and human computation that prevent economic actors in the real world from behaving in ways that approximate the predictions of classical and neoclassical theory." Economic actors suffer cognitive and informational constraints that make it impossible to achieve optimal decisions. Boundedly rational agents cope with a problem by adopting simplifying strategies for its solution. Simon's theory of organization arises from the empirically adequate theory of individual choice. Boundedly rational agents "satisfice"; the mode of choice leads them to behave in a routine, myopic, but reasonably adaptive manner. Organizational structure is shaped by attempts to reduce the effects of the cognitive and informational constraints. By assigning corporate goals to different subunits within the firm, managers reduce the amount of information necessary to monitor organizational performance, thereby relieving cognitive and informational overload for a given unit. This process also provides economic actors in the subunits with clear cognitive outcomes to attain, thus aligning individual behavior with the overall goals of the organization.

In most microeconomic textbooks, the firm is represented as a "black box" (in purely technological terms, as a production function or production set), utilizing inputs to produce outputs in response to changes in prices. The firm is an autonomously created entity in which rational choices are made on the allocation of resources without any specification of the particular processes involved. In the neoclassical framework of complete contingent contracting, initial ownership rights and organizational structures do not matter for economic efficiency. No matter how unrealistic the assumptions may be about the firm, the need for firms to be efficient takes center stage in the neoclassical economic perspective. The neoclassical theory of the firm in market economies has been challenged for many years. The most critical objection was raised by Ronald Coase's famous 1937 paper on the nature of the firm. Coase

(1937) notes that real-world production processes generally involve many transactions among owners of capital, labor, land, specialized knowledge, and other inputs, and that these transactions are costly. Coase's (1937) question is why any transactions take place in firms if markets are so good at allocating resources. According to Coase (1937), a firm will arise when it is cheaper to carry out a transaction in a firm than it is to do so over the market. The driving force is efficiency. That is, economic agents substitute authority relations for market relations, in order to reduce transaction costs and produce more efficiently. Coase (1937) makes an argument that lays the foundation for transaction cost economics; the decision to use a market or, alternatively, a firm depends on the transaction costs involved. One should note that transaction costs are not costs like production costs, but are the costs of economic exchanges. For Coase (1937), the most obvious cost of using the price system is that of discovering what the relevant prices are. The boundaries of the firm will be determined when the cost of organizing a transaction within the firm equals the cost of using markets. That is, a "point must be reached where the costs of organizing an extra transaction within the firm are equal to the costs involved in carrying out the transaction in the open market or to the costs of organizing by another entrepreneur" (Coase, 1937, p. 395). Firm boundaries depend not only on technology, but on organizational considerations (i.e., the costs and benefits of contracting). Coase's (1937) paper helped the revival of interest in the economics of organizations in the 1970s, with the work of Alchian and Demsetz (1972), Williamson (1975), and others. Much of the revival focused on the costs of using markets quite differently: moral hazard, principal–agent problems, asymmetric information, and opportunism. Its major contribution is the introduction of economic conditions that prevent complete contingent contracting under which initial ownership rights and organizational structures do not matter for efficiency.

Oliver Williamson, in the 1970s, made significant progress on understanding the costs of using markets. He focused on a situation in which parties make relationship-specific investments: relation-specific investments are worth more inside a relationship than outside. Relationship-specific investments could be described, including both specialized physical and human capital, along with intangibles such as firm-specific knowledge or capabilities. A transaction involving relationship-specific investments would be ideally governed by a long-term contract. However, in practice, such contracts are hard to write because of the difficulties of anticipating the future. Any such contract will be incomplete, and will have to be renegotiated *ex post*; this renegotiation process is costly. Information within an organization is incomplete, because information available in the environment outside the organization, as well as that already possessed by some members in the organization, cannot be costlessly transmitted to those who use it. For Williamson, internal organization is a source of conflicts and disputes. Williamson's work stresses the combined effects of bounded rationality and opportunism. Members in organizations are subject to bounded rationality; they act under the constraints imposed by their intellectual capacities. While bounded rationality results in contractual incompleteness, opportunism is thought of as "self-interest with guile." The degree of bounded rationality and self-interest varies among individuals.

Foss et al. (2000) argue that we can understand the existence of firms as a consequence of one or two of the following assumptions not holding true:

a. The assumption of symmetry information concerning "states of nature."
b. The assumption of complete contracting: agents can foresee all future contingencies, and can costlessly write contracts which cover all contingencies.

Thus, according to Foss et al. (2000), the theory of the firm in contemporary mainstream economics can be roughly partitioned into two general groups:

1. Principal−agent type models:
   Agents can write comprehensive contracts characterized by *ex ante* incentive alignment under the constraints imposed by the presence of asymmetric information.
2. Incomplete contracts models:
   It is costly to write elaborate contracts, and contracts will have holes or inefficient provisions. Therefore, there is a need for *ex post* governance.

Thus, principal−agent type models emphasize *ex ante* incentive aspects of contracts, while incomplete contracts models focus on *ex post* governance costs of contracts, such as dispute avoidance and ways of processing disputes. The former models are built on the assumption that every possible, economically significant contingency can, at least in principle, be foreseen and included in contracts between parties. The latter ones, on the other hand, share Coase's (1937) basic assumption that firms cannot write complete contracts, precisely because of the inability or the cost of specifying the preferred action of the partner when facing unforeseen contingencies. Some kind of *ex post* adaptation is then needed to specify the action called for, as information becomes available. When circumstances arise which are not accounted for in the original agreement, individuals will need to negotiate revised terms which address the newly uncovered contingency.

In the principal−agent type models, a basic assumption is that some informational asymmetry exists between the two. The principal−agent type models express the agency relationship, in which one party, the principal, considers entering into a contractual agreement with another, the agent, in the expectation that the agent will subsequently choose actions that produce outcomes desired by the principal. The agent has their own interests at heart, and is induced to pursue the principal's objectives only to the extent that the incentive structure imposed in their contract renders such behavior advantageous. The principal−agent type models focus on the *ex ante* alignment of economic incentives; getting the economic incentives right at the outset leads to efficient contracting. The theoretical orientation is to design contracts that minimize agency costs, given various constraints. The literature makes a distinction between "hidden information" and "hidden action" models. Adverse selection and moral hazard are general problems whose potential is inherent in all contracting and hierarchical relationships. Adverse selection derives from unobservability of the information on which others' decisions are based. In a standard principal−agent model of moral hazard, the agent's effort is not observable, and, due to the presence of uncertainty in the production process, it cannot be inferred by the principal. The compensation scheme is assumed to be based on observable, realized output. According to Foss et al. (2000, pp. 636−638), the principal−agent type models are classified into three subgroups: (1-a) the nexus of contracts view; (1-b) the firm as a solution to moral hazard in teams approach; and (1-c) the firm as an incentive system view.[16]

First, in Jensen and Meckling's (1976) view, the firm is seen as a fictitious entity created by "a nexus of contracts" of the principal−agent variety. The nexus of contracts model treats firms as coalitions of individuals linked through a set of contracts, either with one another directly or with a fictitious central party called "the firm." Then, the firm does not exist as a separate entity, and it cannot be worth more than the sum of the individual contracts that comprise it. As Jensen and Meckling (1976, p. 311) point out, "the private enterprise is simply one form of legal fiction which serves as a nexus of contracting relationships and which is also characterized by divisible residual claims on the asset and cash flows of the organization which can generally be sold without permission of other contracting individuals." The firm is a legal fiction which serves as a focus for a complex process in which the conflicting individuals are coordinated within a framework of contractual relations. Contractual relations with employees, suppliers, customers, creditors, etc., are an essential aspect of the firm. It is difficult to draw a line between transactions within a firm and those between firms. What distinguishes firms from other forms of market contracting is the continuity of contractual relationships between input owners.

Second, Alchian and Demsetz (1972) and Holmström (1982) develop "the firm as a solution to moral hazard in teams approach." Team production highlights three fundamental features of the theory of the firm: (1) gains from trade stemming from complementarities among heterogeneous resources; (2) team technologies that may function as covers for moral hazard; and (3) governance

---

16. According to Foss (1993, p. 271), "[o]n the level of ontology and conceptualizations of human agency, it [complete contract theory] derives from the implicit assumption that the future does not hide anything that cannot be contractually anticipated, at least in broad outlines."

mechanisms that internalize the externalities from (2). Alchian and Demsetz (1972) note that, for complex production processes, there is typically a gain from cooperation. Teams of input owners can produce more in cooperation with one another than separately, and this gives them an incentive to coordinate their actions. Team production is a technological characteristic of a production process whereby individual marginal products are enhanced by the efforts of others. However, this makes the allocation of rewards more difficult. Because of the complex interdependence of tasks, they cannot reward one another according to individual impacts on output. Team production and rewards fall as a result. The free-rider problem is inherent in team production, where each member's action is not observable. If information could be gained on the marginal products of individual members, they could agree to be rewarded on this basis. The solution to this problem is to appoint a monitor who is given the right to fire and hire members of the team by observing their marginal products. Formally, Holmström (1982) analyzes the incentive problems under which the monitor is uninformed about individual effort levels under team production. In a team production situation with unobservable effort levels, Holmström (1982) proves that a budget-balancing incentive system cannot reconcile Nash equilibrium and Pareto optimality.

Third, Holmström and Milgrom (1991, 1994) contribute to the "firms as an incentive system view." They stress that the firm should be viewed as "a system"; it is a coherent set of complementary contractual relationships that endeavor to mitigate incentive problems. Holmström and Milgrom (1991) develop a multi-task, principal−agent model in which the agent allocates effort across multiple tasks, and the principal observes a performance measure for each of these tasks. Designing reward systems that provide high-powered incentives for multiple tasks is quite difficult. They show that increasing compensation in one task will cause some reallocation of attention away from other tasks, and therefore pay-for-performance contracts may not be optimal.

Incomplete contracts models are, on the other hand, classified into five subgroups: (2-a) the authority view; (2-b) the firm as a governance mechanism; (2-c) the firm as an ownership unit; (2-d) relational contracts; and (2-e) the firm as a communication hierarchy.

In the authority view, the firm is defined as an employment relation. When the total transaction costs are large, it can be beneficial to use employment contracts and managerial authority. For Simon (1951), a contract between autonomous agents specifies an action to be taken in the future along with its price, while an employment contract specifies a set of acceptable instructions that the employee has to accept if asked to carry them out by the employer. The advantage of the employment contract is its flexibility; the employer does not have to precommit to an action, and the action of the employee can be adapted to whatever state of nature will occur. The real cost of contracts may be their inflexibility when faced with unforeseen contingencies. Within a firm, the price mechanism is suppressed, and transactions occur as a result of instructions or orders issued by a boss. The boss has "authority," that is, the right to dictate a decision in any circumstances not spelled out explicitly in the original contract (Tirole, 1988). The authority relation is not characterized by command, as classical organization theorists suggest, but rather is two-way. Thus, the nature of authority relation and whether or how well it works depend on both parties to the agreement.

The "firm as a governance mechanism" approach is mostly associated with the work of Oliver Williamson. Williamson (1975) emphasizes *ex post* inefficiencies that arise in bilateral relationships. For Williamson (1985), three concepts are crucial: bounded rationality, opportunism, and asset specificity. Bounded rationality gives rise to incomplete contracts and the contractual parties are, as a result, at risk of being subject to opportunism, which requires governance structures. Transactions occur between bounded rational and opportunistic agents that are subject to bilateral dependency in the form of asset specificity.[17] Once the investment in assets which are specific to a transaction has been made, the agents are effectively committed to the transaction for some period of time. Assets are highly specific when they have value inside a particular transaction, but have relatively little value outside the transaction. Under these circumstances, the continuity of the relationship between agents is highly valued. If the assets used in a set of activities are highly specific, the scope for opportunism from partners is large and the set of activities should take place within a single firm. Thus, the higher the degree of asset specificity, the smaller the ratio of economic activities that will take place through market relations. The efficiency of alternative governance structures is determined by the relative value of assets in the context where they are employed.

The "firm as an ownership unit" approach is the property rights theory or incomplete contracts theory of the firm due to Grossman and Hart (1986) and Hart and Moore (1990). This approach considers the role of asset ownership with respect to explaining firm boundaries.[18] It is sometimes regarded as a formal version of Williamson's approach. A central assumption is that contracts must be incomplete, in the sense that the allocation of control rights cannot be specified for all future states of the world. The parties' *ex ante* incentives may be misaligned when contracts are incomplete in situations where the parties to a relation have to undertake specific and complementary investments in that relation. Grossman and Hart (1986) and Hart and Moore (1990) contribute to incomplete contracting theory by clarifying the meaning of integration (common asset ownership), and by demonstrating why contractual problems might exist in the absence of common ownership. The incompleteness of contracts

---

17. Williamson (1996) identifies six types of asset specificity: (1) site, (2) physical asset, (3) human asset, (4) dedicated asset, (5) brand name capital, and (6) temporal.

18. See Hart and Moore (2008) for an additional role of long-term contracts. They argue that a contract defines a reference point that shapes the expectations of the contracting parties.

means that there are non-contractible elements due to difficulties in contemplating in advance all possible future contingencies and measuring performance under each contingency. Grossman and Hart (1986) and Hart and Moore (1990) introduce ownership concepts in an incomplete contract setting, and emphasize relation-specific assets (both physical and human asset specificity). They demonstrate how asset ownership changes *ex ante* investment incentives. Contracts are necessarily incomplete; there are some contingencies that are not regulated by the contract. As a result, there is a need to allocate the right to decide in the events not specified by the initial contract. Ownership is considered to be the legally defined and enforced possession of an asset. What makes a difference is how ownership is allocated. Owning an asset is important if one undertakes a non-contractible investment which is specific to the asset. That is, if one does not own the asset, one is subject to the hold-up threat by the owner. "Residual rights of control," as the concept of ownership of assets, give the owner the power to make decisions regarding the use or disposition of an asset in all respects that are not explicitly designated or limited by a written contract. Residual rights of control are obtained through the legal ownership of assets, and imply the right to "decide when or even whether to sell the asset" (Hart, 1995, p. 65).[19] Ownership matters, because it is a source of power over residual control rights under incomplete contracting. Grossman and Hart (1986) take ownership of nonhuman assets as the defining characteristic of firms. If two different assets have the same owner, then we have a single, integrated firm; if they have different owners, then there are two firms and dealings between them are market transactions. Grossman and Hart (1986) explain the specific reallocation of ownership rights with changes of boundaries of the firm: "[f]irm 1 purchases firm 2 when firm 1's control increases the productivity of its management more than the loss of control decreases the productivity of firm 2's management" (Grossman and Hart, 1986, p. 691). The owner of an asset can decide how it should be used and by whom, subject only to the constraints of the law and the obligation implied by specific contracts. In Hart and Moore's (1990) model, each agent makes investments in human capital that are complementary with a set of nonhuman assets. The most common measure of human capital specificity involves some notion of the amount of training that is required to produce or use an input. Investments in human assets are assumed to be nonverifiable by a third party and, for this reason, contracting parties cannot contract over the costs or the outcomes of such investments. The allocation of ownership to assets influences the contracting parties' incentives to make investment in human assets. The ownership of the nonhuman assets affects his or her own human capital.

---

19. Demsetz (1998, p. 449, original emphasis) points out "Hart writes as though he thinks that *asset* ownership is an unambiguous concept." For Demsetz, it is impossible to describe the complete set of rights that are potentially ownable.

As the "relational contracts" approach, Baker et al. (2002) point out that relational contracts are informal agreements between the parties about compensation. Relational contracts cannot be enforced by a third party, and so must be self-enforcing agreements. There are differences in the way relational contracts function within firms and between firms. The difference between them lies in what happens if the relational contract breaks down. The strength of the threat to discontinue the relationship determines the implementability of relational contracts. As an example, consider the situation in which the market for the good is volatile. Then, a relational contract may be unworkable, since the supplier has an incentive to violate the relational contract when the market price is high. To be credible, relational contracts require one part or both to have established some reputation. Thus, a firm is not simply the sum of components readily available on the market, but rather is a unique combination, which can be worth more or less than the sum of its parts.

The "firm as a communication hierarchy" approach exploits the idea that one function of the firm is to adapt to and process new information. Marschak and Radner's (1972) work, a classic contribution on team theory, presents the problem of communication when specialized local knowledge is necessary for effective decisions, and yet some coordination of the dispersed decision-makers is vital. It is important to understand the substitution of explicit management activity for market-guided activity. Bolton and Dewatripont (1994) consider a situation in which subordinates have more information than a boss: the boss can ask the subordinates to transmit their information, but this is costly and takes time. Alternatively, the boss can forego communication and let the subordinates decide themselves. Then, decentralization will be preferred to centralization if decisions in the organization are independent, but not if they are highly interrelated.

In organization theory, there are some studies on the behavioral or evolutionary paradigm, though they constitute a small part of the literature. An essential difference between the behavioral or evolutionary theory and more orthodox theory lies in the treatment of the rationality and uncertainty contained in them. First, Cyert and March launch into an attack on the conventional theory by "an empirically relevant, process-oriented, general theory of economic decision-making by a business firm" (Cyert and March, 1963, p. 3).[20] The behavioral theory of the firm places an explicit emphasis on the actual process of organizational decision-making (Cyert and March, 1963; March and Simon, 1958). Empirical accounts of behavior in organizations suggest that behavior is often cognitive and calculative. In spite of their best efforts to deal with the complexity and unpredictability of the external world, individuals are limited in their ability to plan for the future and to accurately

---

20. Along with Herbert Simon, Richard Cyert and James March are regarded as pioneers of behavioral economics associated with the Carnegie School.

predict for the various contingencies that may arise. Then, organizations can be viewed as collections of individuals with multiple goals who operate in a defined structure of authority. In decision-making, these multiple goals act as constraints or limits on a course of action. For Cyert and March (1963), the "natural theoretical language" for specifying this process is the language of a computer program. That is, their theory of the firm is embedded in the form of flowcharts, algorithms, and computer programs. As an alternative, behavioral conceptualization of the firm, Cyert and March (1963) consider the firm as a site of decision-making involving conflict, uncertainty, search, learning, and adaptation over time. Their model emphasizes the interplay between cognition and action or between thinking and experiencing. An evidence-based theory of the firm is developed in which the firm adapts to its environment through learning and experimentation. The boundedly rational managers think along the lines of simplified representations of their decision problem; these thought processes lead to strategic directions that guide local searches. Suboptimal decision-making can be interpreted in terms of the managers' levels of information, and their ability to use it, coupled with a willingness to accept adequate, rather than maximal, profits.

Second, Leibenstein (1966) proposes the X-efficiency framework underlying the notion that economic agents may not achieve allocative efficiency in their productive decisions and behavior. X-efficiency measures the extent to which the firm fails to realize its productive potential: it is the degree of inefficiency in the use of resources within the firm. Individuals are subject to "inert areas" within which their behavior is unresponsive to changes in external constraints. As Leibenstein (1980, p. 27) suggests, "... X-efficiency may arise for reasons outside the knowledge or capability of management attempting to do the managing ...." For Leibenstein, the firm is an organization of different individuals that have no consensus on their objectives.

X-efficiency theory is based on the following postulates (Leibenstein, 1987):

1. Relaxing maximizing behavior:
   It is assumed that some forms of decision-making, such as habits, conventions, moral imperatives, standard procedures, or emulation, can be, and frequently are, of a non-maximizing nature.
2. Inertia:
   Functional relations are surrounded by inert areas, within which changes in certain values of the independent variables do not result in changes of the dependent variables.
3. Incomplete contracts:
   The employment contract is incomplete, in that the payment side is fairly well specified but the effort side remains mostly unspecified. There is a tension between the employee and the employer over how hard the former should work.

4. Discretion:

Every employee has effort discretion and a free-rider incentive to move to his minimum effort level (even though he might want others to work effectively); the effort convention is thus a coordinated persistence of nonoptimal conventions, and helps to explain the existence and persistence of X-inefficient behavior.

Finally, Nelson and Winter (1982) focus on the firm as a decision-making and change-generating unit. Instead of maximizing the behavior of firms, Nelson and Winter (1982) use the concept of decision rules, and instead of equilibrium they see tendencies. In their approach, economics is broadly understood by using the evolutionary analogy, based on population, variety, selection, and replication. Biological analogies for the evolution of organizational forms are conceptually attractive, because they offer a complete view of the processes of evolution. Evolutionary processes involve the generation of novelty, focusing attention on the importance of creativity and innovation. In a biological context, the term "Lamarckism" implies that genotypical inheritance may occur as a consequence of acquired characteristics.[21] Nelson and Winter define their position as Lamarckian: "our theory is unabashedly Lamarckian: it contemplates both the 'inheritance' of acquired characters and the timely appearance of variation under the stimulus of adversity" (1982, p. 11, original emphasis). Nelson and Winter (1982) distinguish between the processes by which present decisions are made, and the processes by which decision capabilities are developed. They describe the concept of "routines" as the regular and predictable behavioral patterns of firms. Routines are patterned, typically in the form of sequences of individuals' actions; the contents are organization-specific. Individual firms have a kind of genetic endowment in terms of technical routines. However, searching for new routines is itself also a routine. The emphasis is placed on the interactions, rather than the individuals that are interacting: routines are collective rather than individual-level phenomena. Firms compete for market shares on the basis of their specific routines that they built up and improved on in the past. Routines refer to formal as well as tacitly understood rules of behavior. Some routines will be simple habits, easily changed when better techniques become known, while others will be taken-for-granted elements that resist change. Nelson and Winter (1982) describe firms as repositories of knowledge and capacities, embodied in firm-specific routines. These routines are created within the firm and evolve over time in competitive environments. A firm's knowledge and capabilities change as new participants and technologies come and go, and the firm gains experience in the market. Nelson and Winter (1982) examine populations of firms with differing

---

21. Though Darwinism does not preclude the possibility of Lamarckian evolution, it criticizes its weak inheritance mechanism (Hodgson and Knudsen, 2006).

routines by addressing the interplay between changing external environments and changing routines. The organizations whose routines are relatively better fit for coping with environmental conditions will thrive. Variations in organizational activities may be conditioned by the nature of the routines, but there is no way to know which of the variations will prove better adapted to the environment.

1.5   SUMMARY

Institutions are defined as a system of rules, beliefs, norms, and organizations that conjointly generate a regularity of behavior. Institutions are considered as durable rules which govern human interactions. The formal and informal rules constitute the institutional structure within which human interactions occur. Here, a "rule" is fundamentally the others' expected behavior which motivates our behavior, rather than the rule itself. Following rules helps people make decisions with some degree of certainty about which behavior is acceptable and which is not. It reduces the amount of information individuals must collect, and enhances their ability to make plans and to coordinate with each other. Institutions coordinate the individual's actions at a lower cost, because they reduce the volatility in the plans of others. Norms describe the uniformity of behavior that characterizes a group. Norms specify a limited range of behavior that is acceptable in a situation, and facilitate confidence in the course of action. Markets are important mechanisms of coordination. However, in modern economies, organizations perform a more relevant role of coordination than the price system does. Economic agents substitute authority relations for market relations, in order to reduce transaction costs and produce more efficiently. An organization is a set of regularly interacting individuals who are bound by a set of internal rules.

REFERENCES

Acemoglu, D., Robinson, J., 2012. Why Nations Fail: The Origins of Power, Prosperity, and Poverty. Profile Books, London.

Alchian, A., Demsetz, H., 1972. Production, information costs, and economic organization. Am. Econ. Rev. 62, 777–795.

Aoki, M., 2001. Towards a Comparative Institutional Analysis. The MIT Press, Cambridge, MA.

Arthur, W.B., 1989. Competing technologies, increasing returns, and lock-in by historical events. Econ. J. 99, 116–131.

Baker, G., Gibbons, R., Murphy, K.J., 2002. Relational contracts and the theory of the firm. Q. J. Econ. 117, 39–84.

Bolton, P., Dewatripont, M., 1994. The firm as a communication network. Q. J. Econ. 109, 809–840.

Bowles, S., 2004. Microeconomics: Behavior, Institutions, and Evolution. Princeton University Press, Princeton. NJ.

Boyer, R., Orléan, A., 1992. How do conventions evolve? J. Evol. Econ. 2, 165–177.

Caldwell, B., 2014. Introduction. In: Caldwell, B. (Ed.), The Collected Work of F. A. Hayek: The Market and Other Orders, vol. 15. University of Chicago Press, Chicago.

Cialdini, R.B., Trost, M.B., 1998. Social influence: social norms, conformity and compliance. In: Gilbert, G., Fiske, S., Lindzey, G. (Eds.), The Handbook of Social Psychology, vol. 2. McGraw-Hill, New York, pp. 151–192.

Cialdini, R.B., Reno, R.R., Kallgren, C.A., 1990. A focus theory of normative conduct: Recycling the concept of norms to reduce littering in public places. J. Pers. Soc. Psychol. 58, 1015–1026.

Coase, R.H., 1937. The nature of the firm. Economica 4, 386–405.

Coleman, J., 1990. Foundations of Social Theory. Harvard University Press, Cambridge, MA.

Cyert, R.M., March, J.G., 1963. A Behavioral Theory of the Firm. Prentice-Hall, Englewood Cliffs, NJ.

David, P.A., 1985. Clio and the economics of QWERTY. Am. Econ. Rev. 75, 332–337.

Demsetz, H., 1998. Review: Oliver Hart's Firm, Contracts, and Financial Structure. J. Political Econ. 106, 446–452.

Denzau, A.T., North, D.C., 1994. Shared mental models: ideologies and institutions. Kyklos 47, 3–31.

Elster, J., 1989. Social norms and economic theory. J. Econ. Perspect. 3, 99–117.

Foss, N.J., 1993. More on Knight and the theory of the firm. Manag. Decis. Econ. 14, 269–276.

Foss, N.J., Lando, H., Thomsen, S., 2000. The theory of the firm. In: Bouckaert, B., Dr Geest, G. (Eds.), Encyclopedia of Law and Economics, vol. III. Edward Elgar Publishing, Cheltenham, UK, pp. 631–658.

Greif, A., 1994. Cultural beliefs and the organization of society: a historical and theoretical reflection on collectivist and individualist societies. J. Political Econ. 102, 912–950.

Greif, A., 2006. Institutions and the Path to the Modern Economy. Lessons from Medieval Trade. Cambridge University Press, Cambridge, MA.

Greif, A., Kingston, C., 2011. Institutions: rules or equilibria? In: Schofield, N., Caballero, G. (Eds.), Political Economy of Institutions, Democracy and Voting. Springer, Berlin, pp. 13–43.

Greif, A., Mokyr, J., 2017. Cognitive rules, institutions, and economic growth: Douglass North and beyond. J. Inst. Econ. 13, 25–52.

Grossman, S.J., Hart, O.D., 1986. The costs and benefits of ownership: a theory of vertical and lateral integration. J. Political Econ. 94, 691–719.

Hardin, G., 1968. The tragedy of the commons. Science 162, 1243–1248.

Hart, O.D., 1995. Firms, Contracts, and Financial Structure. Clarendon Press, Oxford.

Hart, O.D., Moore, J., 1990. Property rights and the nature of the firm. J. Political Econ. 98, 1119–1158.

Hart, O.D., Moore, J., 2008. Contracts as reference points. Q. J. Econ. 123, 1–48.

Hayek, F.A., 1952. The Sensory Order: An Inquiry into the Foundations of Theoretical Psychology. University of Chicago Press, Chicago.

Hayek, F.A., 1967. Studies in Philosophy, Politics, and Economics. Routledge & Kagan Paul, London & Henley.

Hayek, F.A., 1973. Law, Legislation, and Liberty, Vol. 1, Rules and Order. University of Chicago Press, Chicago.

Hayek, F.A., 1976. Law, Legislation, and Liberty, Vol. 2, The Mirage of Social Justice. University of Chicago Press, Chicago.

Hayek, F.A., 1952/1979. The Counter-Revolution of Science, second ed. Liberty Press, Indianapolis, IN.

Hodgson, G.M., Knudsen, T., 2006. Dismantling Lamarckianism: why descriptions of socio-economic evolution as Lamarckian are misleading. J. Evol. Econ. 16, 343–366.

Holmström, B., 1982. Moral hazard in teams. Bell J. Econ. 13, 324–340.

Holmström, B., Milgrom, P., 1991. Multitask principal-agent analyses: incentive contracts, asset ownership, and job design. J. L. Econ. Org. 7, 24–52.

Holmström, B., Milgrom, P., 1994. The firm as an incentive system. Am. Econ. Rev. 84, 972–991.

Jensen, M.C., Meckling, W.H., 1976. Theory of the firm: managerial behavior, agency costs, and ownership structure. J. Financ. Econ. 3, 305–360.

Katz, M.L., Shapiro, C., 1985. Network externalities, competition, and compatibility. Am. Econ. Rev. 75, 424–440.

Lachmann, L.M., 1970. The Legacy of Max Weber. Heinnemann, London.

Lachmann, L.M., 1976. From Mises to Shackle: an essay on Austrian economics and the Kaleidic society. J. Econ. Lit. 14, 54–62.

Leibenstein, H., 1966. Allocative efficiency vs. 'X-efficiency. Am. Econ. Rev. 56, 392–415.

Leibenstein, H., 1980. Inflation, Income Distribution and X-Efficiency Theory. Croom Helm, London.

Leibensein, H., 1987. Inside the Firm: The Inefficiencies of Hierarchy. Harvard University Press, Cambridge, MA.

Lewis, D., 1969. Convention: A Philosophical Study. Harvard University Press, Cambridge, MA.

Liebowitz, S.J., Margolis, S.E., 1995. Path dependence, lock-in, and history. J. L. Econ. Org. 11, 205–226.

March, J.G., Simon, H.A., 1958. Organizations. John Wiley and Sons, Inc., New York.

Marschak, J., Radner, R., 1972. Economic Theory of Teams. Yale University Press, New Haven.

McAdams, R., Rasmusen, E.B., 2007. Norms in law and economics. In: Polinsky, M., Shavell, S. (Eds.), Handbook of Law and Economics, vol. 2. Elsevier, Amsterdam, pp. 1573–1618.

Milgrom, P., Roberts, J., 1990. Rationalizability, learning, and equilibria in games with strategic complementarities. Econometrica 58, 1255–1277.

Nelson, R.R., Winter, S., 1982. An Evolutionary Theory of Economic Change. Harvard University Press, Cambridge, MA.

North, D.C., 1990. Institutions, Institutional Change and Economic Performance. Cambridge University Press, Cambridge, MA.

North, D.C., 2005. Understanding the Process of Economic Change. Princeton University Press, Princeton, NJ.

North, D.C., Weingast, B.R., 1989. Constitutions and commitment: the evolution of institutions governing public choice in seventeenth-century England. J. Econ. Hist. 49, 803–832.

Ostrom, E., 1990. Governing the Commons: The Evolution of Institutions for Collective Action. Cambridge University Press, Cambridge, MA.

Ostrom, E., 2000. Collective action and the evolution of social norms. J. Econ. Perspect. 14, 137–158.

Ostrom, E., 2005. Understanding Institutional Diversity. Princeton University Press, Princeton, NJ.

Pagano, U., Rossi, M.A., 2004. Incomplete contracts, intellectual property and institutional complementarities. Eur. J. Law Econ. 18, 55–76.

Pagano, U., Vatiero, M., 2015. Costly institutions as substitutes: novelty and limits of the Coasian approach. J. Inst. Econ. 11, 265–281.

Polanyi, M., 1969. In: Grene, M. (Ed.), Knowing and Being: Essays by Michael Polanyi. University of Chicago Press, Chicago.

Roland, G., 2004. Understanding institutional change: fast-moving and slow-moving institutions. Stud. Comp. Int. Dev. 38, 109–131.

Rizzo, M.J., 2013. Foundations of The Economics of Time and Ignorance. Rev. Austrian Econ. 26, 45–52.

Schelling, T.C., 1960. The Strategy of Conflict. Harvard University Press, Cambridge, MA.

Shackle, G., 1972. Epistemics and Economics: A Critique of Economic Doctrines. Cambridge University Press, Cambridge, MA.

Simon, H.A., 1951. A formal theory of the employment relationship. Econometrica 19, 293–305.

Simon, H.A., 1987. Behavioral economics. In: Eatwell, J., Milgate, M., Newman, P. (Eds.), The New Palgrave: A Dictionary of Economics, Vol. 1. Macmillan, London, pp. 221–225.

Simon, H.A., 1991. Organizations and markets. J. Econ. Perspect. 5, 25–44.

Simon, H.A., 1969/1996. The Sciences of the Artificial, 3rd ed. The MIT Press, Cambridge, MA.

Simon, H.A., 2000. Public administration in today's world of organization and markets. PS: Political Science and Politics 33, 749–756.

Sugden, R., 1995. A theory of focal points. Econ. J. 105, 533–550.

Tirole, J., 1988. The multicontract organization. Can. J. Econ. 21, 459–466.

Williamson, O.E., 1975. Markets and Hierarchies: Analysis and Antitrust Implications. The Free Press, New York.

Williamson, O.E., 1985. The Economic Institutions of Capitalism. The Free Press, New York.

Williamson, O.E., 1996. The Mechanisms of Governance. Oxford University Press, New York.

Williamson, O.E., 2000. The new institutional economics: taking stock, looking ahead. J. Econ. Lit. 38, 595–613.

Young, H.P., 1998. Individual Strategy and Social Structure: An Evolutionary Theory of Institutions. Princeton University Press, Princeton, NJ.

Chapter 2

Institutions and the Economics of Behavior I

2.1 INTRODUCTION

This chapter illustrates how and why institutions matter. Institutions are constituted by collectives of people who associate with each other. Institutions constrain group members by forbidding some choices of actions, and empower them by making some alternatives possible. Rewards and penalties may be imposed on people who display certain behavior. Institutions are laws, conventions, and other formal and informal rules that give a durable structure to social interactions between the members of a population. Institutional structure shapes beliefs about how people behave. An institution as a system of rules perpetuates, if it elicits behavior that is consistent with the rules. Individuals are motivated to follow the rules of behavior.

Laws are, after deliberate procedures, promulgated by public institutions, such as legislatures, regulatory agencies, and courts; they are enforced by the power of the state. Law is conceptualized as a top-down mechanism to bring order and control violence through coercive enforcement. In the case of legal compliance, individual incentives most often refer to deterrence (Becker, 1968). Individuals are then deterred from criminal activities by a higher fine and by a higher probability of conviction. This ensures that people respond significantly to the deterrent incentives created by the criminal justice system. If so, punishment of criminals may be the best way to reduce the amount of crime. The choice of crime is less appealing when crime is more punished. In the rational choice approach, legal compliance is accounted for by the standard economic incentives of self-utility maximization; compliance is in one's self-interest. People obey the law only when their expected compliance utility is greater than their expected violation utility. Offenders are assumed to calculate their probability of success when evaluating criminal opportunities. Policy-makers try to raise the perceived costs of crime by increasing the certainty of direction and strengthen the severity of punishment. Thus, the rational choice approach predicts that the frequency of crimes will decrease, if the perceived costs associated with offending decisions are increased.

The Cognitive Basis of Institutions. DOI: https://doi.org/10.1016/B978-0-12-812023-1.00002-8

However, as Hirschman (1985) says:

Economists often propose to deal with unethical or antisocial behavior by raising the cost of that behavior rather than proclaiming standards and imposing prohibition and sanctions. The reason is probably that they think of citizens as consumers with unchanging or arbitrarily changing tastes in matters of civic as well as commodity-oriented behavior.

(Hirschman, 1985, p. 10)

Furthermore, North (1990) modifies the neoclassical view of human behavior as follows:

If our understanding of motivation is very incomplete, we can still take an important forward step by taking explicit account of the way institutions alter the price paid for one's convictions and hence play a crucial role in the extent to which nonwealth-maximizing motivations influence choices.

(North, 1990, p. 26)

Unlike legal rules, social norms are not supported by formal sanctions. In a bottom-up version, social norms emerge from and percolate through human interactions. The norms are social rules which reflect a historical trial-and-error process. Through the bottom-up process, the norm-produced social rules tend to reflect peoples' diverse and specialized situations. As the number of people complying with a norm increases over time, the expectations lead the norm to continue into the future. Ethical individuals make decisions within a self-imposed moral confine. Arrow (1979) argues that individuals will not always act in their own self-interest, and instead tend to conform to ethical codes that are more efficient economically. Akerlof (1980) suggests that social norms can impact the overall economy by affecting the choices that people make. The emergence of social norms generally focuses on the case in which no formal rule or centralized enforcement regime exists. Social norms are enforced privately in a decentralized way. They often direct individuals to undertake actions that are inconsistent with selfish actions. Norm violators are punished through nonlegal sanctions from social networks. The incentive to comply with norms derives not only from the enforcement of costly punishment by others, but also from reputation building for oneself (Teraji, 2013). The fear of punishment has a positive effect on cooperation. Fehr and Gächter (2000) indicate that many individuals are willing to punish unfair behavior at a personal cost.

2.2 OLD INSTITUTIONAL ECONOMICS

Economic approaches associated with Thorstein Veblen, John R. Commons, and Wesley Mitchell are known, together with their followers, as "American institutional economics" or "old institutional economics" (sometimes now "original institutional economics") (Bronfenbrenner, 1985; Rutherford, 2001).

These three founders of the old institutional economics were quite dissimilar, both in terms of their interests and in their style of analysis: Mitchell was mainly concerned with finance and business cycles; Commons with law and labor; and Veblen with the most diverse topics such as the leisure class, the theory of the firm, and Imperial Germany.[1] For them, mainstream economics was too narrow and individualistic: Mitchell felt that economic theory was a hindrance to an understanding of the business cycle, Commons felt that economic theory should be able to deal with "collective economic behavior," not only with individual economic behavior, and Veblen felt that the economic actor was reduced to something of a caricature, or a lightning calculator of pleasures and pains. After World War II, these approaches suffer a decline in interest, mainly because of the lack of solutions by institutional economists to the problems that emerges during the Great Depression.

Within Veblen's framework, institutions are more than merely constraints on individual action: they are embodied in generally accepted ways of thinking and behaving. In his *The Theory of the Leisure Class*, Veblen (1899) implies that individuals consciously seek to "excel in pecuniary standing," and so "gain the esteem and envy of (their) fellow-men" (p. 32). Veblen's concept of conspicuous consumption is described as a pattern of conduct that is intended to realize the goal of maintaining or enhancing an individual's social position. Individuals seek to excel in their manifestation of pecuniary ability or strength, in order to impress others and, thereby, gain their esteem or envy. Conspicuous consumption is a category of intentional actions whose goal is to bring about an improvement in others' opinions of oneself.

Veblen's central attacks on economics in general (marginal utility theory in particular) revolve around its unrealistic assumptions as to human nature, and its lack of an evolutionary approach to economic phenomena. For Veblen, the "economic man" is not intended as a competent expression of fact, but represents an expedient of abstract reasoning. Assumptions as analytical tools must be judged by their efficacy for the purpose at hand, rather than by standards of descriptive adequacy. Furthermore, economics must become an evolutionary science. Social ideas develop under changing economic conditions.[2] Evolutionary economics seeks to understand the changes in the whole social and economic landscape. Economic processes are to be investigated in terms of the cultural background from which they have emerged. Human nature and social structures are mutually constitutive. For Veblen, institutions are "in substance, prevalent habits of thought with

---

1. According to Dorfman (1958), Veblen and Commons seem to have had little or no personal contact, but Mitchell and Commons corresponded and functioned together professionally for some 25 years.

2. Veblen's criticism of the market is summarized in his vision of the dichotomy between industrial and pecuniary behavior (Veblen, 1904). Veblen viewed competition as the basic obstacle to better societal self-provisioning with goods and services.

respect to particular relations and particular functions of the individual and of the community" (Veblen, 1899, p. 190). To analyze economic behavior, we have to take history into account, and the changes that societies display over time. Ways of thinking and doing become tendencies. Institutions are shared prevalent habits of thought. Habits of thought are viewed as durable but adaptable propensities to think and act in particular ways. Habits of thought are ways and means of thinking and processing information, whereby individuals seek to make sense of, and act capably in, the real world around them (Lawson, 2015). Habits of thought enable us to economize on cognitive capacity and interpret information in a complex environment. Habits reside within individuals. The internalization of institutions through habits as carriers of rules shapes thought and practice. Even though people acquire a habit, they do not necessarily use it all the time. A habit is a disposition to engage in previously adopted or acquired behavior or thought, triggered by an appropriate stimulus or context (Hodgson, 2006).[3] Institutions evolve, with the more suitable ones being selected to deal with a particular environment. Hence, institutional change involves the simultaneous coevolution of both shared prevalent habits of thought and the habits of individuals. The sorts of adaptations or variations in the habits are conditioned by not just traits of human nature, but by the wider institutional context of any such responses (Lawson, 2015). Furthermore, according to Ambrosino (2016), Veblen's instinct—habit theory is close to Hayek's theory of institutions, in which a rule is defined as any behavioral disposition (including instincts and habits) giving rise to regularity in social behavior.

Following Commons (1934), an institution is defined as collective action in the control, liberation, and expansion of individual action. The ultimate unit of economic activity is a transaction. Commons (1934) is a pioneer in recognizing the importance of political and legal institutions in shaping transactions. For Commons, "[i]nstitutional economics openly avows scarcity, instead of taking it for granted, and gives to collective action its proper place of deciding conflicts and maintaining order in a world of scarcity, private property, and the resulting conflicts" (Commons, 1934, p. 7). To maintain order and resolve conflicts, collective action creates a system of government that lays down the working rules guiding individual action. Transactions are not the exchange of commodities in the physical sense of delivery. They are determined by the collective working rules of society. There are three types of transactions: the individual action is participation in bargaining, managing, and rationing transactions. Collective action ranges from unorganized custom to organized "going concerns." A going concern

---

3. For Hodgson (2003, p.164), "[h]abits themselves are formed through repetition of action and thought. They are influenced by prior activity and have durable self-sustaining qualities. Through their habits, individuals carry the marks of their own unique history."

is, for Commons (1934), a larger unit of economic investigation. A going concern is a joint expectation of beneficial bargaining, managerial, and rationing transactions.

2.3 NEW INSTITUTIONAL ECONOMICS

The "new institutional economics" seeks to provide an explanation which focuses on the micro-foundations of the link between institutions and economic outcomes. The new institutional economics tries to account for the emergence and persistence of institutions on the basis of their efficiency. Transaction costs, property rights, and contracts are key concepts of the new institutional economics. The first key concept is transaction costs, which is central to the new institutional economics. The new institutional economics has argued that the primary function of institutions is to reduce transaction costs. Indeed, a synonym for the new institutional economics is transaction cost economics. In neoclassical economic theory, transactions are assumed to be frictionless, and transaction costs are therefore absent. The new institutional economics holds that institutions emerge and persist when the benefits they confer are greater than the transaction costs involved in creating and sustaining them. For Coase (1937), there are costs to transactions in the market; a trader must find someone with whom to trade, get information on price and quality, strike a bargain, draw up a contract, and monitor and enforce the contract. Transaction costs encompass *ex ante* costs (before exchange) associated with search and negotiation, and *ex post* costs (after exchange) of monitoring and enforcement. Under certain circumstances, a firm can reduce these transaction costs by replacing bargaining between owners of the factors of production with coordination by a hierarchy. That is, if the cost of performing the transactions in the market is higher than in firms, then firms "substitute" markets in that task. Firms then emerge in a specialized exchange economy. Thus, transaction costs are defined as the cost of using the price mechanism. If transaction costs are zero, it does not matter for efficiency purposes whether economic activities are organized within firms or through the price mechanism. Firms are contractual entities, the shapes and operations of which are molded by the costs of using the price mechanism. Transaction cost economizing explains why markets are used in some cases, and hierarchies are used in other cases. Markets and firms can be regarded as alternative methods of coordinating production. Firms exist because there are costs of using the price mechanism, and these costs can be reduced by the use of an administrative structure.[4]

---

4. As the analysis of the firm deepens, the distinction between firm and market becomes unclear. As Klein (1983, p. 373) describes, "Coase mistakenly made a sharp distinction between intrafirm and interfirm transactions, claiming that while the latter represented market contracts the former represented planned direction. Economists now recognize that such a sharp distinction does not exist and that it is useful to consider also transactions occurring within the firm as representing market (contractual) relationships."

The assignment of property rights matters, because of positive transaction costs (Coase, 1960).[5] In neoclassical economics, owners whose rights are harmed by others can negotiate payments or transfers with zero transaction costs. Under such circumstances, the airport or factory and its neighbors could negotiate a bargain to allocate rights in a way that would maximize production regardless of the legal assignment of rights. Economic institutions are posited to evolve toward more efficient economic solutions through negotiations between interested parties. If costs of negotiating are not negligible, this transaction process is gradual. With positive transaction costs, the institutional setting (such as ownership, liability, the legal system, and the state) becomes central. The concept is developed by Alchian (1965) in which property rights are defined as a set of rights to take permissible actions to use, transfer, or otherwise exploit the property. Transaction costs are more generally considered to be the resources spent on delineating, protecting, and capturing property rights (Eggertson, 1990). The legal authority or system formally recognizes and enforces a person's right to property, and this generates an expectation that the government will take action if the right is violated. Demsetz (1967, p. 347) argues as follows: "[p]roperty rights are an instrument of society and derive their significance from the fact that they help a man form those expectations which he can reasonably hold in his dealings with others. These expectations find expression in the laws, customs, and mores of a society. An owner of property rights possesses the consent of fellowmen to allow him to act in particular ways. An owner expects the community to prevent others from interfering with his actions, provided that these actions are not prohibited in the specifications of his rights." Thus, in a world of positive transaction costs, governance structures matter for efficiency outcomes according to transaction costs theory (Coase, 1937), and legal rules matter for efficiency outcomes according to property rights theory (Coase, 1960). Property rights may be enforced by law, but are more often enforced by etiquette, social custom, and social ostracism. Ostrom (1990) further develops the concept by analyzing how the damaging effects of poorly-defined and enforced private property rights can be avoided through community governance.

The implications of Coase's concept of transaction costs become fully apparent with the work of Oliver Williamson.[6] Williamson (1975) develops the concept of transaction costs by using what he calls "the lens of contract"

---

5. According to Libecap (1986, p. 228), "[w]here transaction costs are high, as is often the case, the allocation of property rights is more critical, since transfers are less fluid. In those circumstances, the existing property rights arrangement has profound and enduring effects on production and distribution."

6. According to Williamson (2015, p. 223), the essence of Coase is this: (1) push the logic of zero transaction costs to the limit; (2) study the world of positive transaction costs; (3) because hypothetical forms of economic organization are operationally irrelevant, and because all feasible forms of organization are flawed, assess alternative feasible forms of organization in a comparative institutional way; and (4) because the action resides in the details, study the microanalytics of contract, contracting, and organization.

to explore how managers choose between transacting in the market versus transacting within the firm in a world of incomplete contracts and opportunism. Parties to a contract might defect from the spirit of cooperation when the stakes are large. Williamson (1985) provides a "cognitive map of contract" where transaction costs, agency, and property rights theories are placed under a common branch of an economic efficiency theory of contracting. Williamson (1985, p. 41) suggests that the exchange parties mitigate the contractual hazard by "assigning transactions (which differ in their attributes) to governance structures (which are the organizational frameworks within which the integrity of a contractual relation is decided) in a discriminating way." Transactions involve costs which are incurred by finding a counterpart, drawing up a contract, or monitoring task completion. In these circumstances, vertical integration results from economizing on transaction costs. It might be more efficient for the parties to carry out the transaction under the umbrella of a single firm. Bargaining is replaced by authority within a single firm. Transaction cost economics presupposes that organizations continuously search for the most efficient governance structure.

Transaction cost economics focuses on the following three characteristics:

1. the degree of uncertainty of transactions;
2. the frequency with and scale on which transactions take place; and
3. the specificity of assets needed for production.

Within a hierarchal organization, the uncertainty of transactions relates to such aspects as the size of the organization, the complexity of its tasks, and the amount of information to be processed. The main characteristic which differentiates one transaction from another is asset specificity, such as knowledge, sites, and products generated. Williamson (1985) argues that a transaction not governed by market forces should be controlled by, for example, the hierarchical organization if a high degree of asset specificity, high frequency, and high uncertainty coincide at a certain level of bounded rationality and opportunism.

Williamson (1985) introduces the behavioral assumptions of bounded rationality and opportunism in contrast to neoclassical maximizing rationality and simple self-interested behavior with an appeal to the realism of assumptions. Individuals do not behave opportunistically all of the time, but may behave opportunistically some of the time. Bounded rationality, as Herbert Simon (1957) argued, refers to the fact that agents are assumed to be intendedly rational but only limitedly so. Agents have limited possibilities to perceive, store, and process information without error. They also have limited capacities to articulate and communicate their knowledge in a detailed and clear way. These limitations induce some costs in decision-making. Bounded rationality prevents individuals from making complete contracts. Since boundedly rational actors are unable to completely foresee all the future contingencies that affect the execution and fulfillment of contracts, these contracts are inevitably incomplete, and the contractual parties are at risk of being subject to opportunism. Cognitive limitations lead people to implement incomplete contracts in complex environments. With the belief in an economic agent's limited ability to engage in farsighted planning and accurately forecast the impact of potential contractual hazards, there is no reason to believe that the contractual parties in any given transaction will choose the economizing governance structure. Therefore, decision-makers aim to minimize the negative effects of contractual hazards to adopt particular governance structures, such as hierarchy (unified ownership) and long-term contracts, which safeguard against contractual incompleteness.

However, for Hodgson (1988, 2006) in the tradition of the "old institutional economics," the idea of transaction costs is not a sufficient basis for a theory of the nature of the firm. The function of the firm is not simply to minimize transaction costs, but to store and reproduce a large number of gene-like habits and routines. Generally, institutional change can cause changes in habits of thought and action. The old economic institutionalism examines patterns and regularities of human behavior, or habit, as the basis for its institutional analysis. Institutions partially mold and enable individual habits, preferences, and values. For Hodgson, habits represent the key link between institutions and individual behavior:

> By structuring, constraining, and enabling individual behaviours, institutions have the power to mold the capacities and behaviour of agents in fundamental ways: they have a capacity to change aspirations instead of merely enabling or constraining them. Habit is the key mechanism in this transformation. Institutions are the social structures that can involve reconstitutive downward causation, acting to some degree upon individual habits of thought and action.
>
> (Hodgson, 2006, p. 7)

Repeated behavior is important in establishing a habit. Thus, "[h]abits and traditions within the firm are necessarily more enduring because they embody skills and information which cannot always or easily be codified or made subject to rational calculus. If habits become rules, a situation is created in which rules, in turn, foster the emergence of habits. In this self-reinforcing cycle, habits and institutions work symbiotically. What the firm achieves is an institutionalization of these rules and routines within a durable organizational structure" (Hodgson, 1988, p. 208).

Transaction costs provide a different explanation for economic development in Coase (1998):

> [T]he costs of exchange depend on the institutions of a country: its legal system, its political system, its social system, its educational system, its culture, and so on. In effect it is the institutions that govern the performance of an economy, and it is this that gives the "new institutional economics" its importance for economists.
>
> (Coase, 1998, p. 73, original emphasis)

Poorly-constructed institutions are identified as a source of higher transaction costs, increased informational asymmetry, and uncertainty; all of which reduce the potential for and likelihood of trade and exchange. Without the concept of transaction costs, it might be impossible to understand how an economic system works. Transaction costs are not costs like production costs, but the costs of running the economic system of exchange. They are the aggregate of search, negotiation, and enforcement costs. If there are alternative ways to organize transactions, we must compare the costs and benefits of the different arrangements in order to answer the question: which organizes transactions at the lowest possible cost? Different specifications of property rights arise in response to the economic problem of allocating scarce resources, and the prevailing specification of property rights affects economic behavior and economics outcomes. An inefficient initial allocation of property rights may result in fixed bargaining positions that are difficult to reconcile, which leads to persistent suboptimal contracting outcomes.[7]

Nonmarket institutions can be thought of as human devices governing agency relations. Greif (2006) investigates the institutional foundations of medieval economies in Europe and North Africa, and studies the organization of the Maghribi traders, a group of Jewish traders in the Mediterranean in the 11th century. The 11th-century Maghribi traders organized into coalitions to enforce contracts over overseas exchanges. The encounter between delegators (i.e., principals) and delegates (i.e., agents) involved interaction that called for resolution. The Maghribi traders employed agents over long distances. Efficiency was enhanced by letting overseas agents transact business with capital they did not own. Trading was characterized by asymmetric information, slow communication technology, inability to specify comprehensive contracts, and limited legal contract enforceability. Agents are in possession of the Maghribi merchants' capital, and can behave opportunistically to take advantage of this fact. The legal system was not used to mitigate the principal–agent commitment problem, mainly due to the asymmetric information that characterized agency relations; many realizations could not be directly observed, either by the merchant or by the legal system. Without a supporting institution, the Maghribi merchants anticipating opportunistic behavior would not operate through agents, and mutually

---

7. According to Rebitzer (1993), the following three ideas distinguish "radical political economy" from other economics approaches: (1) The role of politics in economics: a "political" dimension is introduced in the sense that key economic processes depend on institutional arrangements that enforce the power and authority of a dominant group. (2) The desirability of institutional change: the institutional arrangements that enforce the power and authority of dominant groups are less efficient than some feasible alternative arrangements. Realizable, superior alternatives to existing social relations will not evolve spontaneously through the market. (3) The historical contingency of economic structures: a notion of path dependency is introduced into the analysis of economic structures that are the contingent result of particular historical developments. These economic structures do not have *a priori* claim to optimality or efficiency.

beneficial exchanges would not be carried out. The resulting exchange took the form of certain norms. The Maghribis were an ethnic and religious community with a trading club whose practice was to hire only member agents, and never to hire an agent who had cheated another member. Information transmitted in correspondence among a commercial and social network of the Maghribi traders supported a reputation mechanism that successfully dissuaded cheating. The Maghribi merchants responded by applying a norm of multilateral punishments against agents who cheated. The coalition was organized through an informal rule indicating that once an agent has cheated another member, he should not be hired in the future. Each merchant in the network was motivated to punish cheaters by the expectation that others would also do so; thus, the punishment was self-enforcing. The reputation-based institutions of the Maghribis were effective in enforcing commercial contracts. Reputation mechanisms alter the likelihood of future transactions with potential trading partners if an agent defaults on an agreement. These mechanisms fundamentally rely on institutions that capture and disseminate information about an agent's performance to a set of potential trading partners (such as coalition members). Thus, the Maghribis relied on collectivist organizations based on multilateral punishment mechanisms (informal codes of enforcement and information sharing).[8] The coalition created a coordination device that promoted efficiency by saving on negotiation costs, allowing flexibility in the horizontal agency relationships.

2.4   INCENTIVES AND EFFICIENCY

Ever since Berle and Means (1932) recognized that the separation of ownership and control impacts firm values, economists have focused on ways to mitigate the agency problems that arise between shareholders and managers. In the 1950s, managerial economics began to analyze how the separation of ownership and control affected the organization of the firm (Baumol, 1959; Marris, 1964). Baumol (1959) postulates that oligopolists maximize sales rather than profits. Marris (1964) argues that managers would pursue growth in sales or assets over profits, in order to increase their own salaries or status. The manager is no longer an automaton who acts in the owners' interests. The conflict of interests between managers and owners is the focus of the principal—agent literature. The principal hires employees to do part of the work. They are paid wages and in exchange usually, though not always, relinquish claims on the profits. The contract to which they agree specifies their duties, their rewards, and the rights of the principal to monitor their

---

8. Milgrom et al. (1990) interpret the role of the merchant guild in the Champagne fairs during the 12th and 13th centuries; the merchant guild may be conceived as an institution to create proper incentives to gather information, make agreements enforceable, secure property rights, and solve commercial contract disputes.

performance. The agency problem is traditionally founded on the principle that managers are egoistic and must be given incentives to act in the best interests of the firm. According to Brennan (1994), however, there is a significant difference between being rational and being self-interested.

Moral hazard is present when one party in a transaction takes actions that a partner cannot observe. In a setting with moral hazard, the informed party has some control over the unobserved attribute. Moral hazard often arises in organizations. It is the problem of agency costs, rather than the law of diminishing returns, that limits the efficient size of firms. Diminishing returns limit merely how much of a single product a firm can produce efficiently. As an organization expands, it will be difficult to correlate the work of a particular employee with the value of the organization's output. Employees have greater scope to engage in behavior that serves their own interests, but not those of the firm. Resources into monitoring of actions are invested in order to use the information in the contract. In order to induce desirable behavior, incentive schemes, which tie rewards and punishments to performance, are often designed.

In team production, output is the joint product of several agents' contributions. Team members interact extensively, and they are accountable as a team for economic outcomes. Alchian and Demsetz (1972) suggest that there is a source of gain from cooperative activity involving working as a team. The combined entity does enjoy opportunities which were impossible for the separate parts to consider. However, team production suffers from a peculiar problem. Compensation must be based on team production, and individual contributions to output cannot be easily identified. Each individual knows that his or her effort has some impact on the team's reward, but that the reward is split among all the members. When the individual shirks by reducing his or her effort expenditures, the savings in effort accrue only to him or her, and the resulting losses in team reward are borne largely by the others. Each member tends to find it in his own best interests to engage in some degree of shirking. Alchian and Demsetz (1972) discuss the metering problem through the example of a university where "the faculty use office telephones, paper and mail for personal uses beyond strict university productivity" (Alchian and Demsetz, 1972, p. 780). For Alchian and Demsetz (1972), organizational control over work can eliminate such practices and increase pecuniary rewards.[9] Holmström (1982) develops the classic team model in which moral-hazard and free-rider problems abound when the effort choices of the team members are unobservable. The output of the team may be readily observable, even though the effort of each member of the team may be completely obscure. Each agent would prefer to cut back

---

9. Williamson (1975) criticizes this conclusion, and underlines that there are interactions between the attitudes of workers with regard to transactions.

effort, while hoping that the other members of the team pick up the slack. The efficient outcome is not a Nash equilibrium, because each would prefer to cut back effort when the other members of the team are working at an efficient level.

Team membership per se may affect the behavior of participants. It is possible that peer norms come about through processes of mutual influence. Peer effects have the potential to internalize externalities in workplaces. Some studies incorporate social concerns into the analysis of behavior inside firms. Kandel and Lazear (1992) stress the role of peer pressure among team members. They argue that teams alleviate costly monitoring of workers by relying on internal motivations through peer pressure. Social ties among co-workers are advantageous, because social sanctions are effective in punishing those who deviate from norms of high effort.

An organization can be inefficient because its outputs lie inside the efficiency frontier. Harvey Leibenstein (1966) dubs this as "X-inefficiency." There is no need to suppose that individuals actually use inputs as effectively as possible. According to Leibenstein (1966), individuals supply different amounts of effort, where effort is a multidimensional variable, under different circumstances. The difference between maximal effectiveness of utilization and actual utilization is considered as the degree of X-inefficiency. For Leibenstein (1976, 1979), the micro–micro problem is the study of what goes on inside a "black box." Leibenstein's approach aims to explain organizations' external activities through the study of their internal processes. In conventional economic theory, effort is assumed to be maximized in its quantity and quality dimensions, irrespective of institutional conditions. According to Leibenstein (1979), the main argument against the maximization postulate is an empirical one—people frequently do not maximize:

> In considering alternatives to cónventional micro theory, one of the questions that arises is how to handle the usual maximization assumption. At present the maximization postulate has an unusually strong hold on the mind set of economists. ... Suffice it to say that in my view the belief in favor of maximization does not depend on strong evidence that people are in fact maximizers.
>
> (Leibenstein, 1979, p. 493)

Economic agents are not all the same and, therefore, are characterized by different objective functions. Many firms remain X-inefficient, even when most of their members know the solutions to perform X-efficiently. X-inefficiency is the failure of a productive unit to fully utilize the resources it commands, and hence attain the familiar production possibility frontier. The introduction of effort variability allows for the existence of the X-inefficient economy. Leibenstein's analysis presents a reasonable vision of human behavior within the organizational context, where X-inefficiency can persist stubbornly. Individual motivations and interactions between individuals are important.

Altman (2000) introduces effort discretion into path-dependent modeling. Given effort discretion, incentives need not exist for agents to adopt superior economic regimes. The introduction of effort variability allows for the existence of suboptimal equilibrium in the long run. The existence of high and low productivity regimes might be a product of history. Furthermore, culture is introduced as a determinant of labor productivity (Altman, 2001). Cultural factors become either facilitators or impediments to X-efficient production, because the quantity and quality of work effort are affected specifically by the cultural milieu of the economic agent.

2.5   TRUST AND PROPERTY RIGHTS

The study of property rights has played a major role in the explanation of economic performance. Property rights are the rights to use, to earn income from, and to transfer or exchange assets and resources (Libecap, 1989). The introduction of property rights in Coase (1960) is accompanied by an emphasis on transaction costs as determinants of how well property rights are delineated and enforced. Property rights are important to make our economic system work, and are a motivating force that determines economic productivity. When property over economic resources is insecure, people typically have to pay transaction costs to enforce their claims on such resources. Transaction costs are incurred by individuals attempting to protect property rights. The presence of positive transaction costs is what makes the study of property rights important. Insecure property rights can be appropriated by others. We need a security mechanism for property rights to support the productive capacities of a society.

How are property rights secured? Many economists and political scientists have emphasized the state's role in creating, defining, and enforcing property rights. They claim that property rights are not well protected in the absence of the state. Thomas Hobbes ([1651] 1991) offers us the intuition of "perpetual war of every man against every man." In the state of nature, there is no property. Political authority is needed, because it gives us the security of property rights.

Property rights are conceptually broad, and we should consider not only the legal aspect of property rights, but the social conventions that govern behavior. Whether a formal institution can achieve high economic outcomes depends, to a large extent, on whether it is supported by informal institutions such as customs, traditions, and codes of behavior. Property rights enforced by the state are not the only motivating force to determine productivity. Property rights can emerge as a result of the efforts of private agents to reduce uncertainty in their socio-economic environment. Demsetz (1969) argues that property rights develop to internalize externalities when the gains of internalization become larger than the cost of internalization. There are many historical and contemporary examples of private enforcement. We focus on the possibility that more secure property rights are conducive to more efficient social allocations in the absence of political institutions of enforcement.

For Hume ([1739−40] 1985), property rights are similar to other useful conventions that are adopted spontaneously.[10] Furthermore, according to Hayek (1983), private property was never "invented," in the sense that people foresaw what its benefits would be, but spread because those groups who by accident accepted them prospered and multiplied more than others. In Hayek's discussion of the emergence of property, the group as a whole enjoys the advantage of property formation. Collectively, societies are better off when their members cooperate with one another to achieve common goals. Individuals, however, face incentives to behave selfishly. Each agent faces a particular situation, in which the action that best serves one's self-interest depends on the actions taken by others. The situation can be viewed as the representation of collective action problems.[11]

The term "social capital" is a comprehensive explanation for why some communities are able to resolve collective action problems cooperatively, while others are unable to bring people together for common purposes. Putnam (2000, p. 19) explains "[w]hereas physical capital refers to physical objects and human capital refers to properties of individuals, social capital refers to connections among individuals." Social capital is the structure of committed relations among individuals that encourages productive activities. It is an aggregate concept that has strong effects on perceptions of the trustworthiness of others. Trust helps determine the effectiveness of committed relations. A lack of trust in relations has negative consequences, because almost all economic transactions have been embedded within relations. By contrast, high levels of social capital will enable agents to deal with each other in a more trustworthy manner. Trust lubricates cooperation and reduces the transaction costs among people. This section argues the notion of trust as a measure of social capital. Trust can be viewed as an expectation or a "cognitive state" that other people will not exploit a vulnerability each agent has created by themselves. Social capital can be thus defined at the individual level as the individual's perceived sense of trust. Furthermore, social capital generates beneficial outcomes through shared trust in a community. Property rights are the endogenous outcomes of collective choices. For property rights to emerge, each agent needs to trust that others will not take away the productive outcome he or she has produced. Then, security must be so strong

---

10. See Sugden (1989) for a game theoretic interpretation of Hume. A convention emerges when individuals believe that others are following the same convention.

11. See Ostrom (2000) for collective action problems. When goods are public, an individual's rational choice is not to contribute to producing the goods. If too many individuals adopt this strategy, then the goods will no longer be available to any individual.

that it would not be worth attacking another person in the community. In the aggregate level, social relationships support informal institutions. The model assumes individual rationality, but it is shown that property rights do not necessarily emerge. The extent to which property rights are perceived to be stable can differ among communities.

In this section, I analyze property rights theory in the behavioral economic model, in which the quantity and quality of effort is a discretionary variable according to the X-efficiency approach. The X-efficiency approach sees inefficiency as a major problem and postulates the existence of inert areas. An inert area is a range of potential effort levels which is set by group or social norms. Agents have discretionary spaces within which to choose their activities, that is, how much they work in production. Then, the effort inputs into the production process are not necessarily maximized. In other words, the model allows for the possibility of suboptimal outcomes where X-inefficiency may be sustainable. Secure property rights have to emerge in order to obtain efficient economic outcomes in human interactions.

As a result, two equilibria are likely to be realized. If it is possible to obtain mutual trust to commitment in the community, secure property rights emerge and the economic activities are efficient. If not, cognitive systems exhibit suboptimal performance within a variety of discretionary spaces to choose activities. One implication of this analysis emphasizes the importance of trust to ensure economic efficiency. Cognition matters in model building, since the perceived framework affects economic performance.

There is a large literature on economic models of conflict and production (Baker, 2003; Grossman, 2001; Grossman and Kim, 1995; Hirshleifer, 1995; Muthoo, 2004). The model presented in this section differs from these models in three points. First, although these models present the basic notion of a state of nature, they do not capture social capital in the community. Social capital affects economic performance via enforcing informal institutions. Through taking the notion of trust as a measure of social capital in the community, this framework explores the origins of property rights. Second, the model emphasizes effort variability in production. Individuals supply different amounts of effort, in which effort is a multidimensional variable under different circumstances. Collective activities become either facilitators or impediments to efficient production, because the quantity and quality of work effort is affected specifically by the level of security in the community. Finally, the model attempts to incorporate the cognitive function into property rights theory about efficiency. Cognition belongs subjectively to one individual agent. Moreover, cognition is closely related to culture. Culture, in the form of beliefs and ideologies, influences the specific way in which individual agents act. Individuals with common cultural backgrounds will share cognitive systems. Some types of mental models, which individual cognitive systems create to interpret the environment, are shared intersubjectively.

2.5.1 Overview of the Model

The model considers a particular community that is composed of a number of agents. The structure of interactions in communities contrasts with that of markets. Market interactions are characterized by anonymity among interacting agents. In communities, on the other hand, the probability that members continue to interact is high. Thus, the model considers communities characterized by non-anonymous relationships among members. In order to consider the emergence of property rights, the model does not assume preexisting political institutions of enforcement. Instead, it emphasizes the role of community governance in providing informal enforcement of property rights.

The economy evolves over four stages, $t = 0$, 1, 2, 3. At Stage 0, all agents simultaneously decide whether or not to commit themselves to securing property rights in the community. At Stage 1, each agent decides whether to be a worker or to be a stealer. If an agent chooses to be a worker, he or she chooses how much effort to exert in production at Stage 2. At Stage 3, payoffs are realized.

At Stage 0, each agent decides whether or not to make an unproductive investment for "community-level" security. In the absence of legal protection, some fraction of property may be taken away. An agent has to commit themself to securing property rights before they pay a cost. The investment is thus specific to each agent. Community-level security will only take place in a trustworthy environment, in which all agents are expected to commit themselves to protect property rights. Generally, commitment is considered as a binding tie between an individual and some other social entity. For a relationship to continue, each agent must incur some cost. The tie of commitment must be strong enough for the relationship to succeed. Mutual commitment is important, because cooperation requires participants to contribute inputs. This framework presents community-based programs of governance. More specifically, community-level security consists of many activities committed, all of which must be successfully completed for property rights to have full value. Effective enforcement in the community is due to strong private protection completed by individual agents. The community as a whole enjoys security of property. Without trust, commitment can waver.

The model permits individual agents to choose the degree to which they trust others' commitment for protection. Trust is closely linked with cooperation. Basically, those who trust others to "do the right thing" (cooperate) will show high rates of cooperation, compared to those who have low trust of others. From a transaction cost perspective, trust is involved in transactions in which an agent faces some loss of opportunity if the transaction is forgone for lack of trust. Mistrust is the possibility that one will receive no payoff from one's actions. Social engagement is considered a public good, and it may be undersupplied. Individuals depend to a large extent on informal norms for security. Social capital generates positive externalities that are achieved through shared trust. This framework shows that trust is a crucial factor for facilitating the creation of security via enforcing informal institutions.

At Stage 1, each agent may choose to be a stealer. Under what conditions is security kept? In the absence of political institutions of enforcement, property rights have to be self-enforcing in order to be secure. This self-enforcing condition requires that each agent has no incentive to be a stealer. Each agent has no incentive to exploit the other in the community, if all agents are expected to commit themselves to protect property rights.

This model focuses on trust in the emergence of property rights in the community. Trust is an expectation or a "cognitive state" that one will not be exploited by another. Trust reduces the transaction costs among people. Instead of monitoring others, individuals are able to trust others to act as expected. Decisions on whether to trust or not are based on the ability to predict what other people are going to do. The model can determine the consequences of different levels of trust on economic performance.

At stage 2, workers supply different amounts of effort in production, where effort is a multidimensional variable. It might be desirable to show how an equation describing effort can be derived from microeconomic principles. As the behavioral economic model attempts to analyze the possibility of suboptimal behavior in production, it does not employ a production function whose dependent variable is maximum output. The introduction of effort variability allows for both efficient and inefficient equilibrium solutions in the economy. This model presents theoretical insights for understanding the importance of trust as a determinant of economic performance. High-trust environments produce more output than low-trust ones. For a sufficient amount of trust, individual agents choose the high effort level in production. A low-trust trap can exist, in which agents choose the low-effort level.

One implication of this analysis emphasizes the importance of belief systems to ensure economic efficiency. Individual belief systems are created to interpret the social environment. Culture, in the form of beliefs and ideologies, influences the specific way in which agents perceive. Trust is cultural, and it is deeply embedded in socially shared understandings. Each agent sees others as predictable, and responds by developing trust in them. If other people are trustworthy, then commitment to the relationship is fostered. Trust facilitates attachments to others, and it influences the development of commitment. Thus, higher levels of trust in the community can achieve collective goals that cannot be accomplished by individuals alone.

2.5.2 The Model

Consider a community consisting of a set of identical agents $N = \{1, 2, \ldots, n\}$, where n is finite. There are no preexisting institutions of enforcement. Agents take actions with individual beliefs as to what the other agents will do. I consider the following four-stage model.

Stage 0: Whether or not to commit oneself to securing property rights.

Protecting property rights is costly, so that unproductive investments are required for their emergence and security in the community. At Stage 0, agent i decides what fraction q_i of his or her resource to use to protect property rights. This fraction shows how much the agent commits himself or herself to protecting property rights in the community. In this model, q_i is assumed to be in an interval [0, 1], where 1 is the highest level that each agent can commit himself or herself to securing property rights. This interval [0, 1] is constant for each agent. Thus, if q_i is 0.98, agent i has a 98% chance of committing to the security activity, and a 2% chance of not committing to it. Each agent simultaneously decides what fraction to commit himself or herself to protecting property rights before he or she pays a cost. The model supposes that each agent incurs the protection cost at Stage 2.

This framework considers an alternative security system of property rights. Costs must be incurred by agents in order to enforce property rights. Collectively, communities are better off when their members cooperate with one another to achieve common goals. In this framework, community-level security will only take place in the trustworthy environment, in which all agents are expected to commit themselves to protect property rights. In the absence of legal protection, some fraction $1 - p$ of productive outcome may be taken away. The fraction that the agent is able to appropriate is the remaining p. The model presents an explicit security mechanism that determines and sustains property rights in the community. The security mechanism permits individuals to choose the degree that they trust others. It reflects common rules that place collective interests above those of individuals. Contributing to security provision jointly generates benefits for members in the community. The security level, $p(q_1, q_2, \ldots, q_n)$, is the fraction of outcome that each agent can appropriate, and it is an increasing function in each q_i, $i = 1, 2, \ldots, n$. More commitment gives individuals the confidence to invest in collective activities. According to people, the investments greatly increase their feeling of security. Effective enforcement in the community is due to strong private protection.

The model permits agents to choose the degree to which they trust others in the community. Trust can be thought of as the subjective probabilities with which an agent assesses that another agent or group of agents will perform a particular action. Specifically, each agent i expects the others' commitment levels. Every agent is supposed to have the same ability for expectation, and to share cognitive systems inter-subjectively. Therefore, the model introduces an identical q_j^*, that is, agent i's expectation or perception of q_j for $j = 1, 2, \ldots, n$ and $j \neq i$. Then, q_j^*, which is also in an interval $[0, 1]$, is the subjective probability that agent i has for agent j's commitment level, or the degree that agent i trusts agent j.

Identical to agent i's "perceived" security level, $p^*(q_1^*, q_2^*, \ldots, q_i, \ldots, q_n^*)$, depends on i's own q_i and i's expectation of q_j for $j = 1, 2, \ldots, n$, and $j \neq i$. It is also increasing in q_i and q_j^* for $j = 1, 2, \ldots, n$ and $j \neq i$. Whether individuals participate in the provision of security services determines a degree of perceived secure property.

More specifically, the perceived security level is captured by:

$$p^*(q_1^*, q_2^*, \ldots, q_i, \ldots, q_n^*) = q_{-i}^* q_i, \tag{2.1}$$

where

$$q_{-i}^* = q_1^* \cdots q_{i-1}^* q_{i+1}^* \cdots q_n^* = \prod_{j \neq i} q_j^*.$$

Each agent is supposed to share the same belief systems about trust, that is, the same degree that he or she trusts others in the community. Thus, each agent has an identical perceived security level. Furthermore, (2.1) represents a particular kind of complementarity, where protection consists of a number of activities all of which have to be completed by agents in the community.[12] Any agent who does not trust others to meet expectations would be foolish to commit to community-level security. If no commitment occurs in any of the activities, the level of security is reduced to zero. The security process consists of many activities committed by different agents. Complexity can be interpreted as the number of activities, n, involved in the process.

Stage 1: To be a worker or to be a stealer.

An agent simultaneously decides whether he or she attempts to steal the other agent's output or not. Let w be the agent's payoff, the wage rate that he or she receives, at stage 3. In the absence of legal protection, some fraction $1 - p^*$ of it may be taken away from an agent. If agent i chooses to be a stealer, his or her expected payoff is $(1 - p^*)w$. If the agent chooses to be a worker, he or she has to exert some amounts of effort in production at Stage 2.

Stage 2: How much effort to exert?

Each worker chooses how much effort to exert in production. Workers' effort levels may depend on wages, because wages affect their morale (Akerlof, 1984). Effort is a continuous variable. It is determined from the utility-maximizing behavior of individual agents. Each agent i chooses to exert effort e at a cost $(1/2)e^2$. Then, agent i's expected payoff is p^*w. Furthermore, each agent i incurs a protection cost for his or her payoff, kp^*w, at Stage 2, where $0 < k < 1$ is a parameter.

---

12. In Cooper and John (1988), a strategic complementarity arises when the optimal strategy of one agent depends (positively) on the strategies of other agents. If there is no way for agents to coordinate their actions, the economy has a low level of activity.

Stage 3: Payoffs are realized.

Production technology is linear. With the number of individuals employed, n, each putting e units of effort into production, output is en. The profit is given by $en - wn$. With the zero-profit condition, we have:

$$e = w. \tag{2.2}$$

Effort per unit of labor input is positively related to the wage rate. Equation (2.2) indicates that a higher wage rate increases the work effort.[13] The model allows the level of work effort to be varied in a continuous manner. Effort discretion as a choice variable plays an important role in the relationship of material incentives to productivity. There are no monitoring problems, thereby inducing individual agents to pay high efficiency wages to promote work effort in production.

2.5.3 Analysis

This section analyzes the agent's commitment problem for securing property rights in steps. In each case, the model is solved by backward induction. The analysis goes from the Stage 2 effort decision in production, to the Stage 1 decision whether or not to work, to the Stage 0 commitment problem. Having done that, we can infer under what circumstances each agent respects and protects property rights.

At Stage 2, each agent chooses the effort intensity in production. Agent i chooses to exert his or her effort, e, at a cost $(1/2)e^2$. Then agent i's expected payoff is p^*w. Furthermore, agent i incurs a protection cost, kp^*w. Thus, the agent's welfare is $p^*w - (1/2)e^2 - kp^*w$. Using (2.2), each agent i chooses e to maximize $(1 - k)p^*e - (1/2)e^2$ for a given p^*. Hence, we have:

$$e = (1 - k)p^*. \tag{2.3}$$

At Stage 1, an agent decides whether to be a worker or to be a stealer. The stealer's welfare is then $(1 - p^*)w = (1 - p^*)e$ from (2.2). Property rights have to be self-enforcing in order to be secure. That is, each agent has no incentive to exploit the others in the community. The self-enforcing condition can be written as:

$$(1 - k)p^*e - (1/2)e^2 \geq (1 - p^*)e, \tag{2.4}$$

for a given p^*.

---

13. The effect of the wage on effort can be justified with familiar efficiency wage arguments like the one in Shapiro and Stiglitz (1984).

At Stage 0, each agent chooses his or her commitment level for security in the community. Each welfare is given by $(1 - k)p^*e - (1/2)e^2 = (1 - k)$ $q^*_{-i} q_i e - (1/2)e^2$ from (2.1). Then every agent i chooses q_i to maximize his or her welfare $U = (1 - k)q^*_{-i} q_i e - (1/2)e^2$ subject to the self-enforcing constraint (2.4).

Using (2.1) and (2.3), (2.3)' can be rewritten as:

$$e = (1 - k)q^*_{-i} q_i, \tag{2.3'}$$

for every agent i.

Inserting the effort level into $e = (1 - k)q^*_{-i} q_i$ into the agent's welfare yields $U = \{(1 - k)q^*_{-i} q_i\}^2 (1/2)\{(1 - k)q^*_{-i} q_i\}^2 = (1/2) \{(1 - k)q^*_{-i} q_i\}^2$. Then $dU/dq_i = \{(1 - k)q^*_{-i}\}^2 q_i \geq 0$. The agent's welfare increases, provided that the self-enforcing constraint is satisfied.

Furthermore, using (2.1) and (2.3)', the self-enforcing condition, (2.4), can be rewritten as:

$$\left(\frac{1}{2}\right)\{(1-k)q^*_{-i} q_i\}^2 \geq (1 - q^*_{-i} q_i)(1 - k)q^*_{-i} q_i \quad \text{or}$$

$$(1 - k)q^*_{-i} q_i \left\{ \left(\frac{1}{2}(3 - k)q^*_{-i} q_i - 1\right) \right\} \geq 0. \tag{2.4'}$$

We allow for the possibility of regimes in which the agent can extract a rent. Suppose that the self-enforcing constraint (2.4)' does not bind for $q_i = 1$. This occurs when $(1/2) (3 - k) \cdot q^*_{-i} - 1 > 0$. This condition can be rewritten as:

$$q^*_{-i} > \Lambda \equiv \frac{2}{3 - k},$$

Here, Λ is the lowest value of q^*_{-i}. In this case, agent i will trust others, provided the perceived probability that they will commit to securing property rights is more than some critical level Λ. For $q^*_{-i} > \Lambda$, the agent's self-enforcing constraint (2.4)' is slack for any q_i. Since $dU/dq_i > 0$, $q_i = 1$ is the solution. Furthermore, from (2.3)', $e = (1 - k)q^*_{-i} > 0$ for some $q^*_{-i} > \Lambda$.

On the other hand, for $q^*_{-i} \leq \Lambda$, the agent's self-enforcing constraint, (2.4), binds. Then $q_i = 0$ is the solution. Furthermore, from (2.3)', $e = 0$.

The system has two types of equilibria. If it is possible to obtain mutual understanding about the community-level security, property rights emerge privately and the economic activities are efficient. If not, cognitive systems exhibit suboptimal performance within a variety of discretionary spaces to choose economic activities. Thus, high-trust communities produce more output than low-trust ones. High levels of trust are crucial to sustain positive output. The model shows that a low-trust trap can exist. The model can determine the consequences of different levels of trust on economic performance.

To summarize:

Proposition 2.1: (Teraji, 2008).

(1) *For $q^*_{-i} > \Lambda$, $q_i = 1$ and $e = (1 - k)q^*_{-i}$. That is, for high levels of trust, agents commit them selves to securing property rights. Then, they exert positive amounts of effort in production.*

(2) *For $q^*_{-i} \leq \Lambda$, $q_i = 0$ and $e = 0$. That is, for low levels of trust, agents do not commit themselves to securing property rights. Then, they exert no effort in production.*

The model has characterized the socio-economic environments in which trust is high or low. Agents determine the degree to which they trust other people by expecting the others' commitment levels. Trust can differ in a systematic way with respect to beliefs systems. The analysis identifies two equilibria that are likely to be realized in the community: one with a full protection of property rights and a high level of trust in others, and another with no protection of property rights and a low level of trust in others. One implication of this analysis emphasizes the importance of commitment to protect property rights; high trust on such commitment induces high economic activities. Thus, communities are part of good governance, because they address certain problems that cannot be handled by individuals acting alone (Bowles and Gintis, 2002). Social capital improves efficiency by solving a coordination failure problem. The effort discretion provides individuals with different choices with regard to whether property rights are perceived to be secure or not.

Whether a certain equilibrium is realized depends on social norms. Here, a social norm is based on a commonly shared belief of how others will behave. The norm arises from informal forms of organizations based on social networks and associates. The evolutionary approach to norms starts with individual minds, and then aggregates up to community-level phenomena (Henrich and Henrich, 2006). At the individual level, norms are adopted as the set of mental representations in the brains. At the community-level, these mental representations are shared by members of the community. Similar mental representations are formed through a process of adaptation for acquiring beliefs from other individuals. Culturally transmitted beliefs that enhance mutual trust will facilitate cooperation and increase conformity. This framework describes such beliefs as the result of an inculcated legacy of the past.

Trust is based on the cognitive systems about how other agents will behave in the community. Cognition includes the process used to predict what others are going to do. Cognition is closely related to culture. Culture, in the form of beliefs and ideologies, influences the specific way in which individual agents act. Individuals with common cultural backgrounds will share cognitive systems inter-subjectively. Trust becomes a potentially useful means to achieve cooperation. Where there is high trust, economic activity can flourish.

According to Fukuyama (1996), as a general rule, trust arises when a community shares a set of moral values in such a way as to create expectations of regular honest behavior. Economic activities require a high level of security which can be based on trust. The employment of trust depends on the probability that other agents will behave in a way that is expected. Trust is the mutual faithfulness on which all social relationships ultimately depend. Social capital is the social relations embedded in the social structure, and is described as a feature that communities possess to varying degrees. Trust contributes to social capital, and is essential to the establishment of civil society. Trust may create a sense of community within citizens. At the individual level, the civil society needs a specific set of attitudes and behavioral patterns, including a certain style of interpersonal interaction and collaboration. Behaviorally, to trust others is to act as if the uncertain acts of others were indeed certain in circumstances wherein the violation of these expectations results in negative consequences for those involved.

In North (1990, p. 54), "[t]he inability of societies to develop effective, low-cost enforcement of contracts is the most important source of both historical stagnation and contemporary underdevelopment in the Third World." Informal rules provide legitimacy to formal rules. In the economy, a low-trust trap can exist. Furthermore, it has been argued that levels of trust are low in post-communist societies. Lovell (2001) argues two legacies of communism: first, the mistrust generated by communist rule (the legacy of active mistrust); and second, the communist way of understanding the conduct of public affairs (the ideological legacy). In post-communism, it is difficult to establish a genuine human community based on mutual trust. Civil society cannot function without a high level of trust. The legacies of communism undermine civil society through atomization of individual citizens.

Trust is based on a cognitive process. Agents cognitively choose whether they trust others or not. Individuals attempt to create mental models or internal representations in order to interpret the environment. Ideologies are the shared framework of mental models that groups of individuals possess. As Denzau and North (1994) point out, individuals with common cultural backgrounds and experiences will share reasonably convergent mental models, ideologies, and institutions. Culture, as a system of beliefs and ideologies, provides shared collective understandings in shaping individuals' strategies. Culture is a possible determinant of economic performance. Cognition may have more subjective aspects, while culture enables individuals to develop intersubjectively shared mental models. Culture affects belief systems about trust. The establishment of trust in interpersonal relations depends on shared mental models. Different cultures may imply different ways in which agents apply a specific logic of action. Economic performance can be explained on the basis of different belief systems. Trust arises from the cultural component which affects individual belief systems.

Trust can be conceptualized as an intersubjective or systematic social reality. Trust exists in a social system insofar as the members of that system act according to the presence of each other. Each person trusts on the expectation that others trust. Trust may decline in contexts of an increase of unknown outsiders. Individuals can develop trust because of the quality of the social system. Investments in community associations increase perceptions of security. Social capital generates positive externalities that are achieved through shared trust.

Furthermore, mental models and institutions are closely related. Following North (2005, p.49), "[b]elief systems embody the internal representation of the human landscape. Institutions are the structure that humans impose on that landscape in order to produce the desired outcome. Belief systems therefore are the internal representation and institutions the external manifestation of that representation." There is no understanding a social structure independently of the mental models or belief systems that help constitute that structure.

The behavioral economic model assumes that the quantity and quality of effort is a discretionary variable. The model presented here yields a set of behavioral patterns with regard to shared mental models. Systems of shared mental models can produce "bad" institutions, in which suboptimal economic performance can emerge. In order to foster economic progress, it is necessary to alter mental models that shape the way individuals think about trust. Economic performance can improve with trust. To trust others is one of the elements of a value system that favors economic prosperity. Shared trust confers social capital benefits for individuals by protecting property rights, reducing uncertainty and transaction costs, and building better bridges to economic opportunities. In contrast, mistrust is typical of communities that resist prosperity.

The model in this section demonstrates the emergence of property rights in the absence of political institutions of enforcement. In new institutional economics, institutions matter for economic performance. On the other hand, in social capital analysis, social relationships matter for economic performance. This section considers an alternative security mechanism in which social relationships support informal institutions. The model emphasizes the role of social relationships in providing informal enforcement of property rights. Individual agents decide whether or not to commit themselves to protect property rights. The model takes the notion of trust as a measure of social capital in the community. Trust is important in networked systems that require close attention to shared commitment to safety and reliability. Without trust, commitment can waver. Trust is a cognitive state that one will not be exploited by another. It is shown that trust is a crucial factor for facilitating the creation of community-level security via the informal enforcement of property rights. The model also presents a behavioral economic framework, in which work effort can vary. Effort discretion is introduced as a choice variable. The analysis identifies two equilibrium solutions: one with a full protection of property rights and a high level of trust, and another with no protection of property rights and a low level of trust. High levels of trust are crucial to produce positive economic outcomes. The model shows that a low-trust trap can exist. The effort discretion provides individuals with different choices with regard to whether property rights are perceived to be secure or not. Trust is based on a cognitive process which discriminates among individuals that are trustworthy and distrusted. The establishment of trust in socio-economic relations depends on shared mental models. Culture, as a system of beliefs and ideologies, provides shared mental models. An individual's degree of trust in others can predict cooperative behavior in the community. Trust can differ in a systematic way with respect to belief systems. An agent will trust others, provided the perceived probability that they will commit to securing property rights is more than some critical level. Trust is conceptualized as an inter-subjective social reality. Cognition matters, since economic activity is affected by the perceived security of property rights. Trusting others enables individual agents to operate more efficiently.

2.6   INSTITUTIONAL INEQUALITY

Is the concept of institutions useful in studying inequality? This section presents a theory of institutional inequality to understand the structure of labor market inequality. The theory shows how differences in institutions lead to differences in economic outcomes. Institutions may result in low levels of economic performance. Institutional aspects of poverty encompass the wide range of hierarchical systems and relationships in society, based on factors as varied as social class, gender, race, ethnicity, age, language, and region. A discriminatory regime affects the structure of opportunities open to different social groups.[14] Ethnic minorities are often denied access to lucrative occupations and, thus, are represented in low-paying jobs. The key idea is to relate wage payment differentials to unequalized opportunities on entry into occupations.

Much of the economic literature has studied the effect of technological innovation on the distribution of income. On the contrary, this section proposes an analytical framework for understanding the nature of institutions contributing to the emergence of the divergence in earnings. It aims to understand how economic inequality is structured by demonstrating the effects of the institutional environment on economic performance. Labor

---

14. Becker (1957) is the seminal work on labor market discrimination. He analyzes the relationship between racial prejudice among whites and discrimination against blacks in a perfectly competitive environment. His model represents prejudice as a distance for cross-racial contact. Prejudice causes an employer to behave as if black workers' monetary wages are higher than they actually are.

earnings are the most important component of income for workers. The level of wage inequality plays an important role for understanding poverty and social stratification.

The concept of social exclusion has encouraged scholars to consider simultaneously the economic, social, and political dimensions of deprivation.[15] Originating in France in the 1970s, and diffusing rapidly in Europe, the framework is concerned with full participation in all aspects of social life as an end in itself. Social exclusion is the opposite of social inclusion which reflects the perceived importance of being part of society. A decline of inclusion generally falls most heavily on the economically disadvantaged. The disadvantaged have a right to inclusion in society. Discussions of exclusion tend to focus on the phenomena of poverty, unemployment, low educational attainment, and barriers to social and political systems. Social exclusion is a repute of social bonds. The least skilled have huge difficulties in finding a decent job, which gives them the feeling that they are not useful members of society. They choose from an impoverished opportunity set. Work fulfills the desire to have a place in society. An individual's nonparticipation in society is due to circumstances outside his or her control.

Social exclusion is, however, a contested term, lacking real agreement over either its definition or its causes. Socially excluded individuals suffer generalized disadvantages in terms of education, training, employment, housing, financial resources, etc. Their chances of gaining access to the major social systems are substantially less than those of the rest of the population. The 1990s marked the era when the social exclusion discourse really proliferated in many European countries. A core element of New Labour's approach in Britain, articulated by Blair on his election in 1997, is "education, education, education," which aims to create citizens fit for the jobs that exist. People with low-wage prospects would improve their qualifications through additional education to gain a higher wage. Social exclusion is therefore seen as a product of underachievement, which assumes that poverty is self-induced or the result of personal failure.

In this section, we see social exclusion as a direct consequence of unequalized opportunities. The impoverishment of opportunity sets itself is not seen as the individual's own free choice. There is a substantial minority in the workforce whose expected earnings do not constitute an income sufficient to enable them to take part in the mainstream of their societies. Nonparticipation in society means nonparticipation in the market economy. On the one hand, technological or organizational change strengthens the tendency toward worker participation; on the other hand, there remains a problem of social exclusion for those who are outside. Unequalized opportunities intrude on the internal structures of organizations.

---

15. The concepts of poverty and social exclusion are related and, to some extent, complementary, even though they are not the same (Atkinson, 1998).

All people should get the opportunity to realize their full human potential, or to realize their own goals and aspirations. This relates to human dignity. Social exclusion can be related to Amartya Sen's concept of capability. Individuals purchase commodities with given characteristics and capabilities (Sen, 1985). In Sen's view, individuals are composed of "functionings." Achieved functionings make up a person's well-being, and an individual's capability to achieve functionings, therefore, represents the individual's true freedom. Thus, social exclusion can be defined as a process leading to a state of functioning deprivations.

Economists usually think of discrimination as a set of restrictions on the actions of an oppressed group that prevents members from earning a competitive market return on their abilities. If these artificial barriers are lowered, they expect the standard of living of the discriminated group to converge rapidly with that of the general population. However, some societal discriminatory affects endure once legal barriers are removed. Why have income differentials not been eliminated? The source of persistence in wage differentials lies in the influence of social categories on the individual. At a very basic level, most people are aware of how they are differentiated from their surroundings. People often perceive themselves as members of social groups. Institutions are not only external to individuals. Institutions are internalized by group members as identities and selves, and they are displayed as personalities. In the identity model of social exclusion (Akerlof and Kranton, 2000), legal equality may not be enough to eliminate racial disparities, and there can be a permanent equilibrium of racial inequality.

Stereotype-based expectations affect individual performance in the domain of the stereotype.[16] The stereotypes against blacks are rooted in the history of slavery and continuing discrimination. Loury (2002) argues that the ideological legacy of slavery in the US stigmatizes blacks, and that stigma is a major factor in the persistence of social inequality. The legacy of discrimination is "spoiled" collective identities. According to Hoff and Pandey (2006), social identity creates a pronounced economic disadvantage for a group through its effect on individual expectations. In their investigation, the public revelation of social identity (e.g., caste) affects cognitive task performance. Indian society is divided into groups called castes.[17] The caste system is a form of social stratification that satisfies a given number of features. There is a broad ranking of castes based on occupations. Occupation choices are restricted, and members of a caste usually follow occupations that the caste has a monopoly over.

---

16. Economists view stereotypes as a manifestation of statistical discrimination (Phelps, 1972; Arrow, 1973). In the conventional economic approach, stereotypes are based on rational expectations.

17. The caste system is formally modeled in Akerlof (1976). People may have beliefs such that breakers of the caste code will be out-casts. Within this structure, there exists a caste equilibrium where jobs are allocated according to caste.

According to Hoff and Pandey (2006), publicly revealing individual membership in a group that has been or is being discriminated against impedes the group's ability to respond to economic opportunities. Social identity affects behavior, largely because it affects expectations. Belief systems that are the legacy of historical conditions of inequality may give rise to behavior that reproduces the inequality. The aggregate effect on the society of expectations associated with caste can be viewed as negative.

2.6.1 Background

The phenomenon of increasing nonparticipation in the workforce by the less-skilled is worthy of serious consideration. For example, in the United States, despite the huge increase in the number of jobs created since 1980, the proportion of less-skilled men who are holding jobs has declined in the 1980s−90s (Freeman, 1995).

The economic approach to the problem of explaining wage inequality focuses on relative demand for and supply of different types of labor. Looking at the demand side, the growth of wage inequality in the United States reflects technological change, which has rendered more-educated workers more valuable to employers than less-educated workers. With regard to the supply side, the compression of wage differentials prior to the 1970s coincided with an increase in the relative supply of more-educated labor. The supply of more-educated labor subsequently failed to keep up with demand, giving rise to sharply increasing returns to education.

The increase in skill differentials in American manufacturing over the 1980s has been well-documented (Murphy and Welch, 1992; Katz and Murphy, 1992). The rise in the wage premium for skilled labor is typically understood with a conventional supply and demand model, as the result of a large rightward shift in the demand function (Johnson, 1997). The literature shows that the rise in earnings inequality in American manufacturing during the 1980s is attributable to a skill-biased technological change. Skill-biased technological change means technological progress that shifts demand toward more highly-skilled workers relative to the less skilled.

According to studies on skill-biased technological change, the decline in the wage and employment prospects of less-educated workers is due to recent advances in technologies. These studies suggest that new technologies are complementary to skills by nature. However, not all technological revolutions increase the demand for skilled labor. Without the organizational and skills infrastructure, technology alone is not enough. In Caroli and Van Reenen's (2001) study, skills appear to complement organizational change (skill-biased organizational change). Organizational changes are characterized by delegation and delayering within organizations. Organizational changes have an independent role, and they reduce the demand for less-skilled labor.

The Great Compression is the term for wage narrowing in the United States during the 1940s, which followed the Great Depression of the 1930s and produced a wage structure more equal than that experienced since (Goldin and Margo, 1992). The relative demand for less-educated workers increased in the 1940s in comparison with later decades, and a rising minimum wage continued to pull up the bottom of the wage distribution. The impact was to compress the wage structure. Along with short-run events affecting the demand for labor, institutional changes, brought about by the war and the command economy, were the primary driving forces behind the Great Compression. The movement toward equality in the 1940s was reversed in the post-1970 period.

The US experience of rapidly rising wage inequality was relatively weak (Fortin and Lemieux, 1997). Some countries, such as France and Germany, saw almost no change in the distribution of wages during the 1980s. The focus on institutional forces differs from explanations that are based on the simple supply and demand model. In fact, the gap between the mean wages of black men and white men in the United States narrowed substantially between 1940 and 1950. Following Maloney (1994), between 1940 and 1950, the black—white wage ratio among non-farm workers rose from 0.48 to 0.61. Those changes in the relative wage can be explained by "race-specific factors," which includes both black workers' characteristics and the degree of discrimination in the labor market. Most of the gains in the 1940s were caused by race-specific factors, including increasing relative wages controlling for worker characteristics.

While the 1960s may be characterized as a period of uniform gain, the 1970s and the 1980s were periods of growing wage inequality in the United States. It is well-established that the real earnings opportunities of less-skilled men deteriorated substantially during the 1980s and 1990s. In late 1980, the fall in labor market activity is especially severe among less-educated and low-wage workers (Juhn, 1992).[18] According to Freeman (1996), the collapse of the job market for less-skilled men may have contributed to their increased rate of criminal activity. He concludes that crime has become an economic activity occupying a proportion of young American men, particularly the less-educated. Many low-skilled individuals are not succeeding in holding down even jobs which require no particularly specialized knowledge. They are not responsible for their low-skill level.

Even in a period of technological change, the creation of low-productivity jobs must be subsidized. Low-productivity job subsidies aim at encouraging firms to increase the relative demand for low-skilled labor, and giving new opportunities to low-skilled unemployed. Phelps (2007) argues

---

18. Among white male high school dropouts, the employment-to-population ratio fell 14 points from 0.89 in 1967, to 0.75 in 1987. In contrast, the employment-to-population ratio for white college graduates fell by only three points from 0.97 to 0.94.

that rising inequality should be of great concern, and that the remedy for this predicament is to choose employment subsidies to lift the wages of low-skilled workers. The problem involves unequal treatment of persons of unequal marketability. Where people face a very low relative wage over the long term, a culture of poverty is apt to grow up. The subsidies would have the "multiplier effect" of alleviating the pessimism that tends to spread like a contagion in poor communities. According to Phelps (2007), employment subsidies would enhance participation in the community and increase social inclusion. Social inclusion is a basic ingredient of any description of a good society. However, eligibility for employment subsidies would have a "stigmatizing effect" on low-wage workers.

This section makes theoretical advances on how economic inequality is structured by demonstrating the effects of institutions on economic performance. First, the institutional environment in which an organization operates is shown to affect the structure of that organization. Second, the institutional environment affects the economic outcomes of individuals employed by organizations. In particular, the institutional environment affects the employment and pay practices of organizations. Thus, it is considered as the coordination on a particular equilibrium in a situation of multiple equilibria. Convergence in per capita income need not take place through the workings of competitive markets.[19]

The institutional environment structures an individual's response to economic opportunities, and shapes work norms. Individuals view the world through the lens of social identities. Generally, economists neglect the social and psychological context in which economic agents make decisions. Different identities have different shared experiences. Indeed, issues of identity and social relations seem to play a more prominent role in low-income situations than they do in high-income ones. Stereotypes are beliefs about some trait of an individual based on his or her membership. Negative stereotype-based expectations can lower economic performance in the domain of the stereotype. Social exclusion is an aspect of how an individual perceives his or her limited access to resources or opportunities in the decision-making environment. Mistrust undermines motivation. The treatment that individuals anticipate in the labor market does influence their perceived earnings. This section shows that there is a threshold level below which negative stereotypes can influence economic performance in ways that cause the beliefs to become self-fulfilling. Economic outcomes depend on whether individuals hold negative stereotypes about their capabilities. Institutions matter for economic performance, because they shape the incentives of economic agents in society. These differences play a key role in shaping economic performance.

---

19. Altman's (2009) behavioral—institutional framework allows for at least two sustainable paths (a low wage path and a high wage path) to economic growth, even in competitive markets.

2.6.2   The Model

Consider a closed economy that is populated by a continuum of workers of Measure 2. The population consists of two different groups of equal size, qualified workers and unqualified ones. All people are assumed to be classified as qualified or unqualified in the economy. Group membership is costlessly observable.

Production takes place either in a traditional organization (T) or in a modern organization (M). The production technology in both organizations exhibits constant returns to scale. L^T and L^M denote the respective amounts of labor employed in the two organizations. The economy is competitive, and all factors are remunerated according to their marginal products.

More specifically, total output in the traditional organization, Y^T, is given by:

$$Y^T = (L^T)^{1-\alpha}, \tag{2.5}$$

where $0 < \alpha < 1$.

In addition to the amount of labor, production in the modern organization is affected by a productivity index A (>1), which reflects the level of knowledge available for production in the modern organization. Thus, total output in the modern organization, Y^M, depends on productivity, A, and the amount of labor working in the organization, L^M:

$$Y^M = A^\alpha (L^M)^{1-\alpha}. \tag{2.6}$$

The supply of skills is a key factor to introduce "organizational changes" from the traditional organization to the modern one. Skills are complementary to organizational design. Therefore, L^M can refer to the quantity of "highly skilled" labor.

There are three relevant dates in the economy, $t = 0$, 1, and 2:

At $t = 0$, firms choose either the traditional organization or, alternatively, the modern organization.

At $t = 1$, the production process starts with an initial stage, during which firms hire a certain number of workers.

At $t = 2$, wages are paid to workers.

Let w^T and w^M be the wage in the traditional organization and the wage in the modern organization, respectively. At $t = 1$, firms hire labor in a competitive market taking the wage rate, w^T or w^M, as given.

The payoff in the traditional organization, π^T, is given by:

$$\pi^T = \max_{L^T \geq 0} \left\{ (L^T)^{1-\alpha} - w^T L^T \right\}. \tag{2.7}$$

Solving this maximization problem, demand for labor in the traditional organization becomes:

$$L^{\mathrm{T}} = \left(\frac{1-\alpha}{w^{\mathrm{T}}}\right)^{\frac{1}{\alpha}}. \tag{2.8}$$

Substituting back into (2.7) gives the payoff in the traditional organization as a function of w^{T}.

$$\pi^{\mathrm{T}} = \alpha \left(\frac{1-\alpha}{w^{\mathrm{T}}}\right)^{\frac{1-\alpha}{\alpha}}. \tag{2.9}$$

Likewise, the payoff in the modern organization, π^{M}, is given by:

$$\pi^{\mathrm{M}} = \max_{L^{\mathrm{M}} \geq 0} \left\{ A^{\alpha}(L^{\mathrm{M}})^{1-\alpha} - w^{\mathrm{M}}L^{\mathrm{M}} \right\}. \tag{2.10}$$

Then, demand for labor in the modern organization becomes:

$$L^{\mathrm{M}} = \left(\frac{1-\alpha}{w^{\mathrm{M}}}\right)^{\frac{1}{\alpha}} A. \tag{2.11}$$

Substituting back into (2.10) gives the payoff in the traditional organization as a function of w^{M}.

$$\pi^{\mathrm{M}} = \alpha \left(\frac{1-\alpha}{w^{\mathrm{M}}}\right)^{\frac{1-\alpha}{\alpha}} A. \tag{2.12}$$

For qualified workers, there are three options: (1) working in the modern organization; (2) working in the traditional organization; and (3) subsistence. Qualified workers possess economic advantages. It is assumed that, in the institutional environment, only qualified workers can work in the modern organization. On the other hand, there are barriers to economic advancement for unqualified workers. Unqualified workers are not candidates for holding down jobs that require highly skilled labor. Unqualified workers are then assumed to have two options: (1) working in the traditional organization and (2) subsistence. They are not responsible for the impoverished set of opportunities from which they choose, and may perceive social segmentation in the labor market. Furthermore, each worker must eat a certain amount of food, c, to live. Thus, the return of subsistence is given exogenously by c.

Suppose that firms choose the modern organization at $t = 0$. Then, there exists no traditional organization in the economy. Qualified workers choose to work in the modern organization if $w^{\mathrm{M}} \geq c$. Here, average economic performance by qualified workers in the modern organization is assumed to be

1. On the other hand, if $w^M < c$, the supply of labor is zero in the modern organization. Thus, the supply of (highly skilled) labor in the modern organization is given by:

$$L^M = \begin{cases} 1 & \text{if} \quad w^M \geq c \\ 0 & \text{if} \quad w^M < c. \end{cases} \qquad (2.13)$$

Suppose that firms choose the traditional organization at $t = 0$. Then, there exists no modern organization in the economy. All workers earn the same wage rate, w^T, from working in the traditional organization. They choose to work in the traditional organization if $w^T \geq c$. However, average economic performance can differ between qualified workers and unqualified ones. Average economic performance by qualified workers in the traditional organization is assumed to be 1. On the other hand, average economic performance by unqualified workers is assumed to be lower than that by qualified ones in the traditional organization. Let Q be average economic performance by unqualified workers in the traditional organization, where $0 < Q < 1$. For all workers, if $w^T < c$, they choose subsistence. Thus, supply of labor in the traditional organization is given by:

$$L^T = \begin{cases} 1 + Q & \text{if} \quad w^T \geq c \\ 0 & \text{if} \quad w^T < c. \end{cases} \qquad (2.14)$$

One possible explanation for the decline in average economic performance by unqualified workers ($Q < 1$) is that negative stereotype-based expectations lower economic performance. Only unqualified workers have expectations of institutional inequality. Agents who perceive social segmentation would have lower levels of job satisfaction and higher levels of psychological distress than their counterparts who do not perceive social segmentation. The perceived social segmentation is negatively associated with workers' economic performance. The identity associated with a negative stereotype hurts economic performance.

Next, this section offers an explanation for a mechanism to sustain economic inequality, in which an institutional environment can cause divergence in the economic performance. We focus on earnings inequality between qualified workers and unqualified ones in the economy. Differences in the institutional environments are associated with differences in economic outcomes for workers.

2.6.3 Analysis

Suppose that firms choose the modern organization at $t = 0$. Furthermore, consider a situation in which the wage in the modern organization fully covers subsistence consumption, $w^M > c$. In this situation, only qualified workers supply highly skilled labor. For unqualified workers, there are institutional constraints on opportunities to work in the modern organization. The total supply of highly skilled labor in the modern organization is 1, from (2.13).

Demand for highly skilled labor in the modern organization is given by (2.11). Equating supply and demand for labor in the modern organization:

$$1 = \left(\frac{1-\alpha}{w^M}\right)^{\frac{1}{\alpha}} A,$$

we get the wage rate in the modern organization as:

$$w^M = (1 - \alpha)A^\alpha. \tag{2.15}$$

Next, suppose that firms choose the traditional organization at $t = 0$. Furthermore, consider a case in which the wage in the traditional organization fully covers subsistence consumption, $w^T > c$. Then, all workers supply labor. The total supply of labor in the traditional organization is $1 + Q$, from (2.14). Demand for labor is then given by (2.8). Equating supply and demand for labor in the traditional organization:

$$1 + Q = \left(\frac{1-\alpha}{w^T}\right)^{\frac{1}{\alpha}},$$

we get the wage rate in the traditional organization as:

$$w^T = (1 - \alpha)\left(\frac{1}{1+Q}\right)^\alpha. \tag{2.16}$$

Since $A > 1$ and $0 < Q < 1$, we have $A > 1 > 1/(1 + Q)$. So, from (2.15) and (2.16), it follows that:

$$(1 - \alpha)A^\alpha > (1 - \alpha)\left(\frac{1}{1+Q}\right)^\alpha.$$

Thus, the wage in the modern organization fully covers the wage in the traditional organization, that is, $w^M > w^T$.

Using (2.12) and (2.15), the payoff in the modern organization is given by:

$$\pi^M = \alpha A^\alpha. \tag{2.17}$$

Using (2.12) and (2.15), the payoff in the traditional organization is given by:

$$\pi^T = \alpha(1+Q)^{1-\alpha}. \tag{2.18}$$

At $t = 0$, firms choose either the traditional organization or the modern organization. If $\pi^M \geq \pi^T$, firms choose the modern organization; and if $\pi^M < \pi^T$, firms choose the traditional organization. Using (2.17) and (2.18), if

$$\alpha A^\alpha \geq \alpha(1+Q)^{1-\alpha},$$

or

$$A^{\frac{\alpha}{1-\alpha}} - 1 \geq Q, \tag{2.19}$$

then firms choose the modern organization at $t = 0$.

For a given A in (2.19), let Q^* denote the value of Q that satisfies the following equation:

$$A^{\frac{\alpha}{1-\alpha}} - 1 = Q^*. \tag{2.20}$$

If $Q \leq Q^*$ in (2.20), firms choose the modern organization. Then only qualified workers supply labor, $L^M = 1$. On the other hand, if $Q > Q^*$ in (2.20), firms choose the traditional organization. Then, all workers supply labor, $L^T = 1 + Q$. There is a threshold level of average performance, Q^*, below which firms have no incentive to choose the traditional organization.

For a lower level of average performance by unqualified workers ($Q \leq Q^*$), there exists an equilibrium with an unequal distribution of earnings. That is, only qualified workers who supply labor, are able to earn the wage rate w^M, while unqualified workers have no chance to find a job. Some lower level of average performance by unqualified workers causes "unemployment" for them.

On the other hand, for a higher level of average performance by unqualified workers ($Q > Q^*$), all workers who supply labor, earn the wage rate w^T in the traditional organization. In this equilibrium, all workers earn the same wage rate, and we have an equal distribution of earnings. Which of these two alternative equilibria will materialize depends on a key behavioral parameter, that is, average performance by unqualified workers in the traditional organization.

To summarize:

Proposition 2.2: (Teraji 2011). *Suppose that productivity in the modern organization is given. In the economy with some lower level of average performance by unqualified workers, only qualified workers are able to earn the higher wage rate in the modern organization. On the other hand, in the economy with some higher level of average performance by unqualified workers, all workers earn the same wage rate in the traditional organization.*

This section demonstrates that the institutional environment affects not only organizational structures, but also the economic outcomes of individuals employed by organizations. The institutional environment affects the employment and pay practices of organizations. An exogenous decrease in the behavioral parameter could increase economic inequality. The higher the average performance by unqualified workers employed in the traditional organization, the lower the level of inequality in earnings. Inequality in earnings is thus dependent on economic performance. A stereotype can influence economic behavior in ways that cause the beliefs to become self-fulfilling. Institutions that take the form of beliefs will display inertia, and transitions among them will occur infrequently. Institutional environments will vary in the degree to which work norms are present. The degree to which norms are present can affect actions. The opportunities for all the individuals in society are equally valuable. Equal treatment can eliminate negative stereotypes. Social exclusion shapes the structure of incentives, and thereby establishes the constraints within which individuals hold the beliefs and attitudes. Thus, social exclusion can, in itself, be the cause of differences in economic performance.

Productivity in the modern organization, A, and average performance by unqualified workers in the traditional organization, Q, determine which equilibrium dominates the other. For $A > 1$ and $0 < Q < 1$, the state space can be separated into the following two sets:

$$S^1 = \left\{ (A, Q) : A^{\frac{\alpha}{1-\alpha}} - 1 < Q \right\},$$

$$S^2 = \left\{ (A, Q) : A^{\frac{\alpha}{1-\alpha}} - 1 > Q \right\}.$$

Decision-makers in the firm have an incentive to choose the traditional organization when (A, Q) is in S^1, where $A > 1$ and $0 < Q < 1$. Then, an equilibrium with an equal distribution of earnings dominates. On the other hand, the decision-makers in the firm have an incentive to choose the modern organization when (A, Q) is in S^2, where $A > 1$ and $0 < Q < 1$. Then, an equilibrium with an unequal distribution of earnings dominates.

This can be illustrated as shown in Fig. 2.1. The border is given by (2.20). For a given A, the larger Q (in the region S^1), the greater is the payoff in the traditional organization, π^T. Then, all workers can earn the same wage rate in the traditional organization. On the other hand, for a given Q, the larger A (in the region S^2), the greater is the payoff in the modern organization, π^M. Then,

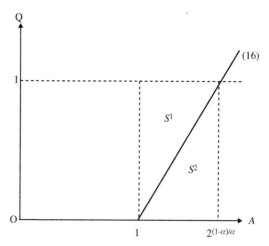

FIGURE 2.1 The state space separated into two sets.

only workers who supply highly skilled labor are able to earn the higher wage rate in the modern organization. In this case, the rise in economic inequality is attributable to skill-biased technological change.

To make the model analytically tractable, this section has assumed that unqualified workers cannot work in the modern organization. However, this assumption is not critical to the above results. Consider what happens as some fraction of unqualified workers is retrained to work in the modern organization. The other unqualified workers remain in the traditional organization, and have lower economic performance, $Q < 1$. There still exists an equilibrium with an unequal distribution of earnings for a lower level of average performance by unqualified workers.

So far, the analysis has assumed that negative stereotype-based expectations lower economic performance for unqualified workers ($Q < 1$). We view differences in norms tied to social identity as differences in economic outcomes. Consider next what happens as workers in the traditional organization are identical in terms of economic performance ($Q = 1$). Unqualified workers do not have expectations of institutional inequality, and institutional inequality does not cause the divergence in economic performance. In this case, there is no relationship between the behavioral parameter (Q) and productivity in the modern organization (A). Then, only skill-biased technological change becomes important for economic inequality. Social identity does not affect economic performance. The divergence in earnings would emerge if productivity in the modern organization is sufficiently high ($A > 2^{(1 - \alpha)/\alpha}$). On the other hand, an equal distribution of earnings would dominate if productivity in the modern organization is sufficiently low ($A < 2^{(1 - \alpha)/\alpha}$).

2.6.4 Discussion

Why does economic performance differ among individuals? A critical assumption of the model is that economic agents are X-inefficient, and do not necessarily maximize the quantity and quality of effort per unit of time contributed to the process of production (Leibenstein, 1979). X-inefficiency can be defined as a difference in economic performance between what can be achieved when economic agents supply effort in an ideal work environment, and what is supplied in a suboptimal work environment. For X-inefficiency to exist, effort must be a discretionary variable.

Traditionally, human capital literature focuses on a single dimension of skill, which is usually measured by cognitive skills. Focusing on labor market outcomes, Murnane et al. (2000) estimate the effect of the General Educational Development (GED) on the earnings of school dropouts.[20] They

---

20. High school dropouts in the United States have the opportunity of achieving high school certification by taking the GED exam. Thus, the GED program is a second-chance program that administers cognitive tests to high school dropouts.

show substantial earning returns to cognitive skills as measured by GED test scores for all groups except white male dropouts. A recent empirical literature highlights the role of a broader set of noncognitive skills for economic performance (Heckman and Rubinstein, 2001; Heckman et al., 2006). GED participants have higher cognitive skills than other dropouts, but exhibit at least as strong problems of self-control and discipline as other dropouts. GED recipients lack skills such as discipline, patience, or motivation, and, as a result, are penalized in the labor market.

Cognitive ability determines both educational attainment and later earnings. Due to the investments that better educated parents can make in their children, cognitive skills contribute to intergenerational persistence. In studies of the intergenerational transmission of economic status, cognitive skills and education have been overstudied, while wealth, race, and noncognitive skills have been understudied (Bowles and Gintis, 2002). Cognitive and noncognitive skills are strongly influenced by parental behavior and abilities. If noncognitive skills are related to family background, they provide another mechanism driving intergenerational persistence.

Economic inequality can be said to be "socially embedded" when individuals are distinguished by their language, ethnicity, race, or other social characteristics. Social embeddedness can impede performance improvements. The average skill differences among social groups may persist across generations. Unequal institutions can persist, due to the nature of these arrangements, and implement income inequality for the populations which they affect.[21] Social capital greatly affects individual's human capital formation (Coleman, 1988). Obviously, social capital is not a homogeneous resource equally available to all individuals or social groups. Only associated individuals have access to more information enabling them to coordinate activities for mutual benefit. Inequality of social capital occurs when a certain group clusters at relatively disadvantaged socio-economic positions. Social capital appears to reproduce itself, giving rise to path dependency in certain social positions.[22] Social networks have value because they provide access to resources and information. Situated in different positions in the social hierarchy, members of a disadvantaged group may find themselves deficient in social capital. They have fewer ties to members of resource-rich classes, and thus have less bridging social capital or contact with the mainstream. In fact, ghettos separate poor blacks from middle-class society. Ghetto residents may learn few skills and acquire norms that are in conflict with mainstream society. Cutler and Glaeser (1997) argue that blacks are significantly worse off in more segregated communities than they are in less segregated

---

21. Esteban and Ray (2011) focus on the role of within- and between-group income differences in determining conflict intensity.

22. For Bénabou (1996), social stratification is endogenous; the rich wish to isolate themselves from the poor because education is a local public good.

communities. Greater intergroup contact, such as the integration of schools, neighborhoods, and workplaces, can serve as an alternative to eliminate negative stereotypes.

Economic success can be influenced by characteristics of the group of individuals, including discrimination, conformist effects on behavior, differential access to information, and complementarities in production. Social relations also have value, because those relationships help to define an individual's sense of self, or the identity, on which individual behavior is based. Behavior has social and psychological foundations rooted in one's identities. Identity is accompanied by stereotypes and beliefs about how an identity should behave in social situations. Different identities have different shared expectations. Individual identity establishes behavioral norms. Furthermore, identity is inherently exclusive, distinguishing in-group members from outgroup members. Unqualified workers may anticipate that they are judged prejudicially. Social exclusion is an aspect of how one perceives his or her limited access to resources or opportunities in the decision-making environment. Mistrust undermines motivation. The multiplicity of equilibria can be interpreted as the possibility of differential treatment of workers based on their observable characteristics. There is a threshold level below which negative stereotypes become sufficiently strong. Movements to an equal income equilibrium can occur when workers break out of *ex ante* beliefs and attitudes. However, social exclusion shapes the structure of incentives, and establishes the constraints within which individuals hold the beliefs and attitudes. As demonstrated above, social exclusion can in itself be the cause of differences in economic performance, which leads to differences in economic outcomes.

This section proposes an analytical framework for understanding the nature of institutions, contributing to the emergence of the divergence in earnings. Institutions shape belief systems about how people behave. Much of the economic literature has studied the effect of technological innovation on the distribution of income. In contrast, the present model shows how differences in institutions can lead to differences in economic outcomes. We view differences in norms tied to social identity as differences in economic outcomes.

The phenomenon of increasing nonparticipation in the workforce by the less-skilled is worthy of serious consideration. The focus on institutional forces differs from explanations that are based on the simple supply and demand model. In this section, social exclusion is considered as a direct consequence of unequalized opportunities. On the one hand, technological change strengthens the tendency toward worker participation; on the other hand, there remains a problem of social exclusion for those who are outside.

The source of persistence in wage differentials lies in the influence of social categories on the individual. Institutions are not only external to individuals. Institutions are internalized by group members as social identities.

Different institutions generate different belief systems, and influence equilibrium outcomes. Social identity may create a pronounced economic disadvantage for a group through its effect on individuals' belief systems. The treatment that individuals anticipate in the labor market does influence their perceived benefits. A stereotype can influence economic behavior in ways that cause the beliefs to become self-fulfilling.

In the model, the population consists of two different groups, qualified workers and unqualified ones. In the economy with some lower level of average performance by unqualified workers, only qualified workers are able to earn the higher wage rate in the modern organization. The decline in average performance by unqualified workers implies negative expectations of institutional inequality. An exogenous decrease in the behavioral parameter can increase economic inequality. Social exclusion shapes the structure of incentives, and, thereby, can in itself be the cause of differences in economic performance. On the other hand, in the economy with some higher level of average performance by unqualified workers, all earn the same wage rate in the traditional organization. The higher the average performance by unqualified workers employed in the traditional organization, the lower the level of inequality in earnings. Thus, there is a threshold level of average performance by unqualified workers below which the divergence in earnings emerges.

2.7 SUMMARY

The new institutional economics seeks to account for the emergence and persistence of institutions on the basis of their efficiency. Groups adopt new institutions if they increase net gains relative to the existing institutions. One of the key concepts of the new institutional economics is transaction costs. Institutions emerge and persist when the benefits they confer are greater than the transaction costs involved in creating and sustaining them. Transaction costs encompass *ex ante* costs (before exchange) associated with search and negotiation, and *ex post* costs (after exchange) of monitoring and enforcement. Without the concept of transaction costs, it might be impossible to understand how an economic system works. Poorly constructed institutions are identified as a source of higher transaction costs. Individuals shape institutions based on their desirability to the group, selecting those institutions that lower aggregate transaction costs. The assignment of property rights matters, because of positive transaction costs. Property rights are defined as a set of rights to take permissible actions to use, transfer, or otherwise exploit the property. Transaction costs are more generally considered to be the resources spent on delineating, protecting, and capturing property rights. An inefficient initial allocation of property rights may result in fixed bargaining positions that are difficult to reconcile, which leads to persistent suboptimal contracting outcomes.

However, the idea of transaction costs is not a sufficient basis for a theory of the nature of the firm. The function of the firm is not simply to minimize transaction costs, but to store and reproduce a large number of gene-like habits and routines. The old economic institutionalism examines patterns and regularities of human behavior, or habit, as the basis for its institutional analysis. Institutions partially mold and enable individual habits, preferences, and values. Institutions evolve, with the more suitable ones being selected to deal with a particular environment. Hence, institutional change involves the simultaneous coevolution of both shared prevalent habits of thought, and the habits of individuals.

REFERENCES

Alchian, A.A., Demsetz, H., 1972. Production, information costs, and economic organization. Am. Econ. Rev. 62, 777–795.

Akerlof, G.A., 1976. The economics of caste and of the rat race and other woeful tales. Q. J. Econ. 90, 599–617.

Akerlof, G.A., 1980. A theory of social custom, of which unemployment may be one consequence. Q. J. Econ. 94, 749–775.

Akerlof, G.A., 1984. Gift exchange and efficiency-wage theory: four views. Am. Econ. Rev. 74, 79–83.

Akerlof, G.A., Kranton, R.E., 2000. Economics and identity. Q. J. Econ. 115, 715–753.

Alchian, A.A., 1965. Some economics of property rights. Il Politico 30, 816–819.

Altman, M., 2000. A behavioral model of path dependency: the economics of profitable inefficiency and market failure. J. Behav. Exp. Econ. 29, 127–145.

Altman, M., 2001. Culture, human agency, and economic theory: culture as a determinant of material welfare. J. Behav. Exp. Econ. 30, 379–391.

Altman, M., 2009. A behavioral–institutional model of endogenous growth and induced technical change. J. Econ. Issues 43, 685–713.

Ambrosino, A., 2016. Heterogeneity and law: toward a cognitive legal theory. J. Inst. Econ. 12, 417–442.

Arrow, K.J., 1973. The theory of discrimination. In: Ashenfelter, O., Rees, A. (Eds.), Discrimination in Labor Markets. Princeton University Press, Princeton, NJ.

Arrow, K.J., 1979. Business codes and economic efficiency. In: Beauchamp, T.L., Bowie, N.E. (Eds.), Ethical Theory and Business. Prentice Hall, New Jersey.

Atkinson, A.B., 1998. Social exclusion, poverty, and unemployment. In: Atkinson, A.B., Hills, J. (Eds.), Exclusion, Employment and Opportunity. Center for Analysis of Social Exclusion, London School of Economics, London, CASE Paper 4.

Baker, M.J., 2003. An equilibrium conflict model of land tenure in hunter-gatherer societies. J. Polit. Econ. 111, 124–173.

Baumol, W.J., 1959. Business Behavior, Value, and Growth. Macmillan, London.

Becker, G.S., 1957. The Economics of Discrimination. University of Chicago Press, Chicago.

Becker, G.S., 1968. Crime and punishment: an economic approach. J. Polit. Econ. 76, 169–217.

Bénabou, R., 1996. Heterogeneity, stratification, and growth: macroeconomic implications of community structure and school finance. Am. Econ. Rev. 86, 584–609.

Berle, A., Means, G., 1932. The Modern Corporation and Private Property. Macmillan, New York.

Bowles, S., Gintis, H., 2002. Social capital and community governance. Econ. J. 112, F419–F436.

Brennan, M.J., 1994. Incentives, rationality, and society. J. Appl. Corp. Finance 7, 31–39.

Bronfenbrenner, M., 1985. Early American leaders: institutional and critical traditions. Am. Econ. Rev. 75, 13–27.

Caroli, E., Van Reenen, J., 2001. Skill-biased organizational change? Evidence from a panel of British and French establishments. Q. J. Econ. 116, 1449–1492.

Coase, R.H., 1937. The nature of the firm. Economica 4, 386–405.

Coase, R.H., 1960. The problem of social cost. J. Law Econ. 3, 1–44.

Coase, R.H., 1998. The new institutional economics. Am. Econ. Rev. 88, 72–74.

Coleman, J., 1988. Social capital in the creation of human capital. Am. J. Sociol. 94, S95–S120.

Commons, J.R., 1934. Institutional Economics: Its Place in Political Economy. Macmillan, New York.

Cooper, R.W., John, A., 1988. Coordinating coordination failures in Keynesian models. Q. J. Econ. 103, 441–463.

Cutler, D.M., Glaeser, E.L., 1997. Are ghettos good or bad? Q. J. Econ. 112, 827–872.

Demsetz, H., 1967. Toward a theory of property rights. Am. Econ. Rev. 57, 347–359.

Demsetz, H., 1969. Information and efficiency: another viewpoint. J. Law Econ. 12, 1–22.

Denzau, A.T., North, D.C., 1994. Shared mental models: ideologies and institutions. Kyklos 47, 3–31.

Dorfman, J., 1958. The mutual influence of Mitchell and Commons. Am. Econ. Rev. 48, 405–408.

Eggertson, T., 1990. Economic Behavior and Institutions. Cambridge University Press, Cambridge, MA.

Esteban, J.M., Ray, D., 2011. A model of ethnic conflict. J. Eur. Econ. Assoc. 9, 496–521.

Fehr, E., Gächter, S., 2000. Cooperation and punishment in public goods experiments. Am. Econ. Rev. 90, 980–994.

Fortin, N.M., Lemieux, T., 1997. Institutional changes and rising wage inequality: is there a linkage? J. Econ. Perspect. 11, 75–96.

Freeman, R.B., 1995. The limits of wage flexibility to curing unemployment. Oxf. Rev. Econ. Pol. 11, 63–72.

Freeman, R.B., 1996. Why do so many young American men commit crimes and what might we do about it? J. Econ. Perspect. 10, 25–42.

Fukuyama, F., 1996. Trust: The Social Virtues and the Creation of Prosperity. Penguin, London.

Goldin, C., Margo, R.A., 1992. The great compression: the wage structure in the United States at mid-century. Q. J. Econ. 107, 1–34.

Greif, A., 2006. Institutions and the Path to the Modern Economy. Lessons from Medieval Trade. Cambridge University Press, Cambridge, MA.

Grossman, H.I., 2001. The creation of effective property rights. Am. Econ. Rev. 91, 347–352.

Grossman, H.I., Kim, M., 1995. Swords or plowshares? A theory of the security of claims to property. J. Polit. Econ. 103, 1275–1288.

Hayek, F.A., 1983. Knowledge, Evolution, and Society. Adam Smith Institute, London.

Heckman, J.J., Rubinstein, Y., 2001. The importance of noncognitive skills: lessons from the GED testing program. Am. Econ. Rev. 91, 145–149.

Heckman, J.J., Stixrud, J., Urzua, S., 2006. The effects of cognitive and noncognitive abilities on labor market outcomes and social behavior. J. Labor Econ. 24, 661–700.

Henrich, J., Henrich, N., 2006. Culture, evolution and the puzzle of human cooperation. Cogn. Syst. Res. 7, 220–245.

Hirschman, A.O., 1985. Against parsimony: three easy ways of complicating some categories of economic discourse. Econ. Philos. 1, 7–21.

Hirshleifer, J., 1995. Anarchy and its breakdown. J. Polit. Econ. 103, 26–52.

Hobbes, T., 1651/1991. Leviathan. Cambridge University Press, Cambridge, MA.

Hodgson, G.M., 1988. Economics and Institutions: A Manifesto for a Modern Institutional Economics. Polity Press, Cambridge.

Hodgson, G.M., 2003. The hidden persuaders: institutions and individuals in economic theory. Camb. J. Econ. 27, 159–175.

Hodgson, G.M., 2006. What are institutions? J. Econ. Issues 40, 1–25.

Hoff, K., Pandey, P., 2006. Discrimination, social identity, and durable inequalities. Am. Econ. Rev. 96, 206–211.

Holmström, B., 1982. Moral hazard in teams. Bell J. Econ. 13, 324–340.

Hume, D., 1740/1985. A Treatise of Human Nature. Penguin, London.

Johnson, G.E., 1997. Changes in earning inequality: the role of demand shifts. J. Econ. Perspect. 11, 41–54.

Juhn, C., 1992. Decline of male labor market participation: the role of declining market opportunities. Q. J. Econ. 107, 79–121.

Kandel, E., Lazear, E.P., 1992. Peer pressure and partnerships. J. Polit. Econ. 100, 801–817.

Katz, L., Murphy, K.M., 1992. Changes in relative wages, 1963–1987: supply and demand factors. Q. J. Econ. 107, 35–78.

Klein, B., 1983. Contracting costs and residual claims: the separation of ownership and control. J. Law Econ. 26, 367–374.

Lawson, T., 2015. Process, order and stability in Veblen. Camb. J. Econ. 39, 993–1030.

Leibenstein, H., 1966. Allocative efficiency vs. "X-efficiency. Am. Econ. Rev. 56, 392–415.

Leibenstein, H., 1976. Beyond Economic Man. Harvard University Press, Cambridge, MA.

Leibenstein, H., 1979. A branch of economics is missing: micro-micro theory. J. Econ. Lit. 17, 477–502.

Libecap, G.D., 1986. Property rights in economic history: implications for research. Explor. Econ. Hist. 23, 227–252.

Libecap, G.D., 1989. Contracting for Property Rights. Cambridge University Press, Cambridge, MA.

Loury, G.C., 2002. The Anatomy of Racial Inequality. Harvard University Press, Cambridge, MA.

Lovell, D.W., 2001. Trust and the politics of post communism. Communist Post-Communist Stud. 34, 27–38.

Maloney, T.N., 1994. Wage compression and wage inequality between black and white males in the United States, 1940–1960. J. Econ. Hist. 54, 358–381.

Marris, R., 1964. The Economic Theory of Managerial Capitalism. Macmillan, London.

Milgrom, P.R., North, D.C., Weingast, B.R., 1990. The role of institutions in the revival of trade: the law of merchant, private judges, and the champagne fairs. Econ. Polit. 2, 1–23.

Murnane, R.J., Tyler, J.H., Willett, J.B., 2000. Do the cognitive skills of school dropouts matter in the labor market? J. Hum. Res. 35, 748–754.

Murphy, K.M., Welch, F., 1992. The structure of wages. Q. J. Econ. 107, 35–78.

Muthoo, A., 2004. A model of the origins of basic property rights. Games Econ. Behav. 49, 288–312.

North, D.C., 1990. Institutions, Institutional Change and Economic Performance. Cambridge University Press, Cambridge, MA.

North, D.C., 2005. Understanding the Process of Economic Change. Princeton University Press, Princeton, NJ.

Ostrom, E., 1990. Governing the Commons: The Evolution of Institutions for Collective Action. Cambridge University Press, Cambridge, MA.

Ostrom, E., 2000. Collective action and the evolution of social norms. J. Econ. Perspect. 14, 137–158.

Phelps, E.S., 1972. The statistical theory of racism and sexism. Am. Econ. Rev. 62, 659–661.

Phelps, E.S., 2007. Rewarding Work: How to Restore Participation and Self-Support to Free Enterprise. Harvard University Press, Cambridge, MA.

Putnam, R., 2000. Bowling Alone. Simon & Schuster, New York.

Rebitzer, J.B., 1993. Radical political economy and the economics of labor markets. J. Econ. Lit. 31, 1394–1434.

Rutherford, M., 2001. Institutional economics: then and now. J. Econ. Perspect. 15, 173–194.

Sen, A., 1985. Commodities and Capabilities. North Holland, Amsterdam.

Shapiro, C., Stiglitz, J., 1984. Equilibrium unemployment as a worker discipline device. Am. Econ. Rev. 74, 433–444.

Simon, H.A., 1957. Models of Man: Social and Rational. Wiley, New York.

Sugden, R., 1989. Spontaneous order. J. Econ. Perspect. 3, 85–97.

Teraji, S., 2008. Property rights, trust, and economic performance. J. Behav. Exp. Econ. 37, 1584–1596.

Teraji, S., 2011. An economic analysis of social exclusion and inequality. J. Behav. Exp. Econ. 40, 217–223.

Teraji, S., 2013. A theory of norm compliance: punishment and reputation. J. Behav. Exp. Econ. 44, 1–6.

Veblen, T.B., 1899. The Theory of the Leisure Class: An Economic Study in the Evolution of Institutions. Macmillan, New York.

Veblen, T.B., 1904. The Theory of Business Enterprise. C. Scribner's Sons, New York.

Williamson, O.E., 1975. Markets and Hierarchies: Analysis and Antitrust Implications. Free Press, New York.

Williamson, O.E., 1985. The Economic Institutions of Capitalism. The Free Press, New York.

Williamson, O.E., 2015. Ronald Harry Coase: institutional economist and institution builder. J. Inst. Econ. 11, 221–226.

FURTHER READING

Henrich, J., 2000. Does culture matter in economic behavior? Ultimatum game bargaining amongthe Machiguenga of the Peruvian Amazon. Am. Econ. Rev. 90, 973–979.

Henrich, J., Boyd, R., Bowles, S., Camerer, C., Fehr, E., Gintis, H., McElreath, R., 2001. In search of Homo economicus: behavioral experiments in 15 small-scale societies. Am. Econ. Rev. 91, 73–78.

Henrich, J., Boyd, R., Bowles, S., Camerer, C., Fehr, E., Gintis, H., 2004. Foundations of Human Sociality: Economic Experiments and Ethnographic Evidence from Fifteen Small-scale Societies. Oxford University Press, Oxford.

Markowitz, H., 1952. The utility of wealth. J. Polit. Econ. 60, 151–156.

Chapter 3

Institutions and the Economics of Behavior II

3.1 THE ECONOMICS OF BEHAVIOR

Over the last several decades, the conventional economic prediction of human behavior has been subject to challenges from empirical results qualified as anomalies (Thaler, 1988). Anomalies are often labeled as decision-making failures or mistakes. Behavioral economics is the study of how individuals behave, and how thinking and emotions affect individual decision-making. Behavioral economics encourages economists to be more receptive to the results of experimental research. By adding insights from a wide range of disciplines (psychology, sociology, neuroscience, and evolutionary biology), behavioral economics tries to modify the conventional economic approach, and to analyze how "flesh-and-blood" people act in social contexts. According to Mullainathan and Thaler (2000), behavioral economics deviates from the standard economic model in three ways. First, under bounded rationality conditions, humans are faced with limited cognitive abilities that constrain their problem-solving abilities. Second, bounded willpower illustrates that people sometimes make choices that are not in their long-term interest. Third, bounded self-interest shows that humans are often willing to sacrifice their own interests to help others.

Psychological research teaches conventional economists about ways to describe preferences more realistically, about biases in belief formation, and about ways it is misleading to conceptualize people as attempting to maximize stable, coherent, and accurately perceived preferences. In Kahneman's (2003, p. 1449) words, "[o]ur research attempted to obtain a map of bounded rationality, by exploring the systematic biases that separate the beliefs that people have and the choices they make from the optimal beliefs and choices assumed in rational-agent models." The idea of behavioral and cognitive modularity helps explicate a variety of otherwise anomalous observations. How closely we wish to interweave economics with psychology depends both on the range of questions we wish to answer, and on our assessment of how far we may trust the assumptions of conventional economic theory as approximations.

The Cognitive Basis of Institutions. DOI: https://doi.org/10.1016/B978-0-12-812023-1.00003-X

Dual process models have their origins in the 1970s and 1980s (Evans, 1989). They have become one of the most important theoretical developments in the understanding of human behavior (Kahneman, 2003, 2011). The human condition is marked by two, often conflicting, cognitive systems. These are labeled System 1 and System 2 to represent two reasoning processes. There are two processes the mind uses to interpret the world. System 1 processes are rapid, automatic, and unconscious, while System 2 processes are slower, deliberate, and conscious. System 1 is driven by habits, emotions, intuitions, and traditions. System 1 drives humans to pursue their wants and satisfactions with an urgency requiring little or no reflection or contemplation. Thus, in the mode of System 1, we are human actors who think about current situations and how to gratify immediate desires. Additionally, System 1 processes are typically thought of as older, shaped by evolution. System 2 processes, by contrast, are typically thought to be more recent. In the mode of System 2, we are economic actors who think more long-range about planning for outcomes. In utilizing System 2, we carefully calculate costs, benefits, and any other aspects salient to the situation before acting. Kahneman (2003, 2011) applies System 1 and System 2 to propose that the human mind consists of both an intuitive system and an evolutionary newer cognitive system. Our thinking is a combination of the two systems, System 1 and System 2.

Since the early 1970s, two psychologists, Daniel Kahneman and Amos Tversky, published a series of articles testing the conventional economic theory of choice under uncertainty. The main contributions of Kahneman and Tversky can be divided into three areas: heuristics and biases, framing effects, and prospect theory.

People rely on simplifying strategies or heuristics. While heuristics are frequently useful shortcuts, they also lead to biases. According to Thaler (1991, p. 4), Kahneman and Tversky have shown that "mental illusions should be considered the rule rather than the exception." The presence of an error of judgment is demonstrated by comparing intuitive inferences and probability judgments to the rules of statistics and the laws of probability:

> *In making predictions and judgments under uncertainty, people do not appear to follow the calculus of chance or the statistical theory of prediction. Instead, they rely on a limited number of heuristics which sometimes yield reasonable judgments and sometimes lead to severe and systematic errors.*

(Kahneman and Tversky, 1973, p. 237)

Heuristics describe how people make judgments and decisions based on approximate rules of thumb. Heuristics simplify the assessment of complex probabilistic hypotheses, and require less effort compared to a rational, calculated choice.

People are often insensitive to sample size. According to Tversky and Kahneman (1971, 1972), people ignore sample size and apply a "law of small numbers" to make predictions in situations where the "law of large numbers" is appropriate. People have much more faith in small samples than they should. An individual's erroneous behavior is the result of false beliefs, for which the individual cannot really be blamed. Then, the "deviations of subjective from objective probability seem reliable, systematic, and difficult to eliminate" (Kahneman and Tversky, 1972, p. 431).

Anomalies arise from the way humans process information to form beliefs. Tversky and Kahneman (1974) describe three heuristics that are employed to assess probabilities and to predict values. First, representativeness is the tendency for people to assess the likelihood of an event based on the similarity of that occurrence to their stereotypes of similar occurrences. For example, people ascribe characteristics to groups or subgroups based on their experiences with, or perceptions of, members of a group. When the experiences with members of a population are not representative of that population, the person might incorrectly ascribe the characteristic to the entire population. Second, availability is the tendency to assess the frequency or likely causes of an event by the degree to which instances or occurrences of that event are readily "available" in memory. That is, if people can readily think of examples of events, they will inflate their probability estimates of the likelihood of their occurrence. Finally, anchoring and adjustment refer to the tendency to assess quantities by starting from an initial value and adjusting to yield a final decision. People are overly influenced by anchors, even arbitrary ones.

According to Tversky and Kahneman (1983, p. 293), "perhaps the simplest and the most basic qualitative law of probability is the conjunction rule: the probability of a conjunction, $P(A\&B)$, cannot exceed the probabilities of its constituents, $P(A)$ and $P(B)$, because the extension (or the possibility set) of the conjunction is included in the extension of its constituents." To test whether decision-makers abide by the conjunction rule, they asked subjects to rank the likelihoods of certain conclusions that can be drawn from hypothetical personality sketches of fictitious individuals. In one version of the experiment, subjects were given the following personality sketch and asked to identify which of the two alternatives was more probable:

Linda is 31 years old, single, outspoken, and very bright. She majored in philosophy. As a student, she was deeply concerned with issues of discrimination and social justice, and also participated in anti-nuclear demonstrations.

1. Linda is a bank teller.
2. Linda is a bank teller and is active in the feminist movement.

In this well-known Linda problem, Tversky and Kahneman (1983) report that 85% of respondents indicated (2) more likely than (1). A majority of respondents (85%) declared the conjunction (bank teller and feminist) to be more probable than its constituent (bank teller), thereby violating the conjunction rule. Tversky and Kahneman (1983) conclude that the conjunction fallacy (assigning higher probability to the conjunction than its constituents) is prevalent in situations where likelihood judgments are mediated by intuitive heuristics, such as representativeness and availability.

People respond differently to gains versus losses when they make choices. Peoples' preferences are affected by whether the decision problems are framed in terms of gains or losses. A framing effect is said to be present when different ways of describing the same choice problem change the choices that people make, even though the underlying information and choice options remain essentially the same. Framing effects are seen as a violation of the axiom of rational choice known as "descriptive invariance." Tversky and Kahneman (1981) discuss framing effects involving the different responses to gains and losses. They illustrate the impact of presentation on framing by showing that simple wording changes (e.g., from describing outcomes in terms of lives saved to describing them in terms of lives lost) can lead to different preferences. In the experiment conducted by Tversky and Kahneman (1981), subjects are given the following choices:

Imagine that the United States is preparing for the outbreak of an unusual Asian disease, which is expected to kill 600 people. Two alternative programs to combat the disease have been proposed. Assume that the exact scientific estimates of the consequences are as follows:

Positive frame:
- If Program A is adopted, 200 people will be saved.
- If Program B is adopted, there is a one-third probability that 600 people will be saved, and a two-thirds probability that no people will be saved.

Negative frame:
- If Program C is adopted, 400 people will die.
- If Program D is adopted, there is a one-third probability that nobody will die, and a two-thirds probability that 600 people will die.

The expected value of Programs A, B, C, and D are the same: 400 lives saved and 200 lives lost. Despite this, subjects receiving the positive frame solidly prefer Program A (72%), and those receiving the negative frame strongly prefer Program D (78%). Thus, by restating the consequences of the alternative programs in terms of potential losses ("will die") rather than the potential gains ("will be saved"), peoples' preferences are reversed, even though the choices are identical.

People evaluate identical facts differently, depending on whether they are presented in a positive or negative manner in terms of a reference point. This implies that values are attached to changes rather than final states. Prospect theory, developed by Kahneman and Tversky (1979), accounts for the framing effect through a "value function," where values are calculated with regard to a reference point. A reference point is a base position from which changes are assessed. It determines whether something is entered as gains or as losses from an endowment or status quo point of view. The value function presents three main characteristics: (1) people implicitly evaluate outcomes in terms of gains and losses; (2) people are more sensitive to variations between outcomes the closer they are to a reference point; and (3) people experience gains and losses with different levels of intensity. Prospect theory suggests that people choose options as if they were evaluating outcomes with reference to a zero point, rather than in terms of total wealth. Prospect theory is based on experimental results that do not confirm the expected utility theory: subjects make decisions based on changes in wealth rather than total wealth.[1] Peoples' assessments of values are suggested to be based on an S-shaped subjective value function of gains and losses. Its S-shaped value function represents changes (i.e., gains and losses) in the objective value levels on the horizontal axis, and the subjective value resulting from these changes on the vertical axis. The function for gains is concave, and the function for losses is convex. There is an asymmetry in the value function under prospect theory. The value function is steeper for losses than it is for gains. A loss would be felt more strongly than a gain of an equivalent amount.

3.2 NUDGES

Human decision-making is prone to phenomena like anchoring, framing, status quo bias, and inertia. Systematic biases and temporally inconsistent motivations lead to poor choices. Policy-makers are asked to design policies in light of these cognitive shortcomings. Then, by changing the choice architecture as the context in which people make decisions, outcomes can be improved in a way that makes choosers better off (Thaler and Sunstein, 2008).[2] That is, the policy maker can steer or "nudge" decisions toward a different, more beneficial outcome. According to Thaler and Sunstein (2008, p. 6), "a nudge . . . is any aspect of the choice architecture that alters people's behavior in a predictable way without forbidding any options or significantly changing their economic incentives." They think that governments can operate smarter by nudging people toward valued outcomes, rather than by forcing them to comply through the promulgation and enforcement of laws. A

1. Markowitz (1952) also proposes a utility function that explains gambling and insurance which differs significantly from the Friedman and Savage (1948) utility function. He claims that not only total wealth, but also change of wealth, may be a factor in the decision-making process.
2. According to Blumenthal-Barby (2013), there are several ambiguities and objections to the claim that choice architecture can serve as a mechanism for making people better off: (1) what is meant by "better off"? (2) who is made "better off"? (3) the problem of cognitive and affective biases and errors affecting the choice architect's judgment; (4) motivational problems such as indifference, conflicts of interest, or malevolence affecting the choice architect's judgment; and (5) whether choice architects are subject to epistemological deficiencies that affect their ability to influence people beneficently.

nudge also has an aspect of a policy intervention on the choice architecture; it is intended by the policy-maker to steer the chooser's behavior away from the behavior implied by the cognitive shortcoming, and toward his or her ultimate goal or preferences. A nudge seeks to bring about beneficial ends by either exploring or preventing systematic biases. Choice architecture reflects the fact that what is chosen often depends on how the choice is presented. Nudges are adopted to help people make better choices in a large variety of areas of human life: retirement, investment, prescription drug coverage, social security, environment, organ donation, school choice, and even marriage. Different architectures affect how people choose. A prominent example used to describe nudging is the cafeteria redesign program: placing healthy food upfront and at eye-level in the cafeteria, while placing snacks and less healthy food alternatives behind the counter. An individual can choose between option A (healthy food) and option B (unhealthy food). The "pro-self" nudge intervention in the cafeteria redesign example aims to make people eat healthy food.

By setting a default rule, a choice architect determines the option that an individual will receive if he or she foregoes making a choice of his or her own. Collections of default settings determine the way individuals initially encounter products, services, or policies. Defaults are choices that apply to individuals who do not take active steps to change them. Default rules may well be the most effective nudges. A substantial increase in money transfer to retirement savings is set as the default rule. Automatic enrollment in retirement plans can increase people's savings significantly. People are more likely to make choices about retirement savings in line with their long-term interests. Why do default rules work? According to Thaler and Sunstein (2008):

> [M]any people will take whatever option requires the least effort, or the path of least resistance. ... All these forces imply that if, for a given choice, there is a default option—an option that will obtain if the chooser does nothing—then we can expect a large number of people to end up with that option, whether or not it is good for them.
>
> (Thaler and Sunstein, 2008, p. 83)

The design of nudges largely relies on results from behavioral economics about the use of heuristics. The nudge approach assumes "somewhat mindless, passive decision makers" (Thaler and Sunstein, 2008, p. 37). Nudges are considered as having a social connection with the System 1 "intuitive thinking" of dual process models. Consider the following illustration of framing as a policy intervention. A government aiming to encourage energy conservation informs the public in two different ways:

- If you use energy conservation methods, you will save $350 per year.
- If you do not use energy conservation methods, you will lose $350 per year.

Thaler and Sunstein (2008) suggest that campaigns employing the "loss" frame are much more effective than those using a "gain" frame. Framing is a powerful nudge, because people tend to be somewhat mindless, passive decision-makers.

As John Stuart Mill ([1869] 1991, p. 91) suggests, ". . . with respect to his own feelings and circumstance, the most ordinary man or woman has means of knowledge immeasurably surpassing those that can be possessed by any one else." An individual knows best about one's own interests, while no benevolent planner can know better. Mill ([1869] 1991) argues against paternalism:

> *All errors which he is likely to commit against advice and warning, are far out-*
> *weighed by the evil of allowing others to constrain him to what they deem his*
> *good.*

(Mill, [1869] 1991, p. 92)

Thus, according to Mill, the harm of interfering with an individual is far greater than having the individual commit errors.

Thaler and Sunstein (2008, p. 144) stress the freedom of individual choice: "[g]overnment should respect freedom of choice, but with a few improvements in choice architecture, people would be far less likely to choose badly." Nudges help people without compulsion. Choices themselves are not affected; people are free to make the same (inferior) choices they would have made without the nudge. Sunstein and Thaler (2003) characterize "libertarian paternalism" as the advocacy of governmental use of pro-self nudges. Libertarian paternalism is said to be "libertarian" because nudges arguably do not interfere with the freedom of choice, and "paternalist" because the interventions in question are "pro-self" in the sense of aiming to steer people's behavior in a private welfare-promoting direction. A nudge steers the paternalized person, but always leaves open the option for the paternalized person to choose another course. Thus, "[i]f no coercion is involved, we think that some types of paternalism should be acceptable to even the most ardent libertarian" (Thaler and Sunstein, 2003, p. 175). Despite the attempt to preserve option-freedom, one of the most prominent arguments against nudging is that it disrespects autonomy (Grüne-Yanoff, 2012; White, 2013). Autonomous agents are self-governing. They have the ability to discern and consider options, and the capacities to act according to these preferences. The influence of nudges may not be easy to resist, even for such agents, especially if nudges are more effective. For example, preserving freedom to choose, default rules are often regarded as prototypical instruments of libertarian paternalism. When consumers are aware that defaults may be set as recommendations in some cases, or manipulation attempts in other cases, they successfully retain autonomy and freedom of choice. However, if default rules have an effect because consumers are not aware that they have choices, or because the transaction costs of changing from the default option are too high, defaults impinge on liberty.

3.3 DISCOUNTING

3.3.1 Hyperbolic Discounting

Most people complain that they do not have enough of self-control. Self-control problems exist when a choice is between a currently available good and an option that requires bearing a current cost associated with a deferred benefit. Agents have a tendency to succumb to short-term impulses at the expense of their long-term interests. They tend to over-consume goods that provide an immediate reward but affect negatively their future welfare, through a deterioration of health (drinking, smoking, overeating, etc.) or a decrease in wealth (compulsive credit card purchases, consumption of luxury goods, gambling, etc.). That is, the individual's current "self" outweighs the present relative to the future.

Inter-temporal choices are decisions with consequences that play out over time. The standard economic theory has analyzed inter-temporal decision-making by using the discounted utility model, which assumes that people evaluate the pleasures and pains resulting from a decision in much the same way that financial markets evaluate the gains and losses. The discounted utility model involves an agent who chooses a feasible consumption plan to maximize his or her present exponentially discounted utility. The critical feature of exponential discounting is that it preserves dynamic consistency. Inter-temporal choices are not different from any other type of choice, except that some consequences are delayed. Experimental psychologists have questioned the validity of the assumption of exponential discounting on the basis of experimental evidence (Ainslie, 1975; Kirby and Herrnstein, 1995). They have studied time preferences by eliciting preferences over various alternative rewards obtained at different times. Representations of such time preferences include a specification of discounting. Rewards obtained in the near future tend to be discounted at a higher rate than rewards obtained in the long-run. Self-control and procrastination are now topics of growing interest to behavioral economic theorists (Laibson, 1997; O'Donoghue and Rabin, 1999a,b). The behavioral economic approach has criticized the traditional economic approach, suggesting an alternative specification of discounting, that is, "hyperbolic discounting."

The reversal of preferences is called time inconsistency. A person may reverse the decisions formulated in the earlier stage. Such preference reversals are not consistent with exponential discounting. Agents have diminishing impatience, and discount the future with a declining discount rate. An individual who is offered a choice today of receiving $50 in 100 days or $60 in 101 days will take the $60, but when the options are moved forward, so that he or she can either have the $50 immediately or $60 tomorrow, he or she may choose the $50. Subjects exhibit a reversal of preferences when choosing between a smaller−sooner reward and a larger−later one in experiments. The former reward is preferred when it offers an immediate payoff, whereas the latter reward is preferred when both rewards are received with delay. Time-inconsistent preferences are proposed by Stroz (1956), and further developed by Phelps and Pollak (1968).[3] Time inconsistency implies that an *ex ante* optimal decision is not carried out because a later reevaluation suggests that it is not optimal anymore.

Since Stroz's (1956) analysis, economists have known that inter-temporal choices are time-consistent only if agents discount exponentially using a discount rate that is constant over time. People are more impatient when they make short-run tradeoffs than when they make long-run tradeoffs. Hence, the rates of time preferences would be very high in the short-run, but much lower in the long-run, as viewed from today's perspective. The contrast between short-run impatience and long-run patience has been modeled using hyperbolic discounting. Hyperbolic discounting rationalizes the preferences for present consumption as a form of time inconsistency. This creates a conflict between an individual agent's preferences at different points in time. Ainslie (2001) explains violations of delay-independence in terms of the steepness of a function which discounts the utility of a reward.[4] In contrast to exponential discount curves, hyperbolic discount curves depict a strong but temporary tendency to prefer a smaller−sooner reward to a larger−later one, in the period just before the smaller−sooner reward is due. The time-inconsistency approach postulates that people may have a variety of selves that become dominant at different points, because of their timing. For example, before the party, one self favors a strict limit of two beers during the evening, but at the party, having already consumed the two beers, the current self feels certain of the desirability of having just one more beer. An agent who discounts rewards hyperbolically is not a straightforward value estimator, but a succession of estimators whose conclusions differ. The hyperbolic discounter is then involved in a decision that has "intrapersonal" strategic dimensions. Individual choices are taken to be the outcome of an intrapersonal game in which the same individual agent is represented by a different player at every date. Intrapersonal games are played between the agent's temporally situated selves. A variety of conflicting reward-getting processes will grow and survive within the individual, sometimes leading to harmful choices in the long-run.

In formal discounting models, a consumer's welfare can be represented as a discounted sum of current and future utility. The model assumes that, at each point in time t, the agent consumes goods c_t. The subjective value to

---

3. "Quasi-hyperbolic" discount functions are introduced in Phelps and Pollak (1968), using a model of imperfect intergenerational altruism.

4. As suggested in Loewenstein (1987), this steepness is related to several emotional factors that are involved in decisions over time. Some of our emotions are caused by expectations about outcomes, and people have goals concerning these emotions, as well as concerning the outcomes themselves.

the consumer is given by an instantaneous utility function $u(c_t)$, which translates the consumption measure, c_t, into a single measure of utility at period t. The discount rate measures the rate of decline of the discount function which measures the value of utility, as perceived from the present, at each future time period. Exponential discount functions have a constant discount rate. Specifically, a utility delayed τ periods is worth δ^τ as much as a utility enjoyed immediately ($\tau = 0$). Here, δ is assumed to be less than one (future utils are worth less than current utils). Exponential discount functions also have the property of not generating preference reversals. For a time-consistent agent, discounted utility at period t takes the familiar form:

$$\sum_{\tau=0}^{\infty} \delta^\tau u(c_{t+\tau}). \tag{3.1}$$

An agent who discounts exponentially maximizes his or her inter-temporal utility in expression (3.1). The standard economic procedure to compare outcomes at different times, introduced by Samuelson (1937), is to discount them at a constant rate and sum them up. The standard economic analysis requires that devaluation occurs as an exponential function of decay. The weight at which future costs and benefits influence today's decisions declines exponentially over time. If delayed outcomes are devalued exponentially, the relative preference of expected future outcomes does not change as the individual moves closer in time to those outcomes. Exponential discounting preserves time consistency. Time consistency means that the future actions required to maximize the current value of utility remain optimal in the periods when the actions are to be taken.

Most experimental evidence suggests that the discount rate declines with the length of the delay horizon. When two rewards are both far away in time, people act relatively patiently. But when both rewards are brought forward in time, preferences exhibit a reversal, reflecting more impatience. Individuals show time-inconsistent preferences in decision-making characterized by discount rates varying over time. This means they have distinct preferences over nearby and distant choices, violating the principle of exponential discounting. The actual discount function declines at a greater rate in the short-run than in the long-run. Individuals employ short-run discount rates which are higher than long-run ones, known as hyperbolic discounting. The agent behaves given the predicted behavior of his or her subsequent selves. Such discounting patterns play a role in generating self-control problems.[5]

---

5. In Gul and Pesendorfer (2001), the agent's preferences do not change between periods (there is no dynamic inconsistency). They propose that "temptation" rather than a preference change may be the cause of a preference for commitment. Their dynamically consistent decision-maker is unambiguously better off when *ex ante* undesirable temptations are no longer available.

An individual is modeled as a sequence of autonomous temporal selves. Temporal selves are indexed by periods $t = 0, 1, 2, \ldots$ Self t's discounted utility from present and future consumption is:

$$u\left(c_t\right) + \beta \sum_{\tau=1}^{\infty} \delta^{\tau} u(c_{t+\tau}), \tag{3.2}$$

where β is assumed to be between zero and one. The parameter $\beta < 1$ represents the "bias for the present." For $\beta = 1$, these preferences are time-consistent. Laibson (1997) calls the discount structure in self t's lifetime utility "quasi-hyperbolic." An agent who discounts quasi-hyperbolically maximizes his or her inter-temporal utility in expression (3.2). The quasi-hyperbolic discount function is a discrete time function with values $\{1, \beta\delta, \beta\delta^2, \beta\delta^3, \ldots\}$ (Phelps and Pollak, 1968). The discount factor between consecutive future periods (δ) is larger than between the current period and the next one ($\beta\delta$). Since β is taken to be less than one, a short-term discount factor is less than a long-term discount factor. In this functional form, the rate of substitution between today and tomorrow is smaller than that between any other pair of successive periods. Self t uses the lifetime utility function to evaluate the stream of payments from t onward. This implies that time inconsistency since δ (the marginal rate of substitution between t and $t + 1$ from the point of view of any previous period) is replaced by $\beta\delta$ at t. The quasi-hyperbolic discount function is only "hyperbolic" in the sense that it captures the key qualitative property of the hyperbolic functions: a faster rate of decline in the short-run than in the long-run. Inter-temporal tradeoffs shift with the mere passage of time, and plans or projects that seemed optimal yesterday need no longer be optimal today. In the original psychological research, hyperbolic discount functions are like $1/\tau$ and $1/(1 + \alpha\tau)$ with $\alpha > 0$, which generalizes simpler hyperbolas.

The dynamic inconsistency forces the hyperbolic consumer to grapple with intrapersonal strategic conflict. Let us simply consider a three-period hyperbolic model, the shortest possible that actually generates a time-inconsistency effect. The periods are labeled 0, 1, 2, and subscripts on c refer to the period in question. In Period 0, Self 0's discounted utility is:

$$u(c_0) + \beta\delta u(c_1) + \beta\delta^2 u(c_2), \tag{3.3}$$

while in Period 1, Self 1's discounted utility is:

$$u(c_1) + \beta\delta u(c_2). \tag{3.4}$$

For Self 0, the discount factor between Period 0 and Period 1 is $\beta\delta$, and between Period 1 and Period 2 it is δ. Self 1's discount factor between Period 1 and Period 2 is $\beta\delta$. Therefore, there is a conflict between different selves about how much to consume in a given period. The consumer is dynamically inconsistent when β differs from one. When $\beta = 1$, we are back to the case of a dynamically consistent consumer, with exponential discounting. This hyperbolic model captures a specific type of self-control problem. Since the discount factor Self 0 applies between periods 1 and 2 (or δ) is greater than the discount factor Self 1 applies between the same two periods (or $\beta\delta$), Self 0 would like to behave more patiently in Period 1 than Self 1 actually does. Self 0 benefits from self-control that brings Period 1 behavior in line with his or her wishes. Since each self consumes too much from earlier selves' point of view, each of them would agree to increase savings a little bit in exchange for later selves doing the same. All inter-temporal selves could be made better off if all of them saved a little bit more. By making the implicit parameter restriction $\beta = 1$, standard economic models assume that no such self-control problem exists.

A single individual is modeled as many separate selves, one for each period. The future selves will control his or her future behavior. There is an important problem as to whether a person has beliefs about how his or her future selves will behave.

A "sophisticated" person knows exactly what his or her future selves' preferences will be. The sophisticated consumer chooses c_0 to maximize (3.3), knowing that in Period 1 the consumption levels c_1 and c_2 will be chosen by Self 1 with the discounted utility given by (3.4). Thus, Self 0 takes actions that seek to constrain Self 1. Sophisticates realize that they have hyperbolic preferences, and are fully aware of their future self-control problems. However, Self 0 might expect Self 1 to carry out the wishes of Self 0. Consumers might make current choices under the false beliefs that future selves will act in the interests of the current self.

A "naïve" person believes that future selves' preferences will be identical to those of current self. Naïves are fully unaware of their future self-control problems. The hyperbolic consumer may or may not be aware that his or her preferences will change over time. O'Donoghue and Rabin (1999a) make a comparison of these two extreme types.[6] Actual behavior is likely best described by something between naïves and sophisticates, not just the two extremes. Self 0 may believe with certainty that the discount factor between periods 1 and 2 is $\beta^*\delta$. The parameter β^* reflects Self 0's beliefs about his or her future β, so that $\beta^* = \beta$ corresponds to sophisticates, and $\beta^* = 1$ corresponds to naïves.

Procrastination occurs when present costs are unduly salient in comparison with future costs, leading individuals to postpone tasks until tomorrow without foreseeing that when tomorrow comes, the required action will be delayed yet again (Akerlof, 1991). Akerlof (1991) implicitly corresponds to a model of time-inconsistent procrastination, highlighting the role of naïve

---

6. O'Donoghue and Rabin (2001) examine behavior for a person who is "partially naïve." A partially naïve person is aware that he or she has future self-control problem, but underestimates their magnitude.

beliefs. Hyperbolic agents procrastinate, because they think that whatever they will be doing later will not be as important as what they are doing now. O'Donoghue and Rabin (1999b) explore procrastination in terms of time-inconsistent preferences. Time-consistent agents do not procrastinate. Time-inconsistent agents, on the other hand, procrastinate.[7]

3.3.2 Addiction

It is often difficult to wait for a delayed reward when an immediately gratifying alternative is available. This is related to theories of addiction. Although people initially decide to take a far-sighted course of action (e.g., quitting smoking), they subsequently succumb to temptation. One element common to addictions is the addict's inability to escape when he or she wants to. The immediacy of reward associated with smoking clearly has some importance in understanding why quitting can be difficult. The rewards from smoking are immediate, and the adverse consequences tend to be delayed. People who become addicted may get greater (or longer) pleasure from their substances, and may also discount future rewards more sharply. Self-control can be defined as efforts made by the individual to avoid or resist behaving inconsistently.

According to rational choice theory, habits are a type of addictive behavior. Becker and Murphy (1988) study consumption of an addictive good in a standard exponential discounting model. Exponential discount functions predict consistency of preferences over time. A single utility function describes both the addict's and the non-addict's behavior. Rational addiction means that the addict's behavior, as well as the non-addict's behavior, maximizes utility in the long-run, according to some particular utility function. In Becker and Murphy's (1988) model, people are aware of the addictive nature of cigarette smoking, and choose to smoke simply because the lifetime benefits are greater than the costs. Habit formation is defined as follows: the more of the product a person has consumed in the past, the more he or she desires the product now. Greater past consumption of addictive goods increases the desire for present consumption (i.e., "reinforcement"). The consumption of addictive goods at different times are complements. Consumption today increases the likelihood of consumption tomorrow, and the fear of addiction may generate conducts that resemble compulsive behavior. An increase in either past or expected future prices decreases current consumption. The price-theoretic framework yields the prediction that

---

7. Fudenberg and Levine (2006) propose a "dual-self" model in which many sorts of decision problems can be viewed as a game between a sequence of short-run impulsive selves and a long-run patient self. In their model, the patient long-run self and a sequence of myopic short-run selves share the same preferences over stage-game outcomes. They differ only in how they regard the future.

addicts will respond much more to permanent than to temporary price changes. Becker and Murphy (1988) show the existence of multiple stable steady state equilibria, where agents either abstain or consume substantially.

Gruber and Köszegi (2001) incorporate time-inconsistent preferences into Becker and Murphy's (1988) rational addiction model. Forward-looking behavior does not imply time consistency. The purpose of Gruber and Köszegi (2001) is to show that a time-inconsistent model, where agents are forward-looking but have time-inconsistent preferences, can generate Becker and Murphy's (1988) prediction that current consumption depends positively on future consumption. An implication of time-inconsistent preferences is that individuals, who realize their self-control problems, have a demand for commitment devices.

Let a_t and c_t denote the consumption of the addictive and the "ordinary" (or non-addictive) goods in period t, respectively. Furthermore, let S_t denote the period t stock of past consumption (or the stock of "addictive capital"). S_t depends on both past consumption of a and life cycle events, and it evolves as follows:

$$S_{i+1} = (1 - \eta)(S_i + a_i),$$

where η is the depreciation rate of the stock. An instantaneous utility is additively separable in these two goods:

$$U_i = U(a_i, c_i, S_i) = v(a_i, S_i) + u(c_i).$$

There are two extreme kinds of hyperbolic discounters: naïves and sophisticates. This can be modeled as a sequential game played by the successive inter-temporal selves. For a naïve agent, self t does not realize that self $t + 1$ will in turn overvalue period $t + 1$. A naïve agent maximizes (3.2), unconscious of the fact that his or her future selves will change the plans. On the other hand, for a sophisticated agent, self t knows that self $t + 1$ will want to do something other than what self t would have him or her do. Sophisticated agents have a demand for self-control devices if they want to quit.[8]

In the context of hyperbolic discounting, players are not separate individuals, but they are different inter-temporal incarnations of the same individual. This form of discounting sets up a conflict between the preferences of different inter-temporal selves. The agent's long-run preferences are taken as those relevant for social welfare maximization. In the standard rational addiction model, the optimal tax on addictive goods should depend only on the externalities that their use imposes on society. Sophisticates try to

---

8. A line of research has shown how the combination of self-control and informational concerns can account for forms of "motivated cognitions" documented by psychologists. For example, in Bénabou and Tirole's (2004) model, a person may achieve self-control through the adoption of personal rules based on self-reputation over one's willpower.

influence their future selves through their current consumption of addictive goods. They feel a need to exert control on their future selves by consuming less. This effect helps the government: smokers who have an intention to quit smoking will support antismoking public policies. In the time-inconsistent alternative model, optimal government policy should depend also on the "internality" that consumers impose on themselves. Due to the existence of internality, there is room for government intervention, for example, a tax hike could be welfare improving.

We can think about price interventions of the following form. The hyperbolic consumer lives for three periods, $t = 0$, 1, 2. Consumption only occurs in Period 1. An adjustment is added to the per-unit price that the consumer faces in Period 1. In that period, the consumer has to make a choice between goods a and c. Good a may be pleasant to consume in Period 1, but it causes harm in Period 2. Consumption of one unit of good a leads to harm $h > 0$ in Period 2. Good c is an ordinary good that is pleasant to consume at the moment, and that has no implications for future utility. Normalize the price of good c in Period 1 to 1.

We consider the per-period utility functions as follows:

$$u_t = \begin{cases} 0 & \text{in period } t = 0 \\ v(a) + c & \text{in period } t = 1 \\ -ha & \text{in period } t = 2 \end{cases}$$

Self 0's discounted utility function is:

$$u_0 + \beta\delta u_1 + \beta\delta^2 u_2,$$

while Self 1's discounted utility function is:

$$u_1 + \beta\delta u_2.$$

The parameters β and δ are between 0 and 1. Then, Self 0 would like Self 1 to discount Period 2 relative to Period 1 by a factor of δ. Hence, Self 0 prefers that Self 1 counts the future implications of consuming each unit of good a at a value of δh. However, Self 1's discount factor between Period 1 and Period 2 is $\beta\delta$. Self 1 only counts the future implications of consuming a at a value of $\beta\delta h$. To align Self 1's interests with those of Self 0, we can impose a price adjustment of $\delta h - \beta\delta h = (1 - \beta) \delta h$ per unit of good a. Due to this change in price, Self 1 is forced to fully internalize the future harm of consumption.

3.4 HERDS

Missing information is ubiquitous in our society. Product alternatives at the store, in catalogs, and on the Internet are seldom fully described, and detailed specifications are often hidden in manuals that are not easily accessible. In fact, which product a person decides to buy will depend on the experience of other purchasers. Learning from others is a central feature of most cognitive and informative activities through which a group of interacting agents deals with environmental uncertainty. The effect of observing the consumption of others is described as the socialization effect. The pieces of information are processed by agents to update their assessments. People may change their preferences as a result of interpersonal contact. Then, information externalities arise that drive towards the emergence of some patterns of influences among individuals.[9]

There has been increasing interest in economic models in the presence of information externalities, which has put great emphasis on the notion of social learning. In social and economic situations, we are often influenced in our decision-making by what others around us are doing.[10] Mimetic contagion was thought of as an irrational behavior responsible for pathological dynamics, such as financial bubbles. Some studies have been seen as a way to formulate rigorously a number of challenges to standard economic doctrine.[11]

Individual behavior may be governed by herd externality. What everyone else is doing is informative, because their decisions may reflect information that they have and we do not. Everyone may do what everyone else is doing, even when his or her private information suggests doing something different. This sort of herding behavior corresponds to human behavior reported by Becker (1991). When faced with two apparently very similar restaurants on either side of a street, a large majority chose one rather than the other, even though this involved waiting in line. The consequences will indeed be different due to the behavior of other agents in a society.[12] An individual's choice will then depend on his or her degree of confidence in the majority view concerning the state of the world. It is important to observe the self-reinforcing nature of confidence in the group opinion in a society.[13] In this section, the quality (or precision) of opinions, which reflects the confidence in the majority opinion, is a central parameter. Several questions arise regarding the relationship between the quality of opinions and herd

---

9. This terminology may be a close parallel to what Leibenstein (1950) called the "bandwagon effect" and "snob effect" in his classic study on the static market demand curve. By the bandwagon (snob) effect, he referred to the extent to which the demand for a commodity is increased (decreased) because others are consuming the same commodity.

10. Note that the decision may not be optimal from the social point of view, since the individual does not take account of the effect of his or her decision on the information of others.

11. See, for example, the analysis of "rumors" in Banerjee (1993) and "fashions" in Karni and Schmeidler (1990).

12. See also Kirman (1993) for an explanation of asymmetric aggregate behavior arising from the interaction between identical individuals.

13. In a situation of uncertainty prevailing in a financial market, the only rational form of conduct is to imitate others. In his study of financial speculation, Keynes (1937) sees the important role of imitation.

externality, specifically by comparing situations where each person's decision is more or less responsive to the majority view. This section considers how the quality of opinions has an impact on the efficiency of the long-run outcome in the social learning.

The social learning model of Banerjee (1992) and Bikhchandani et al. (1992) describes the decision problem faced by exogenously ordered individuals, each acting sequentially under the state of the world.[14] An agent conditions the decision in a Bayes-rational fashion on both one's privately observed information and the ordered history of all predecessors' decisions. The rest of the population is then allowed to choose sequentially, with each agent observing the choices made by all the predecessors. An information cascade occurs when agents ignore their own information completely and simply take the same action that the predecessors have taken. The aggregate information that is available in the population is not correctly revealed by the sequence of decisions. This may eventually lead the whole population to take the wrong decision, and, therefore, to a socially inefficient outcome. However, the formation of a cascade is strongly influenced by the initial decisions.

Orléan (1995) considers the dynamics of imitative decision processes in "nonsequential" contexts in which agents are interacting simultaneously and modifying their decisions at each period in time. Such a framework is better suited for the modeling of herd behavior. In many circumstances, agents are always present and they revise their opinions in a continuous mode, not holding one opinion for all decisions. In his setting, opinions are modified endogenously as a result of interaction between agents. He describes the herd behavior as corresponding to the stationary distribution of a stochastic process, rather than to switching between multiple equilibria. In order to offer a clear justification for possible equilibrium patterns, this section provides an explanation for switches from one to the other in the social learning dynamics.[15] It is worth offering an equilibrium selection criterion of how a particular equilibrium will emerge collectively in a nonsequential context.

This section considers a choice between two competing alternatives with unequal payoffs, and shows how individuals are likely to herd onto a single choice. It specifies a simple model for boundedly rational choice, given the information conveyed by the majority opinion. It considers the society

---

14. Anderson and Holt (1997) induce the emergence of information cascades in a laboratory setting. Their results seem to support the hypothesis that agents tend to decide by combining their private information with the information conveyed by the previous choices made by other agents, in conformity with Bayesian updating of beliefs.

15. Akerlof (1980) states that there are multiple equilibria in the sense that different customs, once established, could be followed in equilibrium. In one of these equilibria, a custom is obeyed and the values underlying the custom are widely subscribed to by members of the community. In the other equilibrium, the custom has disappeared, no one believes in the values underlying it, and it is not obeyed.

consisting of a continuum of agents, each of whom are Bayesian optimizers. They must make a short-run commitment to the actions they choose. Opportunities to switch actions arrive at random, which are identical and independent across agents.[16] This friction reflects uncertainty that leads to "inertia." Following this interpretation, no agent can change his or her choice at every point in time, because of the inability to make an assessment of the current configuration of opinions continuously. It seems to capture a certain aspect of boundedly rational behavior. The social dynamics generate equilibrium paths of behavioral patterns in the presence of herd externality. In some situations, the stability properties of stationary states depend not on the initial decisions, but rather on the degree of quality of opinions defined below. What may be surprising is that the two stationary states, corresponding to the two long-run configurations of opinions, possess different stability properties when people change the quality of opinions. The model shows that sufficient quality of opinions tends to yield inefficient herding in the long-run.

A key element is the multiplicity of equilibrium paths of behavioral patterns.[17] Then, one has to determine which equilibrium actually gets established. The emphasis on the quality of opinions also distinguishes this work from other explanations of equilibrium selection in economic problems, e.g., Keynesian macroeconomics in Cooper and John (1988), and economic development in Murphy et al. (1989).[18] The literature offers very few formal approaches to the process through which interacting agents' beliefs are formed. Most approaches in the literature on equilibrium selection have nothing to say about the self-reinforcing nature of confidence concerning the state of the world.

Many aspects of herd behavior can be explained by equilibrium selection problems based on the quality (or precision) of information conveyed by the group opinion. The selection is based on the stability properties of the stationary states. The stability property, which I call "globally accessible" below, assures us that, for any initial behavioral patterns, there exists an equilibrium path along which the opinions converge. As the quality of opinions is smaller than a certain threshold, the society will reach the state that

---

16. The equilibrium dynamics of this kind is used in Matsuyama's (1992) model. However, the interpretation attached to the dynamics here is quite different.

17. The fundamental argument in the path dependency literature (David 1985; Arthur 1989) is that the free market typically generates suboptimal equilibrium solutions to a variety of economic problems, and the probability of suboptimal equilibrium outcomes increases where increasing returns prevail. It is even possible for efficient and inefficient (suboptimal) solutions to prevail simultaneously in the world of path dependency. For this reason, one cannot expect the free market to force the economy to converge to unique equilibrium.

18. Many studies on evolutionary games also address the question of how a particular equilibrium will emerge in a dynamic context; see, for example, Friedman (1991) and Gilboa and Matsui (1991). These studies do not, however, offer an equilibrium selection criterion, since all strict Nash equilibria share the same dynamic properties in their models.

is globally accessible. The consequence is desirable from the social point of view. On the contrary, as the quality is larger than a certain threshold, the society will tend to reach the state where inefficient herding occurs. This situation corresponds to a self-reinforcing process of confidence in the majority opinion. Then the reduction of confidence in the majority view may be socially beneficial in an *ex ante* welfare sense. Thus, a population exhibits inefficient herding if, in the long-run, everyone uses the inferior product, and it exhibits efficient social learning if, in the long-run, everyone uses the superior choice. We find it useful to present the parameter space, which depends on the agents' degree of confidence, to obtain the social dynamics of their beliefs. Accordingly, the emergence of inefficient herding results from a slight modification in the individual's level of confidence in the group opinion.

3.4.1   The Framework

There is a continuum of identical agents who are faced with a choice between two alternative products, technologies, or practices, labeled by A and B. The decision-making is "nonsequential"; at every point in time t, each agent is drawn randomly from the overall population, and has to make a new choice of either adopting A or adopting B. All agents choose one of the two alternatives to maximize their expected payoffs.

There are many situations in which such a binary restriction seems quite reasonable. For example, when we talk about economic thought, we often think in terms of two alternative schools, such as Monetarist versus Keynesian. Even within the field of economic theory, we often debate the pros and cons of two alternative styles of writing, such as a mathematical versus a nonmathematical approach. At certain times, the generality of a model tends to be values, at other times, simplicity tends to be regarded as a virtue. In many of these situations, pursuing a middle ground may not be a practical option.

The agents, in the model of this section, are assumed to follow boundedly rational behavioral rules that incorporate the notions that there is "inertia" in consumer choices, and that they do not take into account the information they could have observed before.

Inertia, which introduces the sluggishness in the social learning process, is modeled with the assumption that agents cannot switch actions at every point in time. Each agent must make a commitment to a particular choice in the short-run. Opportunities to switch actions arrive randomly; at each point in time, some fraction α, $0 < \alpha < 1$, of the agents decide to reevaluate their choice. Thus, some agents are simultaneously present and can make decisions.

Here, α is the expected frequency of the conscious decision made by an agent per unit of time, and could be interpreted as the planning horizon.

Alternatively, α reflects ease of coordination, since a large α implies that a large fraction of agents can switch to an alternative action over a given time interval. In making decisions, agents do not incorporate the entire history of their observations. It is justified if we suppose that each agent does not want to keep track of historical information, because of the infrequent opportunity to switch.

In each period, all agents using the same brand receive the same payoff. Let u^A or u^B be the payoff to each agent's choice A or B. The model supposes that $(u^A - u^B)$ has two possible values, where $\theta^+ > 0 > \theta^-$. Agents do not know the true value of $(u^A - u^B)$. Each agent assigns common prior probability $q > 0.5$ to the event $(u^A - u^B) = \theta^+$. The value q is assumed to be given. Furthermore, it is supposed that $q\theta^+ + (1 - q)\theta^- > 0$, that is, *ex ante* brand A is better than brand B for each agent. Thus, the prior odds ratio, $q/(1 - q)$, satisfies that $q/(1 - q) > - \theta^+ / \theta^- = k$.

The aggregate behavior of the population at each point in time t can be summarized by a state variable x_t, giving the fraction of the population who are using brand A. This variable will also be called the group opinion. It is a macroscopic datum that aggregates all agents having opinion (A) in the population. Here, the initial state x_0 is taken to be given exogenously, or by "history."

Since agents do not know the true payoff realization, they will value the other source of information about it. Before making a choice, an agent is allowed to observe the group opinion as of t in a society. Then he or she can benefit from the information contained in it, and will find out which alternative other agents have chosen in the population.

I will consider some specifications of the decision rules for each agent, who is drawn randomly from the population at t and observes x_t. An agent receives a signal about the state of the world by observing the group opinion in a society. The information need not, of course, be true, and it may be false. However, each agent believes that the majority in the population, $\{x_t > 0.5\}$ or $\{x_t < 0.5\}$, may convey some implicit information about the realization of $(u^A - u^B)$. Then the agents incorporate observations of the relative popularity of the two choices in their decision-making. The agents weigh observations of others' experience, because others' decisions might reflect the information that they have and he or she does not.

Though the relative popularity of the two choices in the population conveys some information, it is "noisy," and does not give true information about the payoff realization. It is assumed that the relative popularity of the two choices, $\{x_t > 0.5\}$ or $\{x_t < 0.5\}$, at every point in time, is linked to the realization of $(u^A - u^B)$ through the following conditional probabilities:

$$\text{Prob}(x_t > 0.5|\theta^+) = \mu, \quad \text{Prob}(x_t > 0.5|\theta^-) = 1 - \mu,$$
$$\text{Prob}(x_t < 0.5 \,|\theta^+) = 1 - \mu, \quad \text{Prob}(x_t < 0.5|\theta^-) = \mu.$$

Within the present framework, $\{x_t > 0.5\}$ (respectively, $\{x_t < 0.5\}$) is better correlated with the state θ^+ (respectively, θ^-) than with θ^- (respectively, θ^+), that is, $\mu > 1 - \mu$. The closer μ is to 1, the more precise information the majority opinion conveys. Here, $\mu > 1/2$, which I call the quality of opinions, reflects the degree of confidence in the majority view concerning the realization. It is assumed to be public information for the decision-makers. Indeed, by making the quality of opinions large, we can make the precision of the information as high as we like. Furthermore, it is assumed that $\text{Prob}(x_t = 0.5|\theta^+) = \text{Prob}(x_t = 0.5|\theta^-) = 0.5$, that is, the group opinion is such that $\{x_t = 0.5\}$ is uncorrelated with the states. The task is to determine exactly how these factors influence each agent's decision-making.

3.4.2   Behavior

An important characteristic of this social learning process is that the revision of beliefs can be expressed as a simple updating of the agent's prior belief by Bayes' theorem. Once each agent observes the popularity of each choice, he or she updates the prior on the basis of it, and then chooses whichever product has the highest current score, given the posterior. This process of revision of probabilities is called Bayesian, after Bayes' theorem. Thus, we have the posterior probability that an individual should attach to the state of the world after receiving the signal. The agent can estimate the precision of the group opinion and decide whether to follow the majority side of the group. In the present framework, if agents observe the current information such that $\{x_t > 0.5\}$ or $\{x_t < 0.5\}$, they update their prior beliefs according to Bayes' rule. Then they must decide whether to follow the majority side of the population. The posterior probability of the event θ^+ given the group opinion in a society, $\{x_t > 0.5\}$, is:

$$\text{Prob}(\theta^+|x_t > 0.5) = \frac{\text{Prob}(x_t > 0.5|\theta^+)q}{\text{Prob}(x_t > 0.5|\theta^+)q + \text{Prob}(x_t > 0.5|\theta^-)(1 - q)}$$

$$= \frac{(\mu/(1-\mu))q}{(\mu/(1-\mu))q + (1-q)}.$$

Here, a central rule of probabilistic calculus is that the relative weight assigned to collective opinion is controlled by the degree of confidence in the majority opinion. This is a consequence of Bayes' rule. When μ increases, the relative weight that agents assign to the opinion of others increases. This section analyzes the way μ affects the decentralized collective learning process. Similarly, it follows that:

$$\text{Prob}(\theta^- \mid x_t > 0.5) = \frac{1-q}{(\mu/(1-\mu))q + (1-q)},$$

$$\text{Prob}(\theta^+ \mid x_t < 0.5) = \frac{q}{q + (\mu/(1-\mu))(1-q)},$$

$$\text{Prob}(\theta^- \mid x_t < 0.5) = \frac{(\mu/(1-\mu))(1-q)}{q + (\mu/(1-\mu))(1-q)}.$$

The agents, who observe the collective configuration of opinions such that $\{x_t > 0.5\}$ at t, choose brand A when $\text{Prob}(\theta^+ \mid x_t > 0.5)\theta^+ + \text{Prob}(\theta^- \mid x_t > 0.5)\theta^- > 0$. This is equivalent to the posterior odds ratio, $\text{Prob}(\theta^+ \mid x_t > 0.5)$ / $\text{Prob}(\theta^- \mid x_t > 0.5)$, being strictly greater than $-\theta^-/\theta^+$, which is denoted by k defined above. That is, $(\mu/(1-\mu))(q/(1-q)) > k$. Note that the above assumption that brand A is optimal under the prior beliefs (the prior odds ratio $q/(1-q)$ exceeds k) and that $\mu/(1-\mu) > 1$. Thus, the agents who observe the group opinion such that $\{x_t > 0.5\}$ at t will choose A with probability 1.

Next, the agents, who observe the group opinion such that $\{x_t < 0.5\}$ at t, choose A when $\text{Prob}(\theta^+ \mid x_t < 0.5)\theta^+ + \text{Prob}(\theta^- \mid x_t < 0.5)\theta^- > 0$, which implies the posterior odds ratio $(\mu/(1-\mu))(q/(1-q))$ exceeds k defined above. It should be noted that this condition is equivalent to $q/(1-q)k > \mu/(1-\mu)$, or

$$\frac{q/(1-q)k}{1 + q/(1-q)k} \equiv \Lambda > \mu.$$

Then the degree of quality of opinions is smaller than a certain threshold Λ. Thus, when the information conveyed by the majority opinion is imprecise, each agent should always follow his or her own prior belief. The agents are not imitative at all. On the other hand, the agents will choose B when $\Lambda < \mu$, that is, the degree of quality of opinions is larger than a certain threshold Λ. Thus, if the signal is precise, each agent should follow the majority side of the group, not one's own prior belief. The population is then more sensitive to the information conveyed by the group opinion. When μ is exactly Λ, agents are indifferent, and hence randomize between the two choices. (If this happens, there is a chance that the agent will flip a coin to decide either to choose A or B.) Furthermore, the agents, who observe the collective opinion $\{x_t = 0.5\}$ at t, will choose A with probability 1, because it follows that $q\theta^+ + (1-q)\theta^- > 0$ in this setting.

Let $\psi_t \in [0,1]$ be the probability that each agent chooses brand A at t. Then there are two different forms corresponding to $\{x_t \geq 0.5\}$ and $\{x_t < 0.5\}$. The form for $\{x_t < 0.5\}$ becomes complicated in this framework. Namely, we have:

$$\psi_t = \{1\}, \quad \text{where } x_t \geq 0.5 \tag{3.5}$$

and

$$\psi_t = \begin{cases} \{1\}, & \text{if } \Lambda > \mu \\ [0,1], & \text{if } \Lambda = \mu, \quad \text{where } x_t < 0.5. \\ \{0\}, & \text{if } \Lambda < \mu \end{cases} \tag{3.6}$$

The following proposition gives a summary of the results discussed above.

Proposition 3.1: [Teraji (2003)]. *Consider the situation in which* ex ante *brand A is better than brand B for each agent.*

1. *Suppose that the agents, given the opportunity to switch the action, observe the group opinion such that $\{x_t > 0.5\}$ at t. Then, they choose A with probability 1.*
2. *Suppose that the agents, given the opportunity to switch the action, observe the group opinion such that $\{x_t < 0.5\}$ at t. Then, for a certain threshold Λ, if $\Lambda > \mu$, they choose A with probability 1; if $\Lambda < \mu$, they choose B with probability 1.*

An implication of the above result is the following. For the group opinion such that $\{x_t \geq 0.5\}$, each agent adopts the action that is socially efficient. On the other hand, for the group opinion such that $\{x_t < 0.5\}$, each agent will adopt the superior choice with a sufficiently small degree of quality of opinions; and with a sufficiently large degree of it, he or she will adopt the choice that is inefficient in the *ex ante* welfare sense. Thus, the decision-making will depend on the relative estimation of the parameters. When μ increases, the importance of imitation increases in the population.

3.4.3 Collective Configurations of Opinions

I consider the social learning system, where the behavioral patterns in the population evolve continuously over time. I analyze a deterministic dynamical system, in which the system variables are the population fractions that use A in each state of the world. A complete characterization is provided to determine the long-run behavior of the system. In particular, the analysis focuses on an equilibrium path that will converge to one of its endpoints, where the population exhibits "conformity."

In Banerjee (1992) and Bikhchandani et al. (1992), the agents are essentially in a line, the order of which is exogenously fixed and known to all, and they are able to observe the binary actions of all the agents ahead of them. An information cascade is a sequence of decisions where it is optimal for agents to ignore their own preferences and imitate the decisions of all those who have entered ahead of them. This model analyzes nonsequential situations, where agents are interacting simultaneously and modifying their decisions at each period in time. All the agents are always present, and they revise their decisions in a continuous mode, and not once and for all. Such a decision structure is better suited for the modeling of market situations. I consider how herding occurs in the process of collective decision-making.

Let us recall that, at every point in time, a fraction α of the agents decide to re-evaluate their choice. Since a fraction $(1 - x_t)$ of the agents are currently using B, the probability that they will choose A is given by ψ_t $\alpha(1 - x_t)$. Similarly, a fraction x_t of these are currently using A, and the probability that they will choose B is given by $(1 - \psi_t)\alpha x_t$.

The dynamics of the fraction of the population, in continuous time, is then characterized by:

$$\frac{dx_t}{dt} = \psi_t \alpha(1 - x_t) - (1 - \psi_t)\alpha x_t,$$

for all $t \in [0, \infty)$. This is the basic equation for the dynamics of social learning. Hence, for all $t \in [0, \infty)$, it satisfies an equilibrium path from x_0.

From (3.5), the dynamic process, which corresponds to $\{x_t \geq 0.5\}$, is determined by:

$$\frac{dx_t}{dt} = \alpha(1 - x_t), \tag{3.7}$$

for all $t \in [0, \infty)$. It is straightforward to show that this process has a degenerate stationary state $x = 1$, where all agents eventually adopt the same choice, A.

Similarly, from (3.6), the dynamic process, which corresponds to $\{x_t < 0.5\}$, is characterized by:

$$\frac{dx_t}{dt} = \begin{cases} \alpha(1 - x_t), & \text{if } \Lambda > \mu \\ [-\alpha x_t, \alpha(1 - x_t)], & \text{if } \Lambda = \mu, \\ -\alpha x_t, & \text{if } \Lambda < \mu \end{cases} \tag{3.8}$$

for all $t \in [0, \infty)$. It is straightforward to show that this process has a degenerate stationary state $x = 1$ (respectively, $x = 0$), where all agents eventually adopt the same choice A (respectively, B), when $\Lambda > \mu$ (respectively, $\Lambda < \mu$) for a certain threshold Λ.

The goal is to study the stability of the stationary states in the dynamical system defined by (3.7) and (3.8), and to demonstrate that two stationary states have different stability properties. Since there are generally multiple equilibrium paths from a given condition, I must be specific about what stability means. It is necessary to introduce some terminology.

Definitions:

1. $x \in [0, 1]$ is accessible from $x' \in [0, 1]$, if there exists an equilibrium path from x' that reaches or converges to x.
2. $x \in [0, 1]$ is globally accessible if it is accessible from any $x' \in [0, 1]$.

The second stability property, which I call globally accessible, states that, for any initial behavioral patterns, there exists an equilibrium path along which the behavioral patterns converge to a degenerate stationary state.

For the dynamics corresponding to $\{x_t \geq 0.5\}$, it follows that:

$$\frac{\partial (dx_t/dt)}{\partial x_t}\Big|_{x=1} < 0,$$

from (3.7). The path implies that $x_t \to 1$ for any $x_t \in [0.5, 1]$. Thus, $x = 1$ is accessible from any $x_0 \in [0.5, 1)$.

For the dynamics corresponding to $\{x_t < 0.5\}$, it follows that, from (3.8),

$$\frac{\partial (dx_t/dt)}{\partial x_t}\Big|_{x=1} < 0,$$

when $\Lambda > \mu$, and

$$\frac{\partial (dx_t/dt)}{\partial x_t}\Big|_{x=0} < 0,$$

when $\Lambda < \mu$. The induced path is $x_t \to 1$ (respectively, $x_t \to 0$) for any $x_0 \in [0, 0.5)$ when $\Lambda > \mu$ (respectively, $\Lambda < \mu$). Thus, $x = 0$ is accessible from any $x_0 \in [0, 0.5)$ when $\Lambda < \mu$. Then, there are multiple equilibrium paths (which converge to $x = 1$ and $x = 0$) with different equilibrium selection mechanisms. Furthermore, $x = 1$ is globally accessible when $\Lambda > \mu$, because it is then accessible from any $x_0 \in [0, 1]$. Then there is a unique equilibrium path of the behavioral patterns, which converges to $x = 1$. All agents will adopt the superior alternative in the limit.

As a consequence, I can identify classes of environment for which the dynamic in the population is such that an homogeneous population arises in the limit. The following key proposition gives a formal summary of the results discussed above.

Proposition 3.2: [Teraji (2003)]. *Consider the situation in which* ex ante *brand A is better than brand B for each agent.*

1. $x = 1$ *is accessible from any* $x_0 \in [0.5, 1]$.
    And for a certain threshold Λ,
2. $x = 0$ *is accessible from any* $x_0 \in [0, 0.5)$ *if* $\Lambda < \mu$.
3. $x = 1$ *is globally accessible if* $\Lambda > \mu$.

Thus, the system will exhibit two possible patterns of behavior in the long-run. First, there is efficient social learning, allowing convergence to the superior extreme ($x = 1$). Second, there is inefficient herding ($x = 0$) if everyone eventually adopts the same choice, but the common choice is not optimal in the *ex ante* welfare sense. The process leads to inefficient information aggregation. This suggests why herd behavior may be undesirable from the social point of view.

Proposition 3.2 shows how these regions arise. For $\{x_t \geq 0.5\}$ at t, the force toward the superior extreme is overwhelming, and efficient social learning occurs. Then, even if the quality of opinions is at the lower level, the efficient state is globally accessible. A low μ implies that the agent might not have confidence in the majority view. The formation is strongly influenced by the initial conditions. On the other hand, for $\{x_t < 0.5\}$ at t, the two stationary states have different stability properties. The long-run behavior of the system is then determined by how a force pushing all agents toward using the same choice combines with the degree of quality of opinions μ. As a result, the system exhibits conformity toward the superior choice when agents have little confidence in the majority view, and exhibits inefficient herding when they have much confidence. These conditions show the possibilities that a society evolves toward efficient herding or inefficient herding. The equilibrium selection mechanism is then based on the presence of the herding externality.

The equilibrium selection problem that arises is solved by agents who follow a sluggish social learning rule. Then the coordination problem, which may prevent the society from attaining the efficient equilibrium, may be alleviated by the reduction of confidence in the majority view. Thus, mass behavior is often fragile, in the sense that the mere possibility of a value change can shatter herding. There are multiple equilibria for some parameter values, and the social learning system leads to fragility only when agents happen to balance on a knife-edge.

3.4.4 Web Herd Behavior: An Example

Digital auctions on the Internet present not only a new market place for transactions, but also a new domain for consumer decision-making. Perhaps the most important factor is that a digital auction is typically a multi-stage process that involves multiple periods. A consumer decides whether to choose or enter a particular auction, which is often followed by a sequence of bidders. The online auction environment provides particular value cues that bidders can rely on. Consumers can update their value assessments based on others' bids. It is important to note that reliance on others' bids in online auctions may often lead buyers to overestimate the value of the auctioned item.

Dholakia and Soltysinski (2001) provide evidence of the herding bias—many buyers tend to bid for items with existing bids, ignoring more attractive items within the same category. Susceptibility to the herding bias implies that the buyer may end up paying a higher price, or winning a less favorable item than necessary. Furthermore, there is a possibility that an item, competitive in all respects, may remain unnoticed and fail to find buyers. Because of the herding bias, the buyers, who participate in these so-called efficient digital market places, routinely violate principles of consistency and make suboptimal bidding decisions.

Behavioral decision research has shown that, in many instances, consumers are influenced by contextual informational cues when making choices, and exhibit inconsistent preferences in different choice contexts (Simonson and Tversky, 1992). In fact, online bidders tend to have a concern for other persons and the outcomes derived from them. When bidding in a digital auction, others' preceding behavior may provide valuable information, and may be perceived as having greater credibility than seller-originating content, such as descriptions or pictures. The process of social identification with other bidders may further increase the influence of this cue. Such a spiraling escalation may magnify the bias, elevating the inferior item's market value furiously.

This situation corresponds to a self-reinforcing process of confidence in the majority opinion. In this situation, each person's decision is more responsive to the majority opinion. The reduction of confidence in the majority view may be socially beneficial in an *ex ante* welfare sense. The society may be better off by encouraging the consumers to use their own information, not joining the herd.

Many studies clearly explain how increasing returns to adoption can lead potential adopters to a situation of lock-in. The sources of lock-in are well established, and are principally network externalities, informational increasing returns, technological interrelatedness, and evaluation norms. However, the interaction between the agents and the resulting aggregate phenomena is often not clear.

One feature of economic activity is the tendency for agents to form coalitions. The aggregate behavior is the result of the interaction between individuals and coalitions. A good example of this sort of approach is the analysis proposed by De Vany (1996), who examines the emergence of self-organized coalitions among decentralized agents playing a network coordination game. The network evolves to optimal or suboptimal coalition structures for some parameter settings. With some settings, the system freezes on suboptimal states. But noisy evolution of coalitions, which is implemented by a "Boltzman network," can overcome lock-in and reach global optima by leaping to new paths. The "temperature" parameter effectively adds "noise" to the signals reaching the agent, and it exponentially increases the paths over which the system may evolve. We can see it as analogous to the model presented in this section.

The model has identified the long-run properties of herd behavior in the economic environment with multiple equilibria. In this section, there are two equilibrium patterns of choices; one is efficient and the other is inefficient in the *ex ante* welfare sense. The model has found the relation between the degree of quality of opinions and the properties of equilibrium. With the small degree of quality of opinions, the efficient equilibrium is unique and globally accessible in the situation. That is, as long as the signal is imprecise, the system can escape suboptimal states. However, for sufficient ranges of quality of opinions, there are multiple steady states. In one of them, there is inefficient herding in the long-run. High signal accuracy gives too much credibility to the majority opinion in society. Each person's decision is then more responsive to the majority view. This is the self-reinforcing nature of confidence in the majority opinion. A reduction of confidence in the majority view has serious consequences in terms of *ex ante* welfare. This is essential in leading the whole society away from inefficient herding.

Because of the simplicity of this model, I have assumed that the confidence an agent attaches to the other agent's opinions is uniform in society. It may be more realistic to consider the quality of opinions as private information. Furthermore, I have assumed that the opportunities to switch actions arrive at random: a more natural assumption is to assume that agents can decide when to decide. Waiting incurs some costs, but it may allow an agent to make a better decision at a later period of time.[19]

Choice decisions made by an individual depend crucially on the perceptions of choice objects. To the extent that such perceptions are affected by social elements, they cannot be independent of a particular social environment in which decisions are made. An individual participates in the economy not simply as an economic abstract with idiosyncratic tastes, but as a whole person with a variety of social concerns and motivations.

3.5 CULTURE

Culture can be understood as heterogeneous. There is cultural variation in the way people think about themselves and about others. These differences are responsible for differences in the way people behave. Cultural differences are, to a large extent, due to environmental differences. Therefore, patterns of social interactions affect the structure of cultural systems. Henrich et al. (2004) propose as the mechanism reducing "intra-group" differences and maintaining "inter-group" differences, by biasing individuals in favor of copying the common ideas, beliefs, and values. Culturally transmitted ideas, beliefs, and values are important for understanding human behavior. As Henrich et al. (2004) suggest, cultural evolution is likely to proceed much more rapidly than genetic evolution, because cultural transmission can spread

---

19. See, for strategic delays caused by information externalities, Chamley and Gale (1994).

novel behavior, ideas, and practices among populations within a single generation. Cultural transmission involves learning from others. It allows individuals to adapt more quickly to changing environments than is possible under either a strictly genetic mode of transmission, or a system that includes only individual trial-and-error learning. Using socially transmitted information, people can make predictions about the intent of others. Once a cultural trait is acquired and internalized, it tends to direct actions without significant cognitive effort or reflection.

The ultimatum bargaining game involves two parties with payoffs that are monetary and transferrable. This sequential-move bargaining game begins when the first player, the "proposer," proposes a division of a fixed monetary pie (typically provided by the experimenter). This proposed split represents a take-it-or-leave-it offer. The second player, the "responder," may accept or reject the offer. The interaction between the players is anonymous. If the responder accepts the proposed division, then the division is, as proposed by the proposer, implemented. If the responder rejects the proposed division, each player earns nothing. Game theory predicts that a self-interested responder will accept any proposed positive payoff, no matter how small. Anticipating this choice, a self-interested proposer will offer nothing more than the smallest amount possible. The equilibrium offer (i.e., the division for which no player has anything to gain by doing something differently) allocates the smallest positive payoff to the responder. The proposer offers the responder the smallest possible amount of money, and the responder accepts: this is the sub-game perfect equilibrium. The sub-game perfect equilibrium gives all the bargaining power to the proposer. That is, any demand that leaves the second player with anything should be accepted, and consequently, the proposer should get almost all of the pie. Experimental subjects repeatedly violate the theoretical predictions.[20] Results from ultimate game experiments in developed countries show that proposers do not demand nearly this amount, generally demanding between 50% and 60% of the total. Research on ultimate game experiments has consistently found that responders tend to reject low offers and thus behave at odds with the assumption that they simply maximize their self-interest (Camerer, 2003).

In experimental research, games have been used recently in anthropology to investigate the basis of prosocial human behavior such as fairness, altruism, and cooperation (Henrich et al., 2004). Using games, researchers standardize protocols across field sites, and compare behavior and groups to explore the range of cross-cultural variation. The experiment in Roth et al. (1991) is the first cross-country comparison study on ultimatum bargaining in the United States, Israel, Japan, and Slovenia (Yugoslavia). In this

---

20. See Güth et al. (1982) for early experiments.

experiment, subjects were first assigned a role: proposers and responders. They played the game in their respective roles for 10 rounds, each time with a different, anonymous, and randomly selected opponent. The pie was worth the equivalent of $10 in terms of purchasing power in all four countries. Subjects were paid according to their monetary payoffs in one randomly selected round. The responder can choose to accept or reject the offer. If the responder accepts the offer, the proposer gets the demanded amount, and the responder gets the remainder. If the responder rejects, neither player receives anything.

The experimental results in Roth et al. (1991) can be summarized as follows:

- The frequencies of the offers implied by the sub-game perfect equilibrium of the game ($0 or $1) ranged from 1% of the time in Slovenia to 9.3% in Israel.
- Offers were highest in the United States and Slovenia, then in Japan, and lowest in Israel.
- The conditional frequency of rejected offers was inversely related to the fraction of the pie offered. That is, low offers were rejected more frequently than high offers. The conditional frequencies of acceptance of offers were lowest in Slovenia, then in the United States, and highest in Israel and Japan.
- Furthermore, the probability that a given offer is rejected was lower in counties where lower offers were observed.

Thus, Roth et al. (1991) find "small," significant differences in the ultimatum bargaining game. They conclude that these differences can be explained as "cultural differences." Camerer (2003) argues that cross-cultural comparison raises at least four difficult methodological problems: stakes, languages, experimenter interactions, and confounds. First, controlling for stakes requires the experimenter to match the purchasing power of the stake in different cultures. Second, keeping the meaning of instructions as constant as possible is important. Third, the biggest mistake in controlling for identity is to use a different experimenter in each culture. The ideal experimenters speak both languages, and are perceived similarly in both cultures. Fourth, it is extremely difficult to avoid the effects of potentially important variables that are confounded with culture (causing "identification problems" in econometrics terms).

In the ultimatum bargaining game, existing experimental data and analyses have shown that subjects from industrial societies behave almost similarly; the mean offer from proposers averages between 40% and 50% of the total, and responders often reject offers lower than 20% of the total. However, Henrich (2000) shows that the Machiguenga of the Peruvian Amazon behave very differently from subjects drawn from industrial societies in the ultimatum bargaining game. The Machiguenga are described as "socially disconnected," with economic life centering on the individual family and little opportunity for anonymous transactions. For the Machiguenga, the economic unit is the family; families fully produce for their own needs (food, clothing, etc.), and do not rely on institutions or other families for their social or economic welfare. Cooperation above the family level is almost unknown, and the Machiguenga are quite socially disconnected. In Henrich's (2000) test, Machiguenga proposers offered only 26% of the total sum, and responders almost always accept offers less than 20% of the total. Machiguenga proposers seem to possess little or no sense of obligation to provide an equal share to responders, and responders seem to have little or no expectation of receiving an equal share. Cultural differences greatly influence economic behavior.

Henrich et al. (2001) recruited subjects from 15 small-scale societies to expand the diversity of economic and cultural conditions in various games, including the ultimatum bargaining game. Their sample consisted of three foraging groups, six slash-and-burn horticulturists, four nomadic herding groups, and two sedentary, small-scale agricultural societies. The experimental results in Henrich et al. (2001) can be summarized as follows:

- The canonical model of self-interested behavior was not supported in any society studied.
- There was considerably behavioral variability across groups. The mean offers from proposers ranged from 26% to 58%. The most common behavior for the Machiguenga was to offer zero. In some groups, rejections were extremely rare, even in the presence of very low offers, while in others, rejections rates were substantial, including frequent rejections of offers above 50%.
- Group-level differences in economic organization and the degree of market integration explain a substantial portion of the behavioral variation across societies. The higher the degree of market integration, and the higher the payoffs to cooperation, the greater the level of cooperation and sharing in experimental games.
- Individual-level economic and demographic variables do not explain behavior, either within or across groups.
- Behavior in the experiments is generally consistent with economic patterns of everyday life in these societies.

Why would people reject a positive amount of money? The responder would reject the offer because he or she is angry at being treated unfairly, and he or she is willing to pay to hurt the person who perpetrated this unfairness. This is not just a matter of developing a capability for anger, but rather a matter for developing a cognitive understanding of whether an angry response is appropriate. What matters is that the "right" response is elicited in the "right" settings. Prosocial emotions (the empathy people feel toward those who have been kind to them, and the hostility they feel toward whose who have not) thus serve to enhance the individual and group fitness in the social conditions. This reflects strong reciprocity, namely, a propensity to reward those who have behaved cooperatively and correspondingly to punish those who have violated norms of acceptance behavior, even when reward and punishment cannot be justified in terms of outcome-oriented preferences.

3.6   CULTURE AND HIERARCHY

Standard economic models are based on the assumption that preferences are strictly personal. Economists often hesitate to incorporate values, norms, and more generally, cultural aspects. They do not deny the need to inquire about preferences, but culture has not been seen as a problem for economists. However, in real life, cultural aspects matter to most people. Preferences are not simply formed in isolation, but they are affected and shaped by the social categories to which individuals belong.[21] The purpose of this section is to investigate culture as a determinant of individual decision-making in a behavioral model of the firm hierarchy. The model demonstrates how individual actions can determine differences in economic outcomes among firms.

What is the substance of culture? At the present time, there is no consensus on a definition of culture. Culture is complex, and it ranges from underlying beliefs and assumptions to visible structures and practices. The study of culture has been overwhelmingly the province of anthropology. For example, in Harris (1971, p. 144), "a culture is the total socially acquired life-way or life-style of a group of people. It consists of the patterned, repetitive ways of thinking, feeling, and acting that are characteristic of the members of a particular society or segment of society."

Numerous cross-cultural studies have demonstrated how peoples' styles are influenced by cultural factors. As an example, individualism/collectivism describes the relationship between the individual and the prevailing collectivity in a society. Individualism implies a loose social framework in which people focus on their goals, needs, and rights more than community concerns. On the other hand, collectivism is characterized by a tight social framework in which people value in-group goals and concerns. According to Hofstede (1984), Australia and other western nations measured high on individualism, whereas East Asian nations measured high on collectivism.

It is a common view for most economists that globalization would raise the overall level of economic activity. Globalization is identified as a new way that firms organize their production process. Members of the global firms come from different cultural orientations. Global firms face cultural factors that are causal variables affecting economic performance. When

---

21. As an example of the formation of beliefs, see the economic analysis of religion in Iannaccone (1998).

people from different cultural orientations are interacting, cultural factors can cause some conflict in economic environments.

There is no need to suppose that individuals actually use inputs as effectively as possible. According to Harvey Leibenstein, many firms remain X-inefficient, even when most of their members know the solutions to perform X-efficiently. X-inefficiency is the failure of a productive unit to fully utilize the resources it commands, and hence attain the familiar production possibility frontier. There are a large number of studies on X-inefficiency originating from Leibenstein (1966). He presented a reasonable vision of human behavior within the organizational context, where X-inefficiency can persist stubbornly. Individuals supply different amounts of effort, where effort is a multidimensional variable, under different organizational and environmental circumstances. The introduction of effort variability allows for the existence of the relatively X-inefficient economy. Furthermore, in Altman (2001), culture is introduced as a determinant of labor productivity. Cultural factors become either facilitators or impediments to X-efficient production, because the quantity and quality of work effort is affected specifically by the cultural milieu of the economic agent. This section aims to present a more general behavioral model of the firm hierarchy, where effort is not contractable because of nonenforceable *ex ante* contracts.

Defining culture as some notion of shared beliefs, expectations, customs, jargon, and rituals, Lazear (1999a) investigates the role of language and cultural assimilation. If there are more than two cultures in a country, a common culture facilitates trade between individuals, allowing them to have common expectations and customs. Members of a minority group may find it urgent to coordinate with the majority of the society. Minority members may be assimilated in order to survive in the society. Furthermore, in Lazear (1999b), a global firm, which is a multicultural team, is analyzed. Global firms face costs to combine workers who have different cultures, legal systems, and languages. In his model, the idea of exploring language acquisition and cultural assimilation is used to analyze the globalization of firms. However, it is not said so much that individuals must give up some freedom to preserve their own cultures and languages in order to coordinate behavioral patterns. Individuals may place a high value on holding on to their original culture.

The model presented here is also related to Akerlof and Kranton (2000). A person's sense of self is "identity" in their terminology. At a very basic level, most people are aware of how they are differentiated from their surroundings. People often perceive themselves as members of social groups. Such group memberships may affect behavior in ways that cannot easily be reduced to material self-interest. According to Akerlof and Kranton (2000, p. 716), "[b]ecause of its explanatory power, numerous scholars in psychology, sociology, political science, anthropology, and history have adopted identity as a central concept. ... We incorporate identity into a general model of behavior and then demonstrate how identity influences economic outcome." Akerlof and Kranton (2000) propose a utility function that incorporates identity as a motivation for behavior. Identity is based on social difference.

In Akerlof and Kranton's (2000) prototype model, identity depends on two social categories, Green and Red, and the correspondence of one's own and others' actions to behavioral prescriptions for their category. There are two possible activities: Activity One and Activity Two. Each person has a taste for either Activity. They consider an interaction between an individual with a taste for Activity One (Person One), and an individual with a taste for Activity Two (Person Two). If Person One (or Two) undertakes Activity One (or Two), he or she earns some positive utility. A person who chooses the activity that does not match his or her taste earns zero utility. As behavioral prescriptions, Greens should engage in Activity One, and Reds should engage in Activity Two. For example, by choosing Activity One, a person could affirm his or her identity as Green. First, identity changes the payoffs from one's own actions. Person One who chooses Activity Two would lose his or her Green identity. Second, identity changes the payoffs of others' actions (externality). If Person One and Person Two are paired, Activity Two on the part of Person Two diminishes Person One's Green identity. Third, the choice of different identities affects an individual's economic behavior. While Person Two could choose between Green and Red, he or she could never be a "true" Green. Finally, the social categories and behavioral prescriptions can be changed, affecting identity-based preferences. Furthermore, Akerlof and Kranton (2000) show that identity can explain many economic phenomena (e.g., gender discrimination in the labor market, the household division of labor, and the economics of social exclusion and poverty). Differently from their framework, this section introduces effort variability into the firm hierarchy model with different cultures, and shows the co-existence of both efficient and inefficient equilibrium solutions.

Incorporating cultural heterogeneity into inherent preferences, the model constructs a behavioral model of the firm hierarchy which consists of a principal and an agent. It emphasizes two individuals, a principal and an agent, as the basic decision units. In doing so, it is explicitly shown that individuals will supply different amounts of effort under different circumstances. Effort put forth by individuals depends on motivation from both incentives and pressures. Effort reflects the pace at which an individual will carry out the activities, the quality of the activities, and the time spent on the activities. Effort may differ across individuals, and is likely to influence their productivity. In this setting, the organization requires both the principal's effort and the agent's effort to implement a project. The more the principal exerts effort, the higher the agent exerts effort. There are thus effort complementarities between the principal's effort and the agent's effort within the organization. Given effort variability, this complementarity property enforces a multiplicity of possible equilibrium solutions within the firm hierarchy.

Building on the work of Grossman and Hart (1986) and Hart and Moore (1990), there is a growing literature on the economics of organizations in a world of incomplete contracts.[22] The incomplete contracting approach is useful for understanding the nature of the firm and the role of property rights. The parties cannot write enforceable contracts contingent on any aspect that might appear in the environment. Under conditions of incompletely specified obligations, an agent's general job attitude becomes important. It is necessary to specify how the surplus is divided between the parties. The organization then faces the friction of incomplete contracts in arm's-length relationships. However, these studies of economic organizations have not focused on cultural factors as causal variables affecting differentials in economic outcomes.

Culture relates to core organizational values. All organizations have cultures that influence the way people behave in a variety of areas such as the rites, rituals, routines, and the language used. An organizational culture is characterized by members shared ability to understand specific concepts within the organization (Karathanos, 1998). Organizations tend to be effective if they have cultures that are well coordinated. However, cross-cultural interactions bring together people who may have different patterns of behaving and believing. Changing from one's own culture to the alternative organizational culture can be uncomfortable. Focusing on costly cultural assimilation in organizations, this section analyzes culture as a determinant of different economic outcomes among firms. Given effort variability, the effort level supplied can be affected by cultural factors. If there are different cultures, both efficient and inefficient economic outcomes can exist simultaneously in a world where effort discretion exists.

3.6.1   The Setting

The organization is a hierarchical system which is composed of interrelated parts. Organizational capabilities are built through specialization. A firm hierarchy is engaged in a common production process. Activities in the firm hierarchy are coordinated by conflicts and incentive problems. In the model, a hierarchy is assumed to consist of a principal ("she") and an agent ("he") to implement a project.[23] This assumption seems to be unrealistic. Usually, there are some relations between individuals at different levels (employees versus the hierarchy). If there are several agents, each agent cares about the others' views, and he is subject to peer pressure. Peer pressure can encourage

---

22. See Aghion and Tirole (1997) for a model of authority and its delegation within organizations.
23. In the present model, the hierarchical size is fixed. See Williamson (1967) and Calvo and Wellisz (1978) for problems of "loss of control" in expanding hierarchical size.

additional work effort from several agents, which is another problem.[24] To highlight cultural assimilation that occurs in the hierarchy, this model assumes that the hierarchy exists between a principal and an agent.

A principal engages an agent to run the project. Within the organization, the principal develops a blueprint in production. The principal is able to have access to the blueprint by exerting some effort. However, knowledge has no economic impact unless some effort is made to implement it. Thus, to implement the blueprint, the firm needs some effort to be exerted by the agent. It may be difficult to observe an individual's contribution to output. Then, individuals who exert higher levels of effort on the job are expected to exhibit greater productivity. High effort yields high output with some probability, whereas low effort yields low output with some probability. The probability of success increases with effort. In this organizational setting, the non-substitutability of efforts exerted by both parties is crucial.

A principal matches with an agent with no cost. The following setting is one of incomplete contracts. The parties cannot sign *ex ante* enforceable contracts. In this case, the actual supply of effort to the production process is not secured by contract. Although no enforceable contract can be signed *ex ante*, the parties bargain over the surplus from the relationship after the blueprint is implemented. This model considers this *ex post* bargaining as a generalized Nash bargaining game, in which the agent obtains a fraction α, where $0 < \alpha < 1$, of the *ex post* gains from the relationship.

A society has different cultural orientations. For simplicity, the model supposes that there are only two cultures, that is, culture A and culture B. Culture is where preferences come from. Culture is a kind of meta-preference, telling us what we should want. Different cultural orientations correspond to different value systems that individuals obey. One individual belongs to one culture originally, with members of one culture preferring one value, while members of the other culture prefer another. One wishes to maintain the culture of origin in terms of attitudes, behaviors, and ways of life. Different individuals may have different cultural-based attitudes over a set of value systems.

Each firm hierarchy requires coordination on a common organizational culture. Cultural assimilation occurs through interaction, whereby boundaries are reduced between members within organizations. Assimilation seems to be useful, because it allows the organization's members to coordinate activities tacitly without having to reach agreement explicitly in every instance. Several studies have revealed that dominant group members often desire the nondominant group to be assimilated into the dominant society. The "ultimate attribution error" is the tendency for members of dominant groups, especially prejudiced individuals, to discount and explain away the achievements of

---

24. See Kandel and Lazear (1992) for a detailed discussion of peer pressure.

oppressed group members (Pattigrew, 1979). As part of the ultimate attribution error that dominant group members may incur, they are likely to view assimilation attitudes and behaviors of nondominant group members positively. Nondominant members will be assimilated to work, even though they may place a value on holding on to their original culture.

The present model considers the situation in which the agent is required to coordinate on the principal's culture. Individuals wish to maintain their cultural identity. Even though the agent prefers his own culture inherently, he has to adopt the principal's culture to work. Pressure is thus created in the presence of the need to coordinate on the alternative culture. An individual who has to change cultures suffers a utility (or virtue) loss for the self-inflicted loss of cultural identity. Adopting the principal's culture is more costly if the agent has to abandon his own culture to work. Cultural assimilation thus requires an additional cost if the agent has to change cultures in the firm hierarchy.

This framework incorporates culture into a behavioral model in the presence of costly cultural coordination. The approach treats culture as a determinant of economic outcomes, given effort is discretionary. Individuals are assumed to be rational utility maximizers. In the model, however, effort is a discretionary variable. Given effort variability, individuals can supply different amounts of effort in the firm hierarchy. The economic performance depends on the amounts of effort supplied by members of the firm hierarchy. The framework demonstrates how effort is affected by culture in the work environment.

3.6.2   The Framework

Consider a simple firm hierarchy which is composed of a principal ("she") and an agent ("he"). A principal and an agent are matched in pairs. Upon matching, the principal develops a blueprint and the agent implements it. Let π be the (pecuniary) total benefit when the project is implemented. Individual interactions are not governed by complete enforceable contracts. Effort is not subject to contract. Although no enforceable contract can be signed *ex ante*, the two parties bargain over the benefit π after the blueprint is implemented. Let $(1 - \alpha)\pi$, where $0 < \alpha < 1$ denotes the principal's benefit when the project is implemented. Similarly, $\alpha\pi$ is the agent's benefit when the project is implemented. The parameter α is the congruence parameter capturing the degree of conflict between the principal and the agent. Here, π is supposed to be known to each party *ex ante*.

This situation can be also seen as an appropriability problem. Organizations cannot exist if the legal rights of ownership cannot be enforced. It is a possible court challenge that may be initiated by either party after the project is implemented. This appropriability is captured with a parameter α that is the probability a court awards the benefit to the agent, if such a court challenge takes place.[25] Thus, if the agent is involved in the project, he must be paid the *ex post* benefit $\alpha\pi$.

Suppose that there exist two cultures in a society, culture A and culture B. Furthermore, principals are assumed to belong to culture A. On the other hand, each agent may belong to one of two cultures. Thus, one principal may match with an agent from culture A, while the other principal may match with an agent from culture B.

The principal's effort translates into a blueprint with probability E, and into no blueprint with probability $1 - E$. Let e_A (respectively, e_B) denote the effort exerted by the agent from culture A (respectively, culture B). Then the agent from culture A (respectively, culture B) implements the blueprint with probability e_A (respectively, e_B), and does not implement it with $1 - e_A$ (respectively, $1 - e_B$). All blueprints are assumed to yield the same benefit. In the present model, economic outcomes are distinguished only by the amounts of effort supplied by each party. Thus, $e_A E(1 - \alpha)\pi$ (respectively, $e_B E(1 - \alpha)\pi$) is the principal's *ex ante* expected benefit if she hires the agent from culture A (respectively culture B). On the other hand, $e_A E\alpha\pi$ (respectively, $e_B E\alpha\pi$) is the agent's *ex ante* expected benefit if he belongs to culture A (respectively, culture B) originally.

The principal chooses how much effort to devote to developing the blueprint. Developing the blueprint generates some private cost $g(E)$ for the principal. More specifically, this is assumed to generate some convex cost, that is, $g(E) = (1/2)E^2$. Thus, the more the principal exerts effort, the higher is the marginal cost of further effort.

Furthermore, to make the analysis tractable, it is assumed that each agent picks his own effort in some interval $[0, \hat{e}]$, where $0 < \hat{e} < 1$, and that he has linear disutility of effort in this interval. This interval $[0, \hat{e}]$ is constant for each agent. For example, the time spent on the project is limited in a day. The common maximum effort is assumed to be $\hat{e}$. On the other hand, if the agent does not implement the project, his effort level is 0 (the common minimum effort). Then he receives no benefit. Thus, the agent from culture A (respectively, culture B) implements the blueprint with probability e_A (respectively, e_B) by exerting some cost $\gamma_A e_A$ (respectively, $\gamma_B e_B$) in the interval $[0, \hat{e}]$, where γ_A, $\gamma_B > 0$.

Each firm requires an agent to coordinate on the principal's culture, that is, culture A. All agents must adopt culture A to work. However, changing cultures is more costly if the agent belongs to culture B by nature. Coordination on the principal's culture thus requires an additional cost for the agent from culture B. The parameter γ_B is supposed to contain such an additional cost when he exerts effort in the firm hierarchy. On the other hand, γ_A is smaller, since the agent from culture A can work in the identical

---

25. Other transactions, including an employment contract that specifies any particular level of effort, cannot be enforced by the court.

cultural environment. Thus, we suppose that $\gamma_A < \gamma_B$ holds. This inequality takes into account cultural differences, and implies that the cost to work can differ among agents. It is costly for the agent to adopt the alternative culture in the work environment.

3.6.3 Analysis

If a principal matches with an agent from culture A, the two parties' expected relevant utilities are:

$$U = e_A E(1 - \alpha)\pi - \left(\frac{1}{2}\right)E^2$$

and

$$u_A = e_A E \alpha \pi - \gamma_A e_A.$$

The first-order conditions of the two parties with respect to their effort levels E and e_A are:

$$\text{Principal: } E = e_A(1 - \alpha)\pi$$

and

$$\text{Agent: } e_A = \begin{cases} \hat{e} & \text{if } E\alpha\pi \geq \gamma_A \\ 0 & \text{if } E\alpha\pi < \gamma_A \end{cases}.$$

The relationship between the principal's effort E and the agent's effort e_A is critical in understanding this organizational enterprise. The principal exerts more effort the higher the agent's effort e_A, the higher her fraction of the benefit (the higher $1 - \alpha$), and the higher the *ex post* benefit π. The agent, in turn, exerts the highest level of effort the higher the principal's effort E, the higher his fraction of the benefit (the higher α), the higher the *ex post* benefit π, and the lower his γ_A.

There are multiple Nash equilibriums of effort levels, $(e_A{}^*, E^*)$, in the firm hierarchy. That is:

$$e_A^* = \hat{e} \text{ and } E^* = \hat{e}(1 - \alpha)\pi \quad \text{if } k \geq \gamma_A$$

$$e_A^* = 0 \text{ and } E^* = 0 \quad \text{if } k < \gamma_A$$

where

$$k \equiv \hat{e}(1 - \alpha)\alpha\pi^2.$$

We can see there are effort complementarities between the principal's effort and the agent's effort within the firm. Under this complementarity property, an increase (respectively, decrease) in the agent's effort increases (respectively, decreases) the principal's effort. Therefore, the economic system has a multiplicity of possible equilibrium solutions.

Similarly, if a principal matches with an agent from culture B, the two parties' expected relevant utilities are given by:

$$U = e_B E(1 - \alpha)\pi - \left(\frac{1}{2}\right)E^2$$

and

$$u_B = e_B E\alpha\pi - \gamma_B e_B.$$

The first-order conditions of the two parties with respect to these effort levels E and e_B are:

$$\text{Principal: } E = e_B(1 - \alpha)\pi$$

and

$$\text{Agent: } e_B = \begin{cases} \hat{e} & \text{if } E\alpha\pi \geq \gamma_B \\ 0 & \text{if } E\alpha\pi < \gamma_B \end{cases}.$$

Therefore, we can compute the Nash equilibrium effort levels, $(e_B{}^*, E^*)$, in the firm hierarchy.

$$e_B{}^* = \hat{e} \text{ and } E^* = \hat{e}(1 - \alpha)\pi \quad \text{if } k \geq \gamma_B$$

$$e_B{}^* = 0 \text{ and } E^* = 0 \quad \text{if } k < \gamma_B$$

where

$$k \equiv \hat{e}(1 - \alpha)\alpha\pi^2.$$

Thus, if γ_B is so large that $k < \gamma_B$ holds, the agent from culture B supplies the minimum level of effort 0, and the principal supplies the effort level 0 as a result. Then they receive no benefit. On the other hand, if $k \geq \gamma_A$ holds, the agent from culture A supplies the maximum level of effort $\hat{e}$, and the principal supplies $\hat{e}(1 - \alpha)\pi$.

Culture is a kind of meta-preference. If cultural assimilation occurs, it is costly for the agent from culture B to coordinate on culture A. In the utility function, cultural differences reflect γ_A and γ_B. If the agent's disutility to change from culture B to culture A is sufficiently large, he supplies the minimum level of effort, that is, $e_B = 0$. On the other hand, if the disutility for the agent of culture A is sufficiently small, he supplies the maximum level of effort, that is, $e_A = \hat{e} (> 0)$. If cultural differences exist, effort inputs may vary. Economic outcomes are distinguished only by the amounts of effort supplied.

In summary:

Proposition 3.3: (Teraji, 2008). *Suppose that $\gamma_A \leq k < \gamma_B$. If a principal meets with an agent from culture A, the agent's effort level is $\hat{e}$ and the* *principal's effort level is é(1 − α)π. On the other hand, if a principal meets with an agent from culture B, the agent's effort level is 0 and the principal's effort level is also 0.*

Thus, we have both efficient and inefficient equilibrium solutions simultaneously in a world where effort discretion exists. Economic outcomes, which are distinguished only by the effort level supplied by each party, can vary among firms. For some firms, the dominant equilibrium may be suboptimal. This results from changing cultures. Individuals interact in the production process, and they may exert spillovers over each other. The organization confronts coordination problems. The organization is then "cooperative" in the sense that the increase (respectively, decrease) in the agent's effort encourages (respectively, discourages) the principal in supplying her effort. If the agent's disutility to change from culture B to culture A is sufficiently large, he supplies the minimum level of effort, which is 0, in the firm hierarchy. It is discouraging for the principal to expect that the agent would supply the minimum level of effort. Therefore, the principal also supplies the minimum level of effort. The equilibrium solution is inefficient, because of the lowest performance of the firm hierarchy. On the other hand, if the disutility for the agent of culture A is sufficiently small, the equilibrium solution is efficient because of the highest performance.

The circumstances under which culture is a crucial factor affecting economic performance have been presented. It is possible to expect there to be both efficient and inefficient equilibrium solutions potentially. Given effort variability, effort complementarities between the principal's effort and the agent's effort allow the suboptimal equilibrium solution in the firm hierarchy. Which equilibrium solution is chosen depends on the agent's original culture. Changing cultures can generate inefficient economic environments if effort is a discretionary variable.

The presence of culture, which may lead to inefficient economic outcomes, conflicts with the neoclassical economic model. Many aspects of social interactions are not governed by contracts. Will the recognized inefficiency persist under the condition of competitive markets? Will people be willing to pay to bring the improvement if they know the more efficient economic opportunities? Preferences are not simply formed in isolation.[26] "Preferences are learned as an accent or a taste for a national cuisine is acquired, that is, by processes which may but need not be intentional" (Bowles, 1998, p. 80). Culture is the non-market factor. Culture is, was, and will be a major characteristic of the human being. Market forces cannot

---

26. Akerlof (1983) assumes that values are not fixed, but may change as people go through various experiences. Parents' experiences are especially important in shaping what they value for their children. In Akerlof (1983), theses value-changing experiences are called "loyalty filters."

deprive the agent of his cultural identity. Even though the agent knows that his original culture causes the inefficient outcome, he hesitates to abandon it. People have the freedom to preserve their own cultures.

Therefore, organizations are expected to manage cultural diversity effectively. In order to avoid the negative consequences of cultural assimilation, some adaptations must be achieved through a long-term process. When different cultural groups are in contact over a long period of time, they are involved in a process of various changes, denoted as the acculturation process. Acculturation is a process of changes that involve various forms of mutual accommodation, leading to some long-term psychological and sociocultural adaptations between both cultures. Acculturation is distinguished from assimilation. Acculturation takes place in the dominant group as well as in the nondominant group. The dominant group's acculturation orientations indicate whether the dominant group allows the nondominant group members to maintain their own culture. When both groups prefer similar acculturation orientations, the long-term relationship between the groups will be consensual. Then the agent will not suffer a utility (or virtue) loss for the self-inflicted loss of cultural identity in the long-run. Cultural sensitivity is thus needed. People have to be well aware of cultural differences, know why differences exist, and accommodate these differences within organizations.

It is well-known that repeated interactions may create incentives that are absent in one-shot interactions (Axelrod, 1984). However, this situation differs from Axelrod's "tit-for-tat" model. In Axelrod's model, employees have extrinsic incentives to behave cooperatively, because employers can punish them by paying low wages. In this situation, employees react to sociocultural approval in organizations. This sociocultural interaction may create another form of extrinsic incentives to behave cooperatively. If the principal allows the agent to maintain his own culture B through a long-term acculturation process, then γ_B may converge to γ_A. Cultural conflict can be avoided only when the two groups agree that mutual accommodation is the appropriate course to follow. Therefore, if the disutility for the agent from culture B is sufficiently small, he gets to supply the maximum level of effort, $e_B = \hat{e}$, in the long-term process. Effort can be homogeneous in the long-term relationship. Acculturation is thus an efficiency-enhancing device in the sociocultural context.

This section has presented a framework for understanding the importance of culture as a determinant of economic performance in the firm hierarchy. Economic outcomes are distinguished only by the amounts of effort supplied in organizations. It has described a behavioral model of the firm hierarchy which consists of a principal and an agent. In this setting, the organization requires both the principal's effort and the agent's effort to implement a project. However, the actual supply of effort to the production process is not subject to contract. There are effort complementarities between the principal's effort and the agent's effort, and the more the principal exerts (respectively, less) effort the higher (respectively, lower) the agent's effort within the organization.

Specifically, there are two cultural orientations in a society, and one individual belongs to one culture originally. The agent has to coordinate on the principal's culture in order to work. If the agent abandons his own culture, he suffers the self-inflicted loss of cultural identity. Adopting the principal's culture is more costly if the agent has to change cultures. Coordination requires an additional cost if the agent has to adopt the alternative culture.

If the agent's disutility to change cultures is sufficiently large, he supplies the minimum level of effort. The principal then supplies the minimum level of effort, expecting the agent's minimum level of effort (i.e., the complementarity property). The equilibrium solution is inefficient because of the lowest economic performance of the firm hierarchy. Thus, for some firms, the dominant equilibrium may be suboptimal. Given effort variability, different coordination costs can generate different amounts of effort among firms. If effort inputs vary, economic outcomes can vary among firms. The complementarity property enforces a multiplicity of possible equilibrium solutions. However, even though the presence of cultural identity generates inefficient equilibrium outcomes, market forces cannot reduce the freedom to preserve original cultures.

3.7   POSSIBLE SELVES

Typically, the standard economic agents do not care about "who they are." In modern economic theory, the individual has a stable and consistent preference ordering over goods and actions. However, preferences are changing at the normal level of personality development. In psychology, people are likely to learn things about themselves and struggle with their identity. The self-concept is an individual's belief about his or her personal qualities and abilities. Our self-concept or identity has profound effects on the way we behave. Economists have largely neglected the self as a primary force of individual behavior.

According to Leibenstein (1976, 1979), the micro–micro problem is the study of what goes on inside a "black box." He argues that conventional micro theory is based on a theory of single person behavior. His approach aims to explain organizations' external activities through the study of their internal processes. Within the firm and the household, he considers personal motivations and interpersonal relations. This section focuses instead on intrapersonal relations in personality development. Different selves of a single person make up that person.[27] Individual choice is often accompanied by internal conflict.

---

27. Presently-biased preferences lead to a "multiple selves" setting which discusses the issues of self-control (Ainslie 2001). Hyperbolic time discounting implies that people will make relatively short-sighted decisions when some costs or benefits are immediate.

The definitional unity of the self is called into serious question by the empirical inconsistencies of economic choice and by the need to construct multiple-self theories to explain them (Lea and Webley, 2005). As Moldoveanu and Stevenson (2001) point out, there are two models of the self: the self as a unified system, and the self as a fragmented entity. A multiplicity of selves is composed of a wide variety of unrelated and conflicting roles. People with multiple selves may make more rational and more adaptive decisions in economic affairs than people with a single unified self in some situations (Lester, 2003).

Akerlof and Kranton (2000) emphasize the importance of identity, a person's sense of self, for economic decision-making. They treat identity as an argument in a person's utility function. Identity in this function is based on the social categories to which an individual belongs. Each individual possesses one particular collection of social characteristics.[28] Their model offers little to explain how individuals' multiple identities are related (Davis, 2007). The Akerlof–Kranton framework does not explain how an individual can change and remain the same individual. Identity choice is very often limited in their framework.

The self-system is multifaceted. An individual's overall self is typically represented as a set of categories, each of which represents a distinct self (Stets and Burke, 2003). Distinct selves are typically tied to a particular situation. Furthermore, the self-system is highly dynamic. Self-conceptions can be activated by certain features of the situation (Kruglanski, 1996). Although people may have distinct multiple selves, only one of them tends to guide behavior.

This section proposes a mechanism for "intentional self-change" over time. Intentional self-change is an effort to construct a particular kind of self. There is no reason to suppose that individuals always remain the same. Intentional self-change redefines who one is. The model provides a simple theory of the decision to change oneself, considering the capacity of the self for change. This change will motivate individuals to revise their ideas and develop different characteristics. Then, their goals and achievements will change. Optimistic self-views can improve performance: in this sense, the present framework is also related to Bénabou and Tirole (2002, 2011) and Compte and Postlewaite (2004).[29] However, my focus is different from theirs, because the self is considered to be immersed in a variety of motives. People may have the motives to steer self-relevant information processing so as to raise the levels of ability or performance. Self-regulation, the capacity of the self to change, is an important function. On the contrary, people may

---

28. See Bazin and Ballet (2006) for an approach to tensions between various forms of identities.
29. Bénabou and Tirole (2011) develop a general model of identity in which people infer their own values from past choices. In their model, an agent who has built up enough of some (economic or social) asset continues to invest it even when the marginal return no longer justifies it.

have motives to direct self-relevant information processing in favor of the confirmation and validation of existing self-beliefs. There may be a cultural regime that generates an X-inefficient economy (Altman, 2001). In this section, I investigate why people sometimes seek positive feedback, and sometimes seek subjectively accurate feedback.

The self can be considered to be a product of social interaction. People compare themselves with others to evaluate their abilities and characteristics. By modifying their inherent preferences and internalizing external values, individuals may get potential benefits from conformity to their social group. The concept of self is closely intertwined with that of the other in a social context. According to identity theory, an identity is a set of "meanings" applied to the self in a social role or situation, defining what it means to be who one is (Burke, 1991; Burke and Tully, 1977). The meanings that define an identity are the identity standards of any group-, role-, or person-based identity such as American, spouse, or honest. Variation in self-concepts is due to the different roles that people occupy in society. An identity operates as a dynamic, self-regulating control system. People engage in behavior to create meanings that correspond to the meanings of their identity standards. Deviations from the identity standards result in distress. People modify their behavior to achieve a match with their identity standards. This process in turn reduces distress.

People have an array of "possible selves." Possible selves are representations of the self in the future, and supply direction for the achievement of the desired goal. Self-discrepancy theory indicates that a person possesses distinct domains of the self (Higgins, 1987). The actual self refers to a representation of the attributes that we believe we currently possess. The ideal self refers to a representation of the attributes that we wish or hope to possess.[30] The self-state representation concerning the actual self constitutes the self-concept. The remaining self-state representation is the "ideal self-guide," that is, self-motivating standard. Ideal self-guides are associated with promotional goals that we strive for. People compare their self-concept to the ideal self-guides, and they are motivated to achieve a state of congruence between their self-concept and the ideal self-guides. In general, discrepancies between the actual self and the ideal self produce dejection-related emotions (depression, disappointment, or sadness).

People may perceive, seek out, and create events in such a way as to enhance their views of themselves with respect to their environmental surroundings. The basic model starts with a conceptualization of the self that includes self-state representations, the actual self, and the ideal self. The

---

30. Furthermore, the ought self refers to a representation of the attributes that we believe we should possess. Ought self-guides are associated with prevention goals that we strive to avoid. Discrepancies between the actual self and the ought self produce agitation-related emotions (anxiety, fear, or uneasiness).

ideal self places principles and values above practical considerations. Comparison between the actual self and the ideal self will have important motivational consequences. Self-regulating possible selves will have significantly greater chances of success. "Self-enhancement" is then the drive to convince ourselves that we are intrinsically worthwhile. Self-enhancement underlies people's tendency to believe that they can improve relative to the past. Human judgment involves consistent departures from normative rationality, and it is distorted by an array of biases. People are motivated to maintain a certain kind of self-enhancement if their perceptions of the feedback they receive about themselves are distorted.

However, it would be an exaggeration to say that self-enhancement is always the dominant self-motive. People may act to maintain a consistent (and negative) view of themselves. The present model accounts for "self-verification" which is the desire to confirm one's existing self-views. People may desire psychological coherence regarding their self-image (Swann et al., 1992). All people possess flaws and weakness, and it makes sense for them to maintain appropriately negative self-views to reflect flaws and weakness. A form of self-verification may occur when people choose partners who see them as they see themselves, thereby creating social environments that are likely to support their self-views. Self-verification is expected to undermine intentional self-change. An identity is thus a perceptual control system. If one identity is more salient than another, that is, more likely to be activated in a situation, the costs of changing that identity will be greater.

3.7.1 The Setting

Most people have positive as well as negative self-views. Under what conditions does self-enhancement occur? Under what conditions is self-enhancement absent? This section defines distinct domains of the self, and presents a model of the decision to change oneself (Teraji, 2009).

Self-conceptions do not only describe how people are at a certain point in time. People also have an array of possible selves. Internal change encompasses the production of possible selves. Possible selves are future-oriented schemata of what people believe they could potentially become. The self-system is multifaceted and dynamic, with different self-representations activated at different times. Self-discrepancy theory considers an aspect of the self that includes distinct self-state representations (Higgins, 1987). My present focus is on the actual self (how we are currently), and the ideal self (how we would like to be). The ideal self signifies the attainment of positive outcomes when the end-state is desired. It is possible to achieve a desired end-state by approaching matches to it.

Comparison between the actual self and the ideal self has important motivational consequences. The actual−ideal congruency represents the presence of positive outcomes, while the actual—ideal discrepancy represents the absence of these outcomes. How does discrepancy mediate change? The response to perceived discrepancy can be shaped toward desired behavioral change. Significant improvement may occur for agents with plausibly self-regulatory possible selves. In particular, the capacity for self-regulation can lead to an increased likelihood of change. Self-regulation occurs with respect to the ideal self-guide. This reflects the desire to bring oneself closer to what one would ideally like to become. Actual—ideal self-regulation involves a promotion focus, that is, a concern with advancement, growth, and accomplishment. People who have high capacities for self-regulation may be more inclined to decide to change themselves, even in the face of obstacles. In the presence of cognitive inconsistencies that threaten self-concept, individuals often eliminate previous behaviors and adopt specific behaviors to reduce dissonance. Individuals are motivated to match self-concept to self-guides, with dejection-related symptoms resulting from unresolved discrepancies between actual and ideal domains.

Self-motives are mechanisms by which people create or maintain certain self-images in their own minds. Specifically, self-evaluation needs may be satisfied either through the positive enhancement of self-relevant information ("self-enhancement"), or the affirmation of already existing self-conceptions ("self-verification"). Self-enhancement is the motivation to elevate positive aspects of one's self-concept or protect oneself from negative information. Self-enhancement has been identified as underlying people's tendency to believe that they have improved relative to the past, and that their personal improvement has been greater than other people's (Leary, 1995). Some people have a better developed self-enhancement motive than others. Self-enhancers are those individuals who perceive themselves more positively than they perceive others.

However, self-verification is also an important factor in the individual's motivational system. Individuals desire not merely to know how great they are (self-enhancement), but also to confirm that they are the type they already thought they were (self-verification). Hence, people might prefer to maintain a coherent image of who they are. Self-verification is defined as the desire to confirm preexisting self-views (subjectively accurate). According to self-verification theory, a preference for self-verifying evaluations should emerge, even if the self-view happens to be negative. Self-verifying information leads to stability in people's self-concepts and makes people feel that they understand themselves (Swann et al., 1992). Self-verification involves the impulse to act in accordance with the preexisting self-concept, and to maintain it intact in the face of any discordant piece of information about it. For example, Rosenberg (1979) reports that adolescents with inconsistent self-concepts showed greater psychological distress.

Intentional self-change represents an acceptance of a new future self. It is considered a strategic action. Decision-making requires a prediction of future selves. When individuals contemplate self-change, they assess the costs and benefits of change. To promote positive views of self is one option. People then seek favorable feedback about their positive self-views. Unless the advantages of change outweigh the costs, the agent's position is fixed through time. People then seek unfavorable feedback about their negative self-views. However, people may seek positive views of self and adopt strategies which promote such self-views. This section formalizes the implications of biases in the prediction of future selves.

Self-confidence or self-esteem refers to a sense of self-worth. The importance of self-confidence lies in what people believe they need to be worthy and valuable human beings. Self-confidence varies among people. People who have high self-confidence will evaluate themselves more favorably than people with low self-confidence. This may be the result of biased attention. People are not always objective at perceiving themselves. For ambiguous attributes, self-concepts are largely idiosyncratic. People with high self-confidence will show more justification of their behavioral change than people with low self-confidence. For example, an individual with high self-confidence may say to himself or herself, "I may not be doing well now, but I will succeed next year." People's perceptions can be biased by their beliefs. Self-confidence or self-esteem shapes how people view themselves on these attributes. A self-serving bias exists when an individual has optimistic beliefs about their own ability. Individuals with self-serving biases are more likely to present favorable images of themselves, independently of the way they actually think about themselves. Hence, the degree of successful self-change tends to be exaggerated.

People will evaluate their behavior in terms of its fit to their own self-expectations for competent conduct. We can expect persons with high self-confidence to perform more effectively than ones with low self-confidence. People with high self-confidence tend to remember past performances as more successful than they were, and to overestimate their abilities. They are cognitively motivated to change when they view themselves as effective persons. Possible selves can facilitate optimism, and belief that change is possible because they provide the sense that the current self is mutable (Markus and Nurius, 1986). People with self-serving biases seek and use information to enhance self-images. By providing hope for a better future, possible selves may fulfill self-enhancement goals and shape better well-being. Possible selves may supply direction for the achievement of the desired goal that represents aspiration.

Affective components gradually receive a growing amount of attention.[31] Motives and emotions are strongly linked. Most judgments and most choices

---

31. For example, in Loewenstein (1996), people in a "cold" state (i.e., not hungry, angry, in pain, etc.) underestimate the influence of a future "hot" state (i.e., craving, angry, jealous, sad, etc.). He argues that this underestimation of future visceral drives is due to constrained memory for visceral experiences.

are made intuitively (Kahneman, 2003).[32] People experience self-conscious emotions when they become aware that they have lived up to, or failed to live up to, some actual or ideal self-representation. An actual–ideal discrepancy might result in depression, disappointment, or sadness (dejection-related emotions).

People may have a powerful sense of psychological coherence. People then limit the possible selves they generate based on their emotions. A person may underestimate the magnitude of self-change. People with low self-confidence seek and use information in order to maintain consistent self-views. They may have an inert area within which they do not attempt self-change and fail to reach optimal economic outcomes.

The self is an adaptive process. People are then influenced by two factors, self-enhancement and self-verification, within the structure of the self-concept. Self-verification is expected to undermine intentional self-change. People who receive self-verifying feedback appear to have more enduring and stable self-views. Expectations regarding personal abilities are central to the formation of human behavior. Self-discrepancy results in depression. People who hold more positive expectancies for themselves (self-enhancement) can lead to better outcomes by coping effectively with stressful situations. They are more likely to overestimate the actual–ideal discrepancy for desired behavioral change. They are motivated to reduce the discomfort associated with dissonance and change themselves. As a result, the greater the discrepancies perceived by self-enhancers, the less consistent the self-conceptions. In contrast, people who hold less positive expectancies for themselves may underestimate the actual–ideal discrepancy, and they act in accordance with the actual self. Self-verifying feedback leads to more stable self-conceptions among persons with low self-confidence.

3.7.2   The Framework

Consider an agent who faces a sequence of decisions over a horizon of two periods: $t = 0, 1$. In each period, the agent selects some level of effort. These decisions affect the economic outcomes. At $t = 0$, "Self 0," which represents the actual self, is activated. At $t = 1$, "Self 1," which represents the ideal self, can be activated. At the beginning of $t = 1$, the agent has to make a judgment about the relative merits of the options he or she is about to face. Then the agent chooses one of two options: one is to choose Self 0, and the other is to choose Self 1. Judgments are modeled as his or her chances of

---

32. Reasoning is done deliberately, but intuitive thoughts seem to come spontaneously to mind.

activating Self 1 successfully before the decision is made. The agent's self-confidence depends on the prediction of successful self-change. The timing is as follows:

At $t = 0$:
- Self 0 exerts effort in the process of production.
- The payoff is realized.

At $t = 1$:
- Judgments are made based on the agent's chances of activating Self 1.
- The agent chooses Self 0 or Self 1.
- Self 0 or Self 1 exerts effort in the process of production.
- The payoff is realized.

At $t = 0$, the agent starts with Self 0. At $t = 1$, self-concepts may be largely idiosyncratic. Self-confidence or self-esteem refers to the extent to which individuals see themselves as having control over their own outcomes. The decision to change oneself is the outcome of appraisal, in which the agent assesses how effective Self 1 would be, and how successfully he or she could activate it. A self-serving bias exists when an individual's preferences affect his or her beliefs in an optimistic direction. The agent may have some self-serving bias for the emergence of Self 1. In this model, optimism depends on the agent's prediction of how successfully he or she activates Self 1. The agent may be biased in the assessment of the probability of successful self-change. The optimistic (or pessimistic) agents tend to overestimate (or underestimate) its chances of success.

Let λ be the subjective belief that the agent activates Self 1 successfully in Period 1. Thus, with probability λ the agent predicts the emergence of Self 1, and with probability $1 - \lambda$ the agent predicts that Self 0 remains in Period 1. The level of self-confidence thus transforms into his or her subjective beliefs over λ. With this formulation, $\lambda = 0$ means no self-serving bias. The manifestation of self-serving bias is optimism about one's own future prospects to activate Self 1. If $\lambda > 0$, the agent may have some self-serving bias; the larger λ is (or the smaller $1 - \lambda$ is), the stronger is the self-serving bias. People with no self-serving bias stick to Self 0 at $t = 1$. They are not able to choose what is best for them. At $t = 1$, Self 1 can produce output with the following technology:

$$\begin{cases} 1 & \text{with probability } \pi E \\ 0 & \text{with probability } 1 - \pi E_t \end{cases}$$

where E is the level of effort that Self 1 chooses in Period 1, and $0 < \pi < 1$ is a parameter. Thus, the production activity is either a success or a failure. Self 1 will receive 1 in the case of success, and 0 in the case of failure.

At $t = 0, 1$, Self 0 can produce output with the following technology:

$$\begin{cases} 1 & \text{with probability } \theta \pi e_t \\ 0 & \text{with probability } 1 - \theta \pi e_t \end{cases},$$

where e_t is the level of effort that Self 0 chooses at $t = 0, 1$, and $0 < \theta < 1$ is a parameter. In each t, Self 0 will receive 1 in the case of success, and 0 in the case of failure. Self 1 has greater chances of succeeding than Self 0 in Period 1.

Self 1's flow payoff at $t = 1$, V, is given by:

$$V = \pi E - \left(\frac{1}{2}\right) E^2,\tag{3.9}$$

where $(1/2) E^2$ is the cost of effort, which measures the task's difficulty. The first-order condition with respect to the level of effort is $E = \pi$. Thus, at the end of $t = 1$, Self 1 obtains:

$$V = \left(\frac{1}{2}\right) \pi^2.\tag{3.9$'$}$$

Likewise, Self 0's flow payoff at $t = 0, 1$, U_t, is given by:

$$U_t = \theta \pi e_t - \left(\frac{1}{2}\right) e_t^2,\tag{3.10}$$

where $(1/2) e_t^2$ is the cost of effort at $t = 0, 1$. Self 0 supplies the amount of effort as follows: $e_t = \theta \pi$. Self 0 always exerts less effort than Self 1. At the end of $t = 0, 1$, Self 0 obtains:

$$U_t = \left(\frac{1}{2}\right) (\theta \pi)^2.\tag{3.10$'$}$$

Realization that current behavior does not measure up to the relevant standard can trigger negative affect. Discrepancies between Self 0 and Self 1 produce dejection-related emotions (depression, disappointment, or sadness) in Period 1. Let c be a (dejection-related) psychological cost to maintain Self 0 in Period 1. That is, the agent suffers the cost c if he or she chooses Self 0 at $t = 1$. For the following analysis, it is assumed that $(1/2)(\theta \pi)^2 > c$.

On the contrary, incoherent self-concepts might result in distress, because people feel some kind of uncertainty regarding their self-image. People have feelings of coherence. Self-verification involves the impulse to act in accordance with the self-concept. The psychological significance of the stable self-views can be appreciated by considering what happens to individuals whose self-views have been changed. Thus, self-change may increase some disutility among individuals. Let d be a disutility cost that the agent attaches to any self-change at $t = 1$. That is, choosing Self 1 involves the cost d in Period 1. For the following analysis, it is assumed that $(1/2)\pi^2 > d$.

Thus, at the end of $t = 1$, with probability λ the agent expects to obtain $V - d$; with probability $1 - \lambda$ the agent expects to obtain $U_1 - c$.

3.7.3 Self-Confidence and Economic Performance

At $t = 1$, the agent decides whether to choose Self 1 or not, having the subjective beliefs over λ (defining the self-serving bias). Formally, let x denote the probability that the agent (who may have some self-serving bias) chooses Self 1 in Period 1.

From (3.10)', the agent's utility as of Period 0 is denoted by:

$$U_0 = \left(\frac{1}{2}\right)(\theta\pi)^2.$$

The expected utility perceived by the agent (who may have some self-serving bias) at the beginning of $t = 1$, W_1, is:

$$W_1 = x\lambda(V - d) + (1 - x)(1 - \lambda)(U_1 - c). \tag{3.11}$$

By using (3.9)' and (3.10)', the expected utility, W_1, is denoted by:

$$W_1 = x\lambda\{(1/2)\pi^2 - d\} + (1 - x)(1 - \lambda)\{(1/2)(\theta\pi)^2 - c\}. \tag{3.11}'$$

The agent with some self-serving bias makes his or her choice in the same way that a fully rational agent would, except that he or she uses biased predictions for the emergence of Self 1.

The agent (who may have some self-serving bias) at the beginning of Period 1 chooses the probability x so as to solve:

$$\begin{aligned} W_1 &= \max_x x\lambda\{(1/2)\pi^2 - d\} + (1 - x)(1 - \lambda)\{(1/2)(\theta\pi)^2 - c\} \\ &= \max_x x[\lambda\{(1/2)\pi^2 - d\} - (1 - \lambda)\{(1/2)(\theta\pi)^2 - c\}] \\ &\quad + (1 - \lambda)\{(1/2)(\theta\pi)^2 - c\}. \end{aligned}$$

This involves comparing the expected utility of Self 1, $\lambda\{(1/2)\pi^2 - d\}$, with the expected utility of Self 0, $(1 - \lambda)\{(1/2)(\theta\pi)^2 - c\}$. If the former is larger than the latter, the agent decides to choose Self 1, that is, $x = 1$. Changing oneself is thus one option. If the former is smaller than the latter, the agent decides to choose Self 0, that is, $x = 0$. The former is equivalent to the latter; the agent is indifferent between Self 1 and Self 0. Or, equivalently:

$$x = \begin{cases} 1 & \text{if } \lambda > \lambda^* \\ 0 & \text{if } \lambda < \lambda^*, \end{cases}$$

where

$$\lambda^* = \frac{1/2(\theta\pi)^2 - c}{1/2\pi^2 - d + 1/2(\theta\pi)^2 - c} \tag{3.12}$$

Thus, the agent whose prediction is $\lambda > \lambda^*$ decides to choose Self 1, that is, $x = 1$. On the other hand, the agent whose prediction is $\lambda < \lambda^*$ decides to choose Self 0, that is, $x = 0$. If $\lambda = \lambda^*$, x may take any value in [0, 1]. Intentional self-change depends on the agent's self-confidence.

Then, in (3.11)', the expected utility perceived by the individual at the beginning of Period 1 is rewritten as:

$$W_1 = \begin{cases} \lambda\{(1/2)\pi^2 - d\} & \text{if } \lambda > \lambda^* \\ (1 - \lambda)\{(1/2)(\theta\pi)^2 - c\} & \text{if } \lambda < \lambda^* \end{cases},$$

when the activity is undertaken in Period 1. At the beginning of Period 1, the agent, whose prediction λ is above λ^*, expects to obtain the utility $\lambda\{(1/2)\pi^2 - d\}$. On the other hand, the agent, whose prediction λ is below λ^*, expects to obtain the utility $(1 - \lambda)\{(1/2)(\theta\pi)^2 - c\}$.

People are influenced by two factors, self-enhancement and self-verification. Psychological costs then contribute to increasing or decreasing λ^*. In (3.12), c (a psychological cost to maintain Self 0) and d (a disutility cost that the agent attaches to intentional self-change) affect λ^*. The agent suffers the cost c if he or she sticks to Self 0 in Period 1 (self-discrepancy). On the contrary, the agent might prefer to have the coherent self-image. Then, choosing Self 1 in Period 1 involves the cost d (self-verification). It follows that:

$$\frac{\partial\lambda^*}{\partial c} < 0, \quad \text{and} \quad \frac{\partial\lambda^*}{\partial d} > 0.$$

Thus, λ^* is decreasing in c, and it is increasing in d. There is an inert area within which a person does not change from Self 0 to Self 1. It is the area of λ below λ^*, that is, the interval [0, λ^*]. The agent with $\lambda < \lambda^*$ sticks to Self 0 in Period 1. Self-discrepancy results in depression. The greater the psychological cost c associated with the actual−ideal discrepancy, the greater the decrease in the inert area ($\partial\lambda^*/\partial c < 0$). People who hold more positive expectancies for themselves (self-enhancement) can cope effectively with stressful situations. Self-enhancers are more likely to overestimate the actual−ideal discrepancy for desired behavioral change. Overestimating the psychological cost c associated with dissonance, they are more likely to be motivated to change themselves. On the contrary, self-verification undermines intentional self-change. Self-verifying feedback leads to a more stable self-concept. The self-serving bias exists when an individual overestimates the discrepancies between the actual self and the ideal self. Furthermore, the individual with some self-serving bias tends to underestimate the psychological cost d to intentional self-change. Then, the greater the psychological cost d to self-change increases the inert area ($\partial\lambda^*/\partial d > 0$). If people overestimate the psychological cost d, they are less likely to be motivated to change themselves. Motives and emotions are closely linked. Self-conscious emotions emerge from self-reflection and self-evaluation.

The self is thus an adaptive process. It can be affected by two factors, self-enhancement and self-verification. If the self-enhancement effect is sufficiently large, Self 1 will be activated in the future. On the other hand, if the self-verification effect is sufficiently large, Self 0 may persist. When the ideal self-guides are brought to mind, people will evaluate their behavior in terms of its fit to their own self-expectations for competent conduct. The self is motivated to maintain a certain kind of self-enhancement effect or self-verification effect. People with low self-confidence will show less justification of their behavioral change than people with high self-confidence. As in optimism, high self-confidence is associated with high levels of self-serving biases. They are more likely to overestimate the psychological cost c associated with the actual−ideal discrepancy. They perceive the smaller cost d to self-change, and do not prefer to have the coherent self-image. In the presence of cognitive inconsistencies that threaten self-concept, people often eliminate previous behaviors, or adopt specific behaviors to reduce dissonance. As a result, the greater the discrepancies perceived by self-enhancers, the less consistent the self-concept. On the contrary, people with low self-confidence are more likely to underestimate the psychological cost c associated with the actual−ideal discrepancy for desired behavioral change. As in pessimism, they are less likely to be motivated to change themselves. Self-change will increase some disutility among persons with low self-confidence. They subsequently act in accordance with preexisting self-conceptions. Thus, in the current model, self-confidence differences can play a role in the dissonance process.

Standard economic theory is based on the assumption that people can identify and choose what is optimal for them. Can people really choose what is best for them? The inert area idea permits both optimal and suboptimal economic results within the same model. Economic variables are subject to inertia. X-inefficiency is the gap between observed performance and existing best performance (Leibenstein, 1979). The assumption of effort variability is critical to X-efficiency theory. The degree of X-efficiency is linked with λ, the subjective belief that the agent successfully activates Self 1. What level of effort an individual exerts depends on some critical value λ^*. The low-effort level persists for one subset of values of λ, where $(0 \leq) \lambda < \lambda^*$. On the other hand, Self 1 exerts the high-effort level for another subset of values of λ, where $(1 \geq) \lambda > \lambda^*$. X-efficiency is obtained only under high levels of self-confidence. Once such self-confidence is reduced, no intentional self-change is likely to occur. The conventional wisdom assumes that effort inputs must be maximized. However, for X-inefficiency to exist, effort must be a discretionary variable. Economic agents have the capacity to choose the quantity and quality of effort supplied in the process of production. Therefore, they would be producing below their potential, or X-inefficiently.

In this model, people are influenced by two factors, self-enhancement and self-verification. People sometimes seek positive or self-enhancing feedback. On the other hand, people may have a "passion for truth." People sometimes seek subjectively accurate or self-verifying feedback. Self-verification solidifies and strengthens self-views. For self-enhancers, comparison between the actual self and the ideal self has motivational consequences. Here, the ideal self is a representation of the attributes that individuals wish or hope to possess in the pursuit of their own self-interests.

The dual motive theory considers the notion of alternative, possible selves (Cory, 2006). Pioneering evolutionary neuroscientist Paul MacLean described how the interactions among instinct, emotion, and rationality in the human brain influence decision-making. MacLean divided the human brain into three "layers" (i.e., the reptilian brain, the limbic system, and the cerebral cortex) that arrived at different stages of evolution. His tri-level conceptual platform of the human brain is the fundamental concept for a comprehensive understanding of human sociality from the standpoint of physiology and medicine. The conflict-system, neurobehavioral model in Cory (2006) is a simplified representation of the brain showing the two opposing circuits of an "ego/self-interest" part, and an "empathy/other-interest" part. Furthermore, the range of "dynamic balance" works to resolve the inherent conflict between ego and empathy. Individuals in the dual-motive model experience "behavioral stress" when either too self-interested or too other-interested. There are potentially stable preferences arising from the co-evolutionary interaction and feedback between ego/self-interest and empathy/other-interest.

Relationships with other people make possible the formation of one's self-image. Self-verification may occur when people choose partners who see them as they see themselves, thereby creating social environments that are likely to support their self-views. People perceive and form appraisals of other people around them. Appraisals enable individuals to adjust their behavior to the abilities and tendencies of their interaction partners. Behavioral change should foster self-concept change. The experimental evidence supports that individuals may have self-regarding preferences and other-regarding preferences in the structure of social interactions (Camerer and Fehr, 2006). The mixture of heterogeneous preferences is likely to create profound effects on aggregate behavior, because individuals explicitly take heterogeneity and incentive interactions between different types of others into account. Choices are complements if individuals have an incentive to match the choices of others. When choices are complements, then self-regarding (or other-regarding) agents may have an incentive to mimic the behavior of other-regarding (or self-regarding) agents. Therefore, there are situations in which other-regarding (or self-regarding) preferences may dominate the outcome of social interactions. People may seek self-enhancing feedback in the pursuit of their own interests if their interaction partners have highly self-regarding preferences. Then, self-regulation occurs with respect to the ideal self-guide. Actual—ideal self-regulation involves a promotion focus. In some situations, however, interaction partners with both self-regarding preferences and other-regarding ones may encourage people to integrate and balance between ego/self-interest and empathy/other-interest.

This section has proposed an economic model of how intentional self-change occurs, by focusing on the cognitive processes by which people interpret their behavior. Possible selves are representations of the self in the future, and supply direction for the achievement of the desired goal. The self-system is multifaceted and dynamic, with different self-representations activated at different times. Self-confidence or self-esteem shapes how people view themselves on these attributes. People with high self-confidence as compared to low self-confidence are more likely to present favorable self-images, independently of the way they actually think about themselves, and to perform effectively. Comparison between the actual self and the ideal self has important motivational consequences. Self-enhancement underlies people's tendency to believe that they can improve relative to the past. Self-enhancers are motivated to match self-concept to ideal self-guides. The response to self-discrepancies can be then shaped toward desired behavioral change. However, self-verification is also an important factor in the individual's motivational system. Self-verification is expected to undermine intentional self-change. Human judgment involves consistent departures from normative rationality. The agent in the present model may be distorted by a self-serving bias. The individual with little self-serving bias has an inert area within which he or she does not attempt self-change. The inert area idea permits both optimal and suboptimal economic outcomes within the same model. The individual with some self-serving bias may reach the optimal economic outcome.

3.8   SUMMARY

Behavioral economics studies how individuals behave, and how thinking and emotions affect individual decision-making. By adding insights from psychology, behavioral economics tries to modify the conventional economic approach, and to analyze how "flesh-and-blood" people act in social contexts. The human condition is marked by two, often conflicting, cognitive systems. These are labeled System 1 and System 2 to represent two reasoning processes. System 1 processes are rapid, automatic, and unconscious, while System 2 processes are slower, deliberate, and conscious. Anomalies are often labeled as decision-making failures or mistakes. Anomalies arise from the way humans process information to form beliefs. Heuristics describe how people make judgments and decisions based on approximate rules of thumb. Heuristics require less effort compared to a rational, calculated choice. Human decision-making is prone to phenomena like anchoring, framing, status quo bias, and inertia. Systematic biases and temporally inconsistent motivations lead to poor choices. By changing the choice architecture as the context in which people make decisions, outcomes can be improved in a way that makes choosers better off.

REFERENCES

Aghion, P., Tirole, J., 1997. Formal and real authority in organizations. J. Polit. Econ. 105, 1–29.

Ainslie, G., 1975. Specious reward: a behavioral theory of impulsiveness and impulse control. Psychol. Bull. 82, 463–496.

Ainslie, G., 2001. Breakdown of Will. Cambridge University Press, Cambridge, MA.

Akerlof, G.A., 1980. A theory of social custom, of which unemployment may be one consequence. Q. J. Econ. 94, 749–775.

Akerlof, G.A., 1983. Loyalty filters. Am. Econ. Rev. 73, 54–63.

Akerlof, G.A., 1991. Procrastination and obedience. Am. Econ. Rev. 81, 1–19.

Akerlof, G.A., Kranton, R.E., 2000. Economics and identity. Q. J. Econ. 115, 715–753.

Altman, M., 2001. Culture, human agency, and economic theory: culture as a determinant of material welfare. J. Behav. Exp. Econ. 30, 379–391.

Anderson, L.R., Holt, C.A., 1997. Informational cascades in the laboratory. Am. Econ. Rev. 87, 847–862.

Arthur, W.B., 1989. Competing technologies, increasing returns, and lock-in by historical events. Econ. J. 99, 116–131.

Axelrod, R., 1984. The Evolution of Cooperation. Basic Books, New York.

Banerjee, A.V., 1992. A simple model of herd behavior. Q. J. Econ. 107, 797–817.

Banerjee, A.V., 1993. The economics of rumours. Rev. Econ. Stud. 60, 309–327.

Bazin, D., Ballet, J., 2006. A basic model for multiple self. J. Behav. Exp. Econ. 35, 1050–1060.

Becker, G.S., 1991. A note on restaurant pricing and other examples of social influences on price. J. Polit. Econ. 99, 1109–1116.

Becker, G.S., Murphy, K.M., 1988. A theory of rational addiction. J. Polit. Econ. 96, 675–700.

Bénabou, R., Tirole, J., 2002. Self-confidence and personal motivation. Q. J. Econ. 117, 871–915.

Bénabou, R., Tirole, J., 2004. Willpower and personal rules. J. Polit. Econ. 112, 848–886.

Bénabou, R., Tirole, J., 2011. Identity, morals, and taboos: beliefs as assets. Q. J. Econ. 126, 805–855.

Bikhchandani, S., Hirshleifer, D., Welch, I., 1992. A theory of fads fashion custom and cultural change as information cascades. J. Polit. Econ. 100, 992–1026.

Blumenthal-Barby, J.S., 2013. Choice architecture: a mechanism for improving decisions while preserving liberty? In: Coons, C., Weber, M. (Eds.), Paternalism: Theory and Practice. Cambridge University Press, Cambridge, MA.

Bowles, S., 1998. Endogenous preferences: the cultural consequences of markets and other economic institutions. J. Econ. Lit. 36, 75–111.

Burke, P.J., 1991. Identity processes and social stress. Am. Sociol. Rev. 56, 836–849.

Burke, P.J., Tully, J.C., 1977. The measurement of role identity. Soc. Forces 55, 881–897.

Calvo, G.A., Wellisz, S., 1978. Supervision, loss of control, and the optimum size of the firm. J. Polit. Econ. 86, 943–952.

Camerer, C.F., 2003. Behavioral Game Theory: Experiments in Strategic Interaction. Princeton University Press, Princeton, NJ.

Camerer, C.F., Fehr, E., 2006. When dose "Economic Man" dominate social behavior? Science 311, 47–52.

Chamley, C., Gale, D., 1994. Information revelation and strategic delay in a model of investment. Econometrica 62, 1065–1085.

Compte, O., Postlewaite, A., 2004. Confidence-enhanced performance. Am. Econ. Rev. 94, 1536–1557.

Cooper, R.W., John, A., 1988. Coordinating coordination failures in Keynesian models. Q. J. Econ. 103, 441–463.

Cory, G.A., 2006. Physiology and behavioral economics. In: Altman, M. (Ed.), Handbook of Contemporary Behavioral Economics. M.E. Sharpe, New York, pp. 24–49.

David, P.A., 1985. Clio and the economics of QWERTY. Am. Econ. Rev. 75, 332–337.

Davis, J.B., 2007. Akerlof and Kranton on identity in economics: inverting the analysis. Camb. J. Econ. 31, 349–362.

De Vany, A., 1996. The emergence and evolution of self-organized coalitions. In: Gilli, M. (Ed.), Computational Economic Systems: Models, Methods and Econometrics. Kluwer Scientific Publications, Boston, pp. 25–50.

Dholakia, U.M., Soltysinski, K., 2001. Coveted or overlooked? The psychology of bidding for comparable listings in digital auctions. Mark. Lett. 12, 223–235.

Evans, J.S.B.T., 1989. Bias in Human Reasoning: Causes and Consequences. Erlbaum, London.

Friedman, D., 1991. Evolutionary games in economics. Econometrica 59, 637–666.

Friedman, M., Savage, L., 1948. The utility analysis of choices involving risk. J. Polit. Econ. 56, 279–304.

Fudenberg, D., Levine, D.K., 2006. A dual-self model of impulse control. Am. Econ. Rev. 96, 1449–1476.

Gilboa, I., Matsui, A., 1991. Social stability and equilibrium. Econometrica 59, 859–867.

Grossman, S.J., Hart, O.D., 1986. The costs and benefits of ownership: a theory of vertical and lateral integration. J. Polit. Econ. 94, 691–719.

Gruber, J., Köszegi, B., 2001. Is addiction "rational"? Theory and evidence. Q. J. Econ. 116, 1261–1303.

Grüne-Yanoff, T., 2012. Old wine in new casks: libertarian paternalism still violates liberal principles. Soc. Choice Welfare 38, 635–645.

Gul, F., Pesendorfer, W., 2001. Temptation and self-control. Econometrica 69, 1403–1435.

Güth, W., Schmittberger, R., Schwarze, B., 1982. An experimental analysis of ultimatum bargaining. J. Econ. Behav. Organ. 3, 367–388.

Harris, M., 1971. Culture, Man andNature: An Introduction to General Anthropology. Thomas Y. Crowell, New York.

Hart, O.D., Moore, J., 1990. Property rights and the nature of the firm. J. Polit. Econ. 98, 1119–1158.

Henrich, J., 2000. Does culture matter in economic behavior? Ultimatum game bargaining among the Machiguenga of the Peruvian Amazon. Am. Econ. Rev. 90, 973–979.

Henrich, J., Boyd, R., Bowles, S., Camerer, C., Fehr, E., Gintis, H., 2004. Foundations of Human Sociality: Economic Experiments and Ethnographic Evidence from Fifteen Small-scale Societies. Oxford University Press, Oxford.

Henrich, J., Boyd, R., Bowles, S., Camerer, C., Fehr, E., Gintis, H., et al., 2001. In search of Homo economicus: behavioral experiments in 15 small-scale societies. Am. Econ. Rev. 91, 73–78.

Higgins, E.T., 1987. Self-discrepancy: a theory relating self and affect. Psychol. Rev. 94, 319–340.

Hofstede, G., 1984. Culture's Consequences: International Differences in Work-related Values. Sage Publications, Beverly Hills.

Iannaccone, L., 1998. Introduction to the economics of religion. J. Econ. Lit. 36, 1465–1496.

Kahneman, D., 2003. Maps of bounded rationality: psychology for behavioral economics. Am. Econ. Rev. 93, 1449–1475.

Kahneman, D., 2011. Thinking, Fast and Slow. Farrar, Straus and Giroux, New York.

Kahneman, D., Tversky, A., 1972. Subjective probability: a judgement of representativeness. Cogn. Psychol 3, 430–454.

Kahneman, D., Tversky, A., 1973. On the psychology of prediction. Psychol. Rev. 80, 237–251.

Kahneman, D., Tversky, A., 1979. Prospect theory: an analysis of decision under risk. Econometrica 47, 313–327.

Kandel, E., Lazear, E.P., 1992. Peer pressure and partnerships. J. Polit. Econ. 100, 801–817.

Karathanos, P., 1998. Crafting corporate meaning (developing corporate culture). Manage. Decis. 36, 123–132.

Karni, E., Schmeidler, D., 1990. Fixed preferences and changing tastes. Am. Econ. Rev. 80, 262–267.

Keynes, J.M., 1937. The General Theory of Employment, Interest and Money. Macmillan, London.

Kirby, K.N., Herrnstein, R.J., 1995. Preference reversals due to myopic discounting of delayed rewards. Psychol. Sci. 6, 83–89.

Kirman, A., 1993. Ants rationality and recruitment. Q. J. Econ. 108, 137–156.

Kruglanski, A.W., 1996. Goals as knowledge structures. In: Gollwitzer, P.M., Bargh, J.A. (Eds.), The Psychology of Action: Linking Cognition and Motivation to Behavior. The Guilford Press, New York, pp. 599–618.

Laibson, D., 1997. Golden eggs and hyperbolic discounting. Q. J. Econ. 112, 443–477.

Lazear, E.P., 1999a. Culture and language. J. Polit. Econ. 107, S95–S126.

Lazear, E.P., 1999b. Globalization and the market for teammates. Econ. J. 109, C15–C40.

Lea, S.E.G., Webley, P., 2005. In search of the economic self. J. Behav. Exp. Econ. 34, 585–604.

Leary, M.R., 1995. *Self-Presentation: Impression Management and Interpersonal Behavior.* Westview Press, Boulder.

Leibenstein, H., 1950. Bandwagon, snob, and Veblen effects in the theory of consumer demand. Q. J. Econ. 64, 183–207.

Leibenstein, H., 1966. Allocative efficiency vs. "X-efficiency". Am. Econ. Rev. 56, 392–415.

Leibenstein, H., 1976. Beyond Economic Man. Harvard University Press, Cambridge, MA.

Leibenstein, H., 1979. A branch of economics is missing: micro-micro theory. J. Econ. Lit. 17, 477–502.

Lester, D., 2003. Comment on "The self as a problem": alternative conceptions of the multiple self. J. Behav. Exp. Econ. 32, 499–502.

Loewenstein, G., 1987. Anticipation and the value of delayed consumption. Econ. J. 97, 666–684.

Loewenstein, G., 1996. Out of control: visceral influences on behavior. Organ. Behav. Hum. Decis. Process. 65, 272–297.

Markowitz, H., 1952. The utility of wealth. J. Polit. Econ. 60, 151–156.

Markus, H., Nurius, P., 1986. Possible selves. Am. Psychol. 41, 954–969.

Matsuyama, K., 1992. The market size, entrepreneurship, and the big push. J. Jpn. Int. Econ. 6, 347–364.

Mill, J.S., 1869/1991. On Liberty. Routledge, London.

Moldoveanu, M., Stevenson, H., 2001. The self as a problem: the intra-personal coordination of conflicting desires. J. Behav. Exp. Econ. 30, 295–330.

Mullainathan, S. Thaler, R.H. (2000). Behavioral economics. NBER Working Paper 7948.

Murphy, K.M., Shleifer, A., Vishny, R.W., 1989. Industrialization and the big push. J. Polit. Econ. 97, 1003−1026.

O'Donoghue, T., Rabin, M., 1999a. Doing it now or later. Am. Econ. Rev. 89, 103−124.

O'Donoghue, T., Rabin, M., 1999b. Incentives for procrastinators. Q. J. Econ. 114, 769−816.

O'Donoghue, T., Rabin, M., 2001. Choice and procrastination. Q. J. Econ. 116, 121−160.

Orléan, A., 1995. Bayesian interactions and collective dynamics of opinion: herd behavior and mimetic contagion. J. Econ. Behav. Organ. 28, 257−274.

Pattigrew, T.F., 1979. The ultimate attribution error: extending Allport's cognitive analysis of prejudice. Pers. Soc. Psychol. Bull. 5, 461−476.

Phelps, E.S., Pollak, R.A., 1968. On second-best national saving and game-equilibrium growth. Rev. Econ. Stud. 35, 185−199.

Rosenberg, M., 1979. Conceiving the Self. Basic Books, New York.

Roth, A.E., Prasnikar, V., Okuno-Fujiwara, M., Zamir, S., 1991. Bargaining and market behavior in Jerusalem, Ljubljana, Pittsburgh, and Tokyo: an experimental study. Am. Econ. Rev. 81, 1068−1095.

Samuelson, P.A., 1937. A note on measurement of utility. Rev. Econ. Stud. 4, 155−161.

Simonson, I., Tversky, A., 1992. Choice in context: tradeoff contrast and extremeness aversion. J. Mark. Res. 29, 281−295.

Stets, J.E., Burke, P.J., 2003. A sociological approach to self and identity. In: Leary, M.R., Tangney, J.P. (Eds.), Handbook of Self and Identity. The Guilford Press, New York, pp. 128−152.

Stroz, R.H., 1956. Myopia and inconsistency in dynamic utility maximization. Rev. Econ. Stud. 23, 165−180.

Sunstein, C.R., Thaler, R.H., 2003. Libertarian paternalism is not an oxymoron. U. Chi. L. Rev. 70, 1159−1202.

Swann Jr., W.B., Stein-Seroussi, A., Giesler, B., 1992. Why people self-verify. J. Pers. Soc. Psychol. 62, 392−401.

Teraji, S., 2003. Herd behavior and the quality of opinions. J. Behav. Exp. Econ. 32, 661−673.

Teraji, S., 2008. Culture, effort variability, and hierarchy. J. Behav. Exp. Econ. 37, 157−166.

Teraji, S., 2009. The economics of possible selves. J. Behav. Exp. Econ. 38, 45−51.

Thaler, R.H., 1988. Anomalies: the ultimatum game. J. Econ. Perspect. 2, 195−206.

Thaler, R.H., 1991. Quasi Rational Economics. Russell Sage Foundation, New York.

Thaler, R.H., Sunstein, C.R., 2003. Libertarian paternalism. Am. Econ. Rev. 93, 175−179.

Thaler, R.H., Sunstein, C.R., 2008. Nudge: Improving Decisions about Health, Wealth, and Happiness. Yale University Press, New Haven, CT.

Tversky, A., Kahneman, D., 1971. Belief in the law of small numbers. Psychol. Bull. 76, 105−110.

Tversky, A., Kahneman, D., 1974. Judgment under uncertainty: heuristics and biases. Science 185, 1124−1131.

Tversky, A., Kahneman, D., 1981. The framing of decisions and the psychology of choice. Science 211, 453−458.

Tversky, A., Kahneman, D., 1983. Extensional versus intuitive reasoning: the conjunction fallacy in probability judgment. Psychol. Rev. 90, 293−315.

White, M.D., 2013. The Manipulation of Choice: Ethics and Libertarian Paternalism. Palgrave Macmillan, New York.

Williamson, O.E., 1967. Hierarchical control and optimum firm size. J. Polit. Econ. 75, 123−138.

Chapter 4

Why Bounded Rationality?

4.1 INTRODUCTION

Neoclassical economic theory postulates a predetermined set of axioms leading to choices that are independent of social context. In neoclassical economic theory, rational agents are assumed to make their choices in a consistent manner. Neoclassical economic theory identifies individuals as a set of preferences conforming to axioms such as completeness, reflexivity, transitivity, and continuity. It treats a choice as rational if it is the one most likely to satisfy these preferences. That is, in neoclassical economics, preferences are axiomatically required to be "rational" so that choices are rational. Its core has been axiomatized in expected utility theory.[1] More specifically, it treats individuals as choosing under risk, where outcomes of actions have a determined probability. Expected utility theory, first made explicit by von Neumann and Morgenstern (1944), teaches us that so long as an agent's preferences over risky options obey certain axioms, then the agent behaves as if he or she is maximizing the expected value of a utility function.

Expected utility theory measures an agent's valuation of things called "prospects," where a prospect is to be understood as a list of consequences with associated probabilities. In expected utility theory, all consequences and probabilities are assumed to be known to the agent; in choosing among prospects, the agent can be said to confront a situation of risk. Any prospect q can be represented by a probability distribution $q = (p_1, \ldots, p_n)$ over a fixed set of pure consequences $x = (x_1, \ldots, x_n)$, where p_i is the probability of x_i, $p_i > 0$ for all i, and $p_1 + \ldots + p_n = 1$. The ordering axiom requires both completeness and transitivity. First, agents are assumed to have their own preferences, and they rank order in a consistent way the preferences for all conceivable prospects. Agents always know and can express that they prefer one prospect to another, or are indifferent to either of them. A preference relation is then complete. Second, if agents choose one prospect, q, over another prospect, r, and r over s, then they will choose q over s. A preference relation is then transitive. Completeness and transitivity together ensure that the agent

---

1. Expected utility theory was first stated by Daniel Bernoulli to solve the St. Petersburg puzzle. In general, the "value" of a gamble to an individual is not equal to its expected monetary value. Instead, individuals place "utilities" on monetary outcomes, and the value of a gamble is the expectation of these utilities.

The Cognitive Basis of Institutions. DOI: https://doi.org/10.1016/B978-0-12-812023-1.00004-1

has a preference ordering over all prospects. Furthermore, the axiom of continuity is required for all prospects; for any prospect **r**, there is always a prospect, **q**, preferred to **r**, and another prospect, **s**, over which **r** is preferred, such that a probability mixture of **q** and **s** is equal to **r** in value. Together the axioms of ordering and continuity imply that preferences over prospects can be represented by a function which assigns a real-valued index to each prospect. The function is a representation of preferences; an agent prefers the prospect **q** to the prospect **r** if, and only if, the function assigns a higher value to **q** than to **r**. The independence axiom places quite strong restrictions on the precise form of preferences: two prospects, **q** and **r**, mixed with a third prospect **s**, maintain the same preference order as when the two prospects are presented independently of the third one.[2] The expected utility hypothesis can be derived from ordering, continuity, and independence axioms.

The rule to maximize expected utility purports to yield rational decisions when options involve risk. However, it has been criticized for neglecting to take account of a decision-maker's attitude toward risk. In expected utility theory, risk aversion is reflected by a concave utility function for money, that is, diminishing marginal utility. Risk aversion and diminishing marginal utility are the same thing in expected utility theory. However, even if the utility function is linear in money, the agent might prefer the certainty of $5 to the gamble "$10 if a fair coin lands heads, nothing otherwise" if he or she wishes to avoid risk. Allais (1952) insists that utility is psychologically real, rejecting the view that an agent's utility function is only a representation of his or her preferences. Allais' treatment of risk is very natural. Suppose that an agent, having wealth of $a, is faced with a choice between $5 for sure (Option A), and either $9 or $1 on the flip of a fair coin (Option B). By applying the utility function u, Option A yields $u(a + 5)$ for sure, while Option B yields $u(a + 9)$ or $u(a + 1)$ with a probability of 0.5 each. On the expected utility view, the agent chooses the option with the highest expected utility, that is, $\max[u(a + 5), 0.5\ u(a + 1) + 0.5\ u(a + 9)]$. On Allais' view, however, an agent's choice among gambles might be influenced by the variance of the possible utilities. Then, the agent will want to reduce the variation in utility that he or she receives. Since Option A has zero variance and Option B has a positive variance, the agent might prefer Option A to B. The agent's attitude toward risk can be reflected in his or her attention to the variance as well as the expectation of the utilities associated with each option.

With respect to the hypothesis of expected utility maximization, we are not very good at judging probabilities, and do not order our preferences consistently; we systematically violate just about every logical implication of decisions theory. People have great difficulties in working with probabilities, assessing risk, and making inferences where uncertainty is involved. An

---

2. Maurice Allais (1952), who originally attacked the independence axiom, regarded expected utility theory as neither empirically plausible nor normatively compelling.

underlying theme of much decision research is that preferences for objects are often constructed in the generation of a response to a judgment to choice task. James March (1978) attributes the construction of preferences to the limits on information processing capacity:

> *Human beings have unstable, inconsistent, incompletely evoked, and imprecise goals at least in part because human abilities limit preference orderliness.*
>
> (March, 1978, p. 598)

Preferences are not necessarily generated by some consistent and invariant algorithm. The information used to construct preferences appears to be highly contingent on a variety of context factors. People often adopt different options in different situations, potentially leading to variance in preferences. Instability or ambiguity of preferences is not necessarily a fault in human choice to be corrected, but often a form of intelligence to be refined.

Herbert Simon (1957) coins the term "bounded rationality" to refer to choices, given the cognitive limitations of the individual decision-makers in terms of acquiring and processing information. For Simon, the cognitive assumption underlying expected utility theory is too strong. Human beings are capable only of very approximate and bounded rationality. A useful way to view bounded rationality is to think of an agent as an information processor: information flows to the agent, he or she processes it in some fashion, and decisions come out. The procedure is a reasonable compromise between the accuracy of the output and the difficulties involved in processing. Human beings do not know all of the alternatives that are available for action, and they are unable to make the calculations that would support optimization.[3] Decision-making that deviates from neoclassical rationality does not imply irrationality. An agent cannot choose an optimal action, but instead must "satisfice," that is, choose an option that meets some predetermined aspiration level.

Oliver Williamson's work, or the "firm as a governance mechanism" approach, is based on two notions: bounded rationality and opportunism. Bounded rationality results in contractual incompleteness; bounded rational actors are unable to completely foresee all the future contingencies that affect the execution and fulfillment of contracts. The contractual parties are, as a result, at the risk of being subject to opportunism. Opportunism is thought of as "self-interest with guile." Williamson (1985) explicates transaction costs as arising from agents' limited capacities to articulate and communicate their knowledge in a detailed and clear way. People have limited communication competencies. These limitations induce some costs in decision-making. The bounded rationality assumption makes complete contracting infeasible,

---

3. For example, in economic theory (Rubinstein, 1998), the rational decision-making methodology leads to the selection of an alternative after completing the following three steps: analyzing the feasibility of the alternative; pondering the desirability of the alternative; and choosing the best alternative by combining both desirability and feasibility.

because there are limits to the capabilities of decision-makers for dealing with information and anticipating the future. Cognitive limitations lead people to implement incomplete contracts in complex environments. As long as the external environment is simple, people are able to specify the set of feasible actions and to determine the optimal solution. However, as soon as uncertainty becomes significant, people are no longer able to determine the optimal solution, and contracts become incomplete. Transaction cost theory differs from conventional economic theory only in its emphasis on transaction costs as distinct from production costs. In Williamson's approach, firms minimize their total costs, made up of both production and transaction costs. According to transaction cost theory, economizing on transaction costs is a central problem in the adoption of governance structures. Transaction cost economics presupposes that organizations continuously search for the governance structure which is economically the most efficient. The governance structure incurs the lowest possible production and transaction costs. Williamson (1985) cites three dimensions of production transactions that combine with the opportunistic behavior and bounded rationality of economic actors to derive transactions from market to contract: imperfect knowledge, the frequency of transactions, and asset specificity. For Williamson, internal organization is a source of conflicts and disputes, which can be explained by the behavioral hypotheses advocated by him, particularly the bounded rationality preventing individuals from making complete contracts. Where costly assets are specific and substantial, investments of time and money leave particular buyers and sellers little choice but to deal with one another, and the time and effort needed to negotiate terms of trade are potent inducements to contract for both sides. The parties seek the security of a long-term relationship as a way to avoid the costs of recurrent bargaining over specific assets. For Williamson, economic actors are able to remedy dysfunctions. Williamson (2009) describes both cognition and self-interest in a two-part way; cognition combines bounded rationality with feasible foresight, while self-interest joins benign behavior with opportunism. For Williamson, complex contracts are unavoidably incomplete by reason of bounded rationality, but economic actors are assumed to have the capacity to look ahead, recognize hazards, and work out the mechanisms in the contractual design by reason of feasible foresight. Hence, economic agents, who are rationally bounded and have asymmetrical information, do not succeed in achieving Pareto efficiency, but do succeed in selecting the best organizational design for a given period. This results in an improvement of economic performance. Williamson pays little attention to the problem inherent in the measurement of outputs. A low degree of output measurability can increase the uncertainty of transactions and the governance of organizations. Furthermore, Williamson makes no use of the idea of satisficing; his central implication of bounded rationality is that all complex contracts are unavoidably incomplete. For Williamson, the rationality of actors tends to be somewhat less bounded when they act opportunistically.

4.2 BOUNDED RATIONALITY

Rationality has been given a very specific meaning in neoclassical economic theory. In Herbert Simon's (1978) words:

> [T]he rational man of economics is a maximizer who will settle for nothing less than the best.

(Simon, 1978, p. 2)

While the neoclassical approach established a close connection between rationality and utility (or profit) maximization, Simon scrutinized the implications of departures of actual behavior from the neoclassical assumptions.[4] Following Simon, the term "rational" is applied not to substantive outcomes, but to the process from which decisions have evolved. Bounded rationality is an alternative conception of rationality that incorporates the cognitive processes of decision-makers more realistically.

Simon's creativity was spread across distinct areas of empirical sciences: social, behavioral, and computer sciences. For Simon, 1958 was a critical year (Jones, 2002). By 1958, four path-breaking lines of his work were published.

1. Bounded rationality:
   In 1957, Simon published a collection of his papers under the title *Models of Man: Social and Rational.*[5] The volume includes his 1955 paper entitled "A behavioral model of rational choice" in the *Quarterly Journal of Economics.*
2. Cognitive psychology:
   In 1958, Allen Newell, Clifford Shaw, and Simon published their paper, "Elements of a theory of human problem solving," in the *Psychological Review.* The standard theory of problem solving, which focuses on how humans respond when they are confronted with an unfamiliar task, was initially outlined by their work. Although Simon was not the progenitor of the "cognitive revolution," his contributions to it were enormous.[6]
3. Artificial intelligence:
   Beginning in the mid-1950s, Newell, Shaw, and Simon's research on the logic theory of machine, their chess playing program, and the general problem solver played a major role in the early development of artificial

---

4. First, neoclassical economics presupposes that each agent has a well-defined utility or profit function. Second, it supposes that all alternative strategies are known to the decision-maker. Third, in neoclassical economics, all the consequences that follow on each of these strategies are assumed to be determined with certainty. Finally, the comparative evaluation of these sets of consequences is assumed to be driven by a universal desire to maximize expected utility or profit.

5. In the preface (p. vii), Simon explained the choice of this tittle: his essays "are concerned with laying foundations for a science of man that will comfortably accommodate his dual nature as a social and as a rational animal."

6. The cognitive revolution sought to undermine the dominance of behavioralism in psychology, focusing on internal psychological processes.

intelligence. In 1957, they published the joint paper, "Empirical explorations of the logic theory of machine."
4. Behavioral organization theory:
An organizational theory must specify how decisions are made. In 1958, March and Simon published their book *Organizations*. The work linked organization studies and the newly-developed behavioral decision theory.

Simon's (1955) paper "A behavioral model of rational choice" argues that if economists are interested in understanding actual decision behavior, the research would need to focus on the perceptual, cognitive, and learning factors that cause human decision behavior to deviate from that predicted by the normative "economic man" model. Economic man is, for Simon (1955, p. 99), "assumed to have knowledge of the relevant aspects of his environment which, if not absolutely complete, is at least impressively clear and voluminous. He is assumed also to have a well-organized and stable system of preferences, and a skill in computation that enables him to calculate, for the alternative courses of action that are available to him, which of these will permit him to reach the highest attainable point on his preference scale." Simon's (1955) original statement on the notion of "bounded rationality" emphasizes the limits of the information and computational capacities for an economic man:

> *Broadly stated, the task is to replace the global rationality of economic man with a kind of rational behavior that is compatible with the access to information and the computational capacities that are actually possessed by organisms, including man, in the kinds of environments in which such organisms exist.*

(Simon, 1955, p. 99)

Simon (1955) focuses on organisms, including man, in complex environments. Thus, the argument is meant to be general, and to highlight how organisms operate in complex environments. Our understanding of economic activity is strongly influenced by the notion of bounded rationality, and by direct analogies and tools from biological and computational sciences.

For Simon (1985, p. 297), "bounded rationality is not irrationality."[7] Bounded rationality, taking into account the cognitive limitations of the decision-maker, is a weakened form of rationality with regard to the maximizing behavior assumed by expected utility theory. An adequate theory of bounded rationality should describe the real processes that individuals use

---

7. In the "heuristics-and-biases" program (e.g., Kahneman and Tversky, 1973; Tversky and Kahneman, 1974), behavior deviating from the laws of logic or the maximization of expected utility is analyzed. According to this view, people need to rely on shortcuts or on heuristics, which make them vulnerable to systematic cognitive biases. Therefore, the use of heuristics explains why human decisions can be irrational or illogical.

to make actual decisions. A choice is a selection of one, among numerous possible alternatives, to be carried out. A behavior is a process through which the selection is performed. Global rationality, or the rationality of neoclassical theory, requires knowledge of all possible alternatives. However, for actual behavior, just a few of these alternatives are considered. The concept of bounded rationality implies that economic agents are, in practice, incapable of exercising global rationality. Bounded rationality is an alternative conception of rationality that models the cognitive processes of decision-makers more realistically. The human's capacity to formulate and solve problems is, in fact, very small compared with the size of the problems whose solution is required for globally rational behavior. Individuals are presumed to attempt to act rationally, but to be bounded in their ability to achieve rationality. Therefore, actors have to employ alternative cognitive strategies to make decisions. This seems to be consistent with Simon's position in the 1990s:

> Global rationality ... assumes that the decision maker has a comprehensive, consistent utility function, knows all the alternatives that are available for choice, can compute the expected value of utility associated with each alternative, and chooses the alternative that maximizes expected utility. Bounded rationality ... assumes that the decision maker must search for alternatives, has egregiously incomplete and inaccurate knowledge about the consequences of actions, and chooses actions that are expected to be satisfactory (attain targets while satisfying constraints).
>
> (Simon, 1997, p. 17)

Rationality is a property of actions, and does not specify the process by which the actions are selected. Simon (1976) coins the terms "substantive rationality" and "procedural rationality" to describe the difference between the question of what decision to make and the question of how to make it. Behavior is substantively rational when it is adequate to the realization of given ends, subjective to given conditions and constraints. Global rationality is understood as substantive one; it is only concerned with "what the decision-maker chooses." On the other hand, the concept of procedural rationality focuses on "what procedures the decision-maker uses." Procedural rationality is "the effectiveness, in light of human cognitive powers and limitations, of the procedures used to choose actions" (Simon, 1978, p. 9). A choice is a selection among numerous possible behavioral alternatives. A decision is a process through which this selection is performed. Procedural rationality can be seen as the process of finding reasonable solutions, given limited information and computational capacities. Procedural rationality thus requires logic of discovery. This interpretation emphasizes the distinction between the real world and the individual's perception of (or reasoning about) the world. The procedural aspect characterizes the presence of a search process. The choice conditions are the subject of a search process.

The decision-maker is goal-oriented. That is, "the decision-maker wishes to attain goals and uses his or her mind as well as possible to that end" (Simon, 1997, p. 293). The decision-maker is faced with several alternative "means—ends" strategies; each with its own consequences, one strategy that the decision-maker selects is followed by the preferred set of consequences. The decision-maker looks for alternatives that satisfy goals, rather than trying to find the best imaginable solution. If the strategy selected is conducive to the achievement of the given end, the decision-maker has behaved rationally. However, the decision-maker often fails to accomplish the intention, because of the interaction between aspects of his or her cognitive architecture and the essential complexity of the environment he or she faces. That is, the decision-maker must operate under the following two constraints. First, because of cognitive limitations, or limitations of knowledge, it is impossible for the behavior to reach any high degree of rationality. Second, since the decision-maker lives in an "environment of givens," his or her behavior is adapted within the limit set by these givens. A wide variety of cognitive tasks require the processing of information distributed across the internal mind and the external environment. It is the interwoven processing of internal and external information that generates much of a person's intelligent behavior. Human thought is adaptive; it is subject to improvement.

People often act without confidence in the efficacy of their decisions. However, people recognize that, in order to act, they do not have to completely understand what they are doing. Neoclassical economic theory can stipulate some objective utility function to be maximized, subject to constraints, and make predictions about the individual's rational behavior. This seemingly simple calculus of utility maximization actually invokes a set of environmentally contingent thought processes. An empirically grounded theory of bounded rationality should be applicable to situations in which the "conditions for rationality postulated by the model of neoclassical economics are not met" (Simon, 1989, p. 377). How is it possible for boundedly rational individuals, with limited information and computational capacities, to attain the equilibrium outcome described by neoclassical economic theory?

Facing bounded rationality, decision-makers simplify the structure of their decisions. In their *An Evolutionary Theory of Economic Change* (1982), Nelson and Winter develop a sophisticated evolutionary theory of economic change, building on Simon's view of bounded rationality. The notion of "routine" is a key concept in the foundation of their evolutionary theory. As Nelson and Winter (1982) note:

> *Man's rationality is "bounded": real-life decision problems are too complex to comprehend and therefore firms cannot maximize over the set of all conceivable alternatives. Relatively simple decision rules and procedures are used to guide action . . .*

(Nelson and Winter, 1982, p. 35, original emphasis)

Routines refer to the regular and predictable aspects of firm behavior. Routines are persistent features of firms that determine their behavior together with environmental conditions. A firm's capabilities, skills, experience, and knowledge "consist largely of the ability to perform and sustain a set of routines" (Nelson and Winter, 1982, p. 142). Firms arise to create and coordinate the capabilities needed to exploit new opportunities in changing environments. Their boundaries are determined primarily by the relative costs of developing needed capabilities internally. As the basis of their theory of organizational evolution, Nelson and Winter (1982) distinguish among three classes of routines: (1) routines, called operating characteristics, that govern organizational decisions and behavior given a particular stock of resources; (2) routines that augment or diminish the stock of resources in response to changes in the state of the organization or the environment; and (3) search routines, including the organization's own R&D and its investigation of what other firms are doing, that can modify various aspects of the operating characteristics. Routines are stores of empirical knowledge and are, as such, conjectural in nature. When a particular routine is no longer deemed to be satisfactory because of changing market conditions, this triggers a search for a new routine, for example, through increased investment in R&D. Search activity generates the variety which is the basis of evolutionary change. Organizations will present a broad array of routines, some long-established and some recently innovated. In Nelson and Winter's (1982) evolutionary theory, there are three basic concepts: routine, search, and selection environment. "Routines in general play the role of genes in our evolutionary theory. Search routines stochastically generate mutations" (Nelson and Winter, 1982, p. 400). The nature of realized variations in organizational forms is random. For organizations, a changing environment may lead to the modification of their routines, or the replacement of them. Some organizations whose routines are relatively better fit for coping with environmental conditions will thrive; other organizations will either imitate the more successful routines, or they will decline.

Simon (1959) distinguishes the "objective" information which exists "out there" in the agent's environment from the information which "enters" his or her faculties and is processed by them. For Simon, people are information-processing organisms. The information which first enters an agent's faculties is selected from the total set of objective information. This selection process follows from the individual's "perception" and "attention." The information, selected by the individual's perception and attention, first enters his or her faculties. The information first entering one's faculties is not necessarily representative of the objective information which is out there. As Simon (1959) points out:

Perception is sometimes referred to as a "filter." This term is as misleading as "approximation," and for the same reason: it implies that what comes through into the central nervous system is really quite a bit like what is "out there." In fact, the filtering is not merely a passive selection of some part of a presented whole, but an active process involving attention to a very small part of the whole and exclusion, from the outset, of almost all that is not within the scope of attention.

(Simon, 1959, pp. 272–273, original emphasis)

Thus, the perceived world of the decision-maker is different from the real world. From the set of information which first enters the individual's faculties, a subset of information will be selected contingent on the problem-solving strategy which has been chosen. The individual eventually ends up processing the information through both rounds of selection: perception and attention in the first round, and choice of problem-solving path in the second. We must pay attention to the procedures that individuals use to choose their actions. As Simon (1978, p. 9) argues: "we must give an account not only of substantive rationality—the extent to which appropriate courses of action are chosen—but also procedural rationality—the effectiveness, in light of human cognitive powers and limitations, of the procedures used to choose actions. As economics moves out toward situations of increasing cognitive complexity, it becomes increasingly concerned with the ability of actors to cope with the complexity, and hence with the procedural aspects of rationality." Human beings must use simplifications in their strategies to deal with complexity. There are problems of too much information in the real world. Simon (1997, p. 357) argues: "[h]uman beings handle this difficulty by attending to only a small part of the complexity around them. They make a highly simplified model of the world, and they make their decisions in terms of that model and the subset of variables that enter into it."

March and Simon (1958) point out that the so-called classical model of rational man has the following three difficulties. First, the model postulates as "given" a set of all available alternative courses of action, a set of all possible consequences attached to each alternative, and a utility function completely ranking each consequence according to its values to the actor. However, in fact, these features are typically not given or even obtainable. Second, it grants that an actor may have risky or uncertain information concerning possible consequences, but does not recognize that one's information concerning available alternatives and the value of consequences may be similarly complicated. Third, by admitting that one's information concerning a situation may be less than perfect, the model implies a distinction between substantive and objective rationality. However, so-called substantive rationality seems to be a poor substitute for "real" rationality. Finally, the model fails to agree with "common-sense notions of rationality."

Simon (1957) hypothesizes that agents perform limited searches, accepting the first satisfactory decision. The agents do not have complete and accurate knowledge about alternatives. Alternatives are generally not fixed in advance, but are generated. Therefore, they have to search for information.[8] They search until they find alternatives that are satisfactory enough, rather than until the marginal expected cost of search equals the marginal expected return. Global rationality refers to the ability of economic actors to make optimal decisions based on information regarding the past, the present, and the future. It is the notion that actors are able to identify and assimilate all of the available information that is relevant to the problem they face, processing it in order to maximize their given objective function. Rather than assuming that agents are globally aware of all the possibilities and comparatively commute them, Simon (1957) emphasizes the search for possibilities and the localness and limits of rationality. For boundedly rational agents, information gathering is costly. The number of possibilities and combinations is so large that search costs are not known in advance. Humans are not knowledgeable about all circumstances, not even all of those that are of interest to their own welfare. There is always a tradeoff between allocating time and resources to gathering further information and proceeding to act on the basis of current information. Starting from their internal representation of the environment, economic agents cope with a problem by adopting simplifying strategies for its solution:

> *The capacity of the human mind for formulating and solving complex problems is very small compared with the size of the problems whose solution is required for objectively rational behavior in the world—or even for a reasonable approximation to such objective rationality.*

> (Simon, 1957, p. 198)

March and Simon (1958, p. 169) propose an alternative model of rationality with the following simplifying features:

1. Optimizing is replaced by satisficing—the requirement that satisfactory levels of the criterion variables be attained.
2. Alternatives of action and consequences of action are discovered sequentially through search processes.
3. Repertoires of action programs are developed by organizations and individuals, and these serve as alternatives of choice in recurrent situations.
4. Each specific action program deals with a restricted range of situations, and a restricted range of consequences.
5. Each action program is capable of being executed in semi-independence of the others—they are only loosely coupled together.

---

8. Simon (1969) explains scientific discovery by his information processing approach. The scientists under investigation share one or more of the processes that are heuristic-driven search mechanisms. An important question is why one agent's use of some shared processes would produce a scientific discovery, while another one's use of it would not. Simon (1969) claims that the environment is the essential factor in producing different results with the same processes. That is, the environment is instrumental in producing the particular scientific discoveries.

People possess limited computational ability, and they use alternative cognitive strategies to make decisions. Economic agents "satisfice" rather than optimize; they simplify their decisions, primarily through the replacement of the goal of optimizing with the goal of satisficing, of finding a course of action that is "good enough" (Simon, 1957). Satisficing is often explained as a procedure of decision-making by which agents search for a solution until they find a satisfactory choice, given the information available and their ability to compute the consequences.[9] Alternatives are generally not fixed in advance, but are generated or identified. As Simon (1997, p. 321) points out, "the greater part of the decision-maker's time and effort is devoted to generating or identifying alternatives." The search for alternatives ends when a certain criterion is reached (a "stopping rule"). Then, the "criteria of satisfaction is closely related to the psychological notion of 'aspiration levels'" (March and Simon, 1958, p. 182). The levels of targets—profit, inventory, or sales, in the firm's case—are not optimized, but are set at a fixed level. The basic departure from the conventional economic theory is described, not in terms of optima, but in terms of values of targets which are satisfactory or unsatisfactory. An empirically grounded theory of bounded rationality should extend to situations, in which an agent cannot choose an optimal action but, instead, must satisfice, that is, choose an option that meets some predetermined aspiration level. Satisficing is a means of addressing when an option is good enough, in the sense that its payoff exceeds an aspiration level. Rejecting an option that does not meet or exceed the aspiration level derives its justification from an observation that the option is rejected in favor of an unknown alternative that produces better consequences. Determination of an aspiration level is based on experience-derived expectations of possible consequences. The search for alternatives is compatible with limited computational resources, and it terminates when an option is identified that exceeds the aspiration level. Satisficing is a way "of simplifying the choice problem, to bring it within the powers of human computation" (Simon, 1957, p. 204).

The basic idea of the satisficing approach is that people form aspirations, search for alternatives satisficing them, and choose the first option that is good enough. The satisficing approach can be described as a decision-making strategy of the following form:

1. Set an aspiration level such that any option which reaches or surpasses it is good enough.
2. Begin to search and evaluate the options on offer.
3. Choose the first option which, given the aspiration level, is good enough.

---

9. In a search-theoretic choice experiment, Caplin et al. (2011) find that many decisions can be understood using Simon's satisficing model: most subjects search sequentially and stop the search when an environmentally determined level of reservation utility has been realized.

Optimizing is thus replaced by satisficing. Satisficing is a way "of simplifying the choice problem to bring it within the powers of human computation" (Simon, 1957, p. 204). An aspiration level is the quantity of a particular value the agent would find satisfactory. The aspiration levels of goals are not rigid, but are revised in an upward or downward direction according to whether performance exceeds or falls short of aspiration. For example, sales in a given period greater or less than the target level fixed for that period will lead to higher or smaller targets, respectively, for the next period. The agent is serially evaluating alternatives as to their likelihood of satisfying his or her preferences. Alternatives can be discovered sequentially through search processes. In the search process, an effort is made to satisfy the aspiration level. The agent continues searching for alternatives if no satisfactory result is found yet. The satisficer stops searching for alternatives by choosing the first option to reach or exceed one's aspiration level.

Thus, we can summarize the following four principles about decision-making:

1. The principle of bounded rationality:
   The capacity of the human mind for formulating and solving complex problems rationally is bounded.
2. The principle of satisficing:
   An individual establishes his or her goal as an aspiration level.
3. The principle of search:
   An individual sequentially searches for alternatives, and selects one that meets the aspiration level.
4. The principle of adaptive behavior:
   An individual continually adjusts his or her behavior to changing environments.

The problem of the evolution of complex systems can be understood as a search problem. The complex whole is a configuration of elementary parts. The space of possible configurations increases with the number of elementary parts; the difficulty of the search problem increases with the complexity of the whole being constructed. In this setting, the evolution of complex systems is interpreted as the organization of increasing numbers of elementary parts. In *The Sciences of the Artificial*, Simon (1969) suggests that the evolution of complex systems requires the intermediate stages of stability between the simplest element stage and the complex whole stage. The intermediate stages of stability consist in the existence of systems of simple elements, which later become subsystems of complex wholes. All complex systems share certain structural features. Simon's notion of complexity is associated with the idea of hierarchic systems. A hierarchic system is composed of interrelated subsystems with a hierarchy among them. Hierarchic systems are decomposable into subsystems; the interactions between subsystems are weak, but not negligible. Simon (1969, p. 93) argues that "the time required for the evolution of a complex form from simple elements depends critically on the numbers and distribution of potential intermediate stable forms." Once a subsystem is formed, all of the parts which compose it can no longer disperse.

4.3 ECOLOGICAL RATIONALITY

Since Herbert Simon's concept of bounded rationality, it is accepted that cheap, adequate solutions are often preferred to costly, perfect ones. Along with the notion of bounded rationality, Simon (1956) points out the importance of considering the environment in which economic agents operate. Simon (1969, p. 52) summarizes his approach by using the following metaphor: "[a]n ant, viewed as a behaving system, is quite simple. The apparent complexity of its behavior over time is largely a reflection of the complexity of the environment in which it finds itself." Human beings are also simple systems whose behavior reflects the complexity of the environment they live in. Human thought is adaptive and subject to improvement. Thus, human behavior is a function of both cognition and the environment. Simon (1956) proposes to integrate environmental factors into the consideration of rationality:

> [W]e might hope to discover, by careful examination of some of the fundamental structural characteristics of the environment, some further clues as to the nature of the approximating mechanisms used in decision making.
>
> (Simon, 1956, p. 130)

Newell and Simon (1972) deal with the "task environment," defined as an environment coupled with a goal, problem, or task. Human behavior is explained by the nature of the task environment; given enough time, human thought takes on the shape of the tasks facing it. They describe the relationship between the external environment and bounded rationality as follows:

> Just as a scissors cannot cut without two blades, a theory of thinking and problem solving cannot predict behavior unless it encompasses both an analysis of the structure of task environments and an analysis of the limits of rational adaptation to task requirements.
>
> (Newell and Simon, 1972, p. 55)

If these two blades are not closely matched, then decision-making will be ineffectual. Human rationality cannot be understood merely by considering the mental mechanisms that underlie human behavior. Instead, we should elucidate the relationship between the mental mechanisms and the environments in which they work. What their metaphor suggests is that we should expand the bounds of bounded rationality outside the head and into the environment. This does not mean replacing the mind with the environment. The mind is an adaptive system that closely fits the environment. An important aspect of any adaptive system is that the system's behavior is adapted to the requirements of specific tasks. The mind has "responded to the shaping forces of an environment to which it must adapt in order to survive" (Simon, 1990, p. 2). Thus, "[h]uman rational behavior ... is shaped by a scissors whose two blades are the structure of task environments and the computational capabilities of the actor" (Simon, 1990, p. 7). This "ecological" perspective on bounded rationality has guided an important line of research on judgment and decision-making that is closely allied with Simon's conception of bounded rationality.

Gigerenzer, Todd, and the ABC (Center for Adaptive Behavior and Cognition) Research Group (1999) define "ecological rationality" as the property of a heuristic. Heuristics describe how people make judgments based on approximate rules of thumb, when strict logics are too time-demanding or not possible at all. Heuristics require less effort than a rational, information-based, and calculated choice. Heuristics are used to simplify complex tasks. Instead of searching for the universal tool that can solve all tasks, humans take a repertoire of specialized heuristics that can solve specific tasks in specific environments. The heuristics individuals use to generate expectations are adapted to the structure of the environment. The ecological rationality perspective is intended to study the mind in its environmental context. Ecological rationality implies a two-way relationship between simple heuristics and their environment:

> Models of bounded rationality describe how a judgement or decision is reached (that is, the heuristic processes or proximal mechanisms) rather than merely the outcome of the decision, and they describe the class of environments in which these heuristics will succeed or fail.
>
> (Gigerenzer and Selten, 2001, p. 4)

The success of simple heuristics is enabled by their fit to environmental structure. A heuristic is ecologically rational to the degree that it is adapted to the structure of an environment. Ecological rationality takes seriously Simon's idea that the structure of the agent's real environment has a decisive influence on his or her cognitive architecture. A decision mechanism relies on some of its work being done by the external environment. The structure of information in the environment will enable a heuristic to make good decisions. Making use of patterns of structured information in the environment is what allows effective heuristics to be simple. The mechanism takes the environmental structure into account in guiding what pieces of information to search for. It does not require seeking all available cues in any order. Simple heuristics can succeed by exploiting the structure of information in an environment.

The more closely a heuristic reflects important aspects of a particular environment, the more likely it is to succeed in the environment. Decision-making mechanisms may not perform well outside the context in which they are involved. The framework of ecological rationality is provided to study the adaptivity of decision-making. This framework assumes that people possess an "adaptive toolbox" (Gigerenzer and Selten, 2001); the heuristics in the toolbox can yield accurate decisions in the face of limited time, knowledge, and computational power. None of the available strategies is an all-purpose tool that can be successfully applied to every situation. To behave adaptively in the face of environmental challenges, agents must be able to make inferences that are fast and frugal. Fast and frugal heuristics, which are matched to particular environments, allow individuals to be ecologically rational.[10] Heuristics can sometimes be more accurate than complex arithmetical calculus. Different environments can have different heuristics that exploit their particular informational structure to make adaptive decisions.

Vernon Smith (2008) introduces two concepts of rationality—constructivist rationality and ecological rationality. Constructivist rationality is defined as follows:

> Constructivist rationality ... *involves the deliberate use of reason to analyze and prescribe actions judged to be better than alternative feasible actions that might be chosen. When applied to institutions, constructivism involves the deliberate design of rule systems to achieve desirable performance.*
>
> (Smith, 2008, p. 2, original emphasis)

Constructivist rationality involves conscious use of knowledge and reason to make decisions to solve specific problems. Neoclassical economic theory characterizes impersonal social systems by competitive equilibrium among fully informed individuals. Their calculated actions are rational in the constructivist sense. The design may include many choices that are explainable by deliberate reasoning. However, humans interact in complex ways, and their actions arise, beyond the reach of conscious reason, in the environment. Human interactions shape the environment, and the environment provides a feedback mechanism that shapes human actions. A trial-and-error learning process is sufficient to achieve the equilibrium outcome. Ecological rationality is defined as follows:

> Ecological rationality *refers to emergent order in the form of the practices, norms, and evolving institutional rules governing action by individuals that are part of our cultural and biological heritage and are created by human interactions, but not by conscious human design.*
>
> (Smith, 2008, p. 2, original emphasis)

---

10. According to Gigerenzer et al. (2008), research in the fast and frugal heuristics program focuses on three interrelated question: (1) What heuristics do organisms use to make decisions, and when do people rely on which heuristics from the tool box? (2) To what environmental structure is a given heuristic adapted, that is, in what situations does it perform well, say, by being able to yield accurate, fast, and effortless decisions? (3) How can the study of people's repertoires of heuristics and their fit to environmental structure aid decision-making in the applied world?

Thus, for Smith (2008), ecological rationality is an evolutionary-oriented notion of rationality. Through a process of social evolution, institutional arrangements emerge from human interactions that enable individuals to better coordinate their behavior. This evolutionary process takes place despite their imperfect knowledge of the structure of their environment. For Smith (2008), constructivist rationality and ecological rationality are not opposed. Constructivism projects real-world rationality on a continuum of logical reasoning and statistical inference, and its results are used as a benchmark to judge actual behavior in the world or in the laboratory. Once an observation is made in the world, one can ask whether it corresponds to a constructivist model of rationality. One can use constructivist rationality to make sense of an observation. For Smith (2008), two concepts of rationality work together through an incessant back-and-forth movement.

When people decide what to do in a given situation, they rely on routines or rules, instead of deliberately and analytically processing all available information. In fact, people follow rules because they know or believe that those rules will produce acceptable actions. Alternatively, people follow rules without checking the soundness of reasons of inherited actions. In Simon's theory of bounded rationality, simple procedures facilitate decision-making when the environment is too complex relative to a human's mental and computational capabilities. Simon (1993) supposes that human choice is driven by a number of motives, not limited to economic gain. Cheap, adequate solutions are often preferred to costly, perfect ones. Individuals have a tendency to act on advice and respect social norms.

People are socially "docile." That is, most of their beliefs are acquired, not by independent verification of the real-world facts, but from social sources which are regarded as legitimate. Simon (1993) argues as follows:

> In large measure, we do what we do because we have learned from those who surround us, not from our own experience, what is good for us and what is not. Behaving in this fashion contributes heavily to our fitness because (a) social influences will generally give us advice that is "for our own good" and (b) the information on which this advice is based is far better than the information we could gather independently. As a consequence, people exhibit a very large measure of docility.

> (Simon, 1993, p.157, original emphasis)

Docility refers to the capacity for being instructed, or the tendency to accept and believe instructions received through social channels. Docile people often make choices under social advice to do so. Social influences can be seen in the way people deal with uncertainty. They tend to acquire socially approved behavior and beliefs. Beliefs are formed on the basis of information received from qualified sources. Certain kinds of common understandings among individuals are required to produce the common associations of ideas that allow conventions to emerge and to reproduce themselves. It is the transmission mechanism of socially learned behavior and beliefs that can be received by docile individuals. Individuals have uncertain knowledge of the relationship between behavioral patterns and the associated payoffs. Therefore, they prefer to adopt behavioral patterns resembling those adopted in similar situations in the past. Simon's (1993) work comes closer to the view of institutions as patterns of social behavior. While rules of thumb are followed by a single individual, the individual pays attention to the social context in which people act and interact. In any case, following rules of conduct may be a simple and practical way of behaving in a complex social environment. Based on the knowledge of rules used in similar situations, social interaction will lead individuals to behave in a more similar manner.

Knudsen (2003) argues that Simon's theory of altruism is consistent with Hamilton's (1964) work. In Hamilton's (1964) work, population viscosity refers to a tendency for individuals to have a higher rate of interaction with their close relatives than with more distantly related ones. If altruists interact with sufficiently increased probability among themselves, they benefit more often from the cooperative behavior of others and enjoy higher fitness than selfish types. Individuals in a population interact with a higher probability with others of their own type. Individuals expect altruism to be a pervasive feature of a population. Docility gives individuals a possible fitness advantage to be realized in society. Docile people tend to accept most instructions received through social channels. The society is sustained by processes favorable to individuals endowed with some docility in following rules.

4.4   ENTREPRENEURSHIP

An important contribution of Frank Knight's (1921) *Risk, Uncertainty, and Profit* is the distinction between risk and uncertainty. Whereas risk has a probability distribution and allows insurance, uncertainty does not have a probability distribution. Risk is calculable *a priori*, and can be treated as a cost. Experience can teach us what percentage of bottles is going to burst in a champagne factory; these damages can be included in our cost calculations. Uncertainty is unpredictable and non-ergodic and, therefore, cannot be understood through conventional economic theory. Uncertainty is uninsurable; it depends on the exercise of human judgment in the decision-making process. The major difference between risk and uncertainty is whether it is possible to make *ex ante* calculations of the incidence of an event. Risk is understood to denote a situation in which known probabilities might be attached to the occurrence of future events. However, a future event is uncertain, if the likelihood of its occurrence is unknown. Knight (1921) regards uncertainty, rather than risk, as the relevant issue for business decisions. Knight's (1921) analysis is mainly devoted to the cognitive processes by which people attempt to reduce uncertainty. In the absence of uncertainty, as Knight (1921, p. 268) suggests, "it seems likely that all organic readjustments would become mechanical, all organisms automata," and "it is doubtful whether intelligence itself would exist in such a situation." When making important decisions, people rarely know either what options are available, or their possible consequences. People differ in their ability to select relevant similarities, and the ability is restricted to some particular field of knowledge. Then, "the existence of a problem of knowledge depends on the future being different from the past, while the possibility of the solution of the problem depends on the future being like the past" (Knight, 1921, p. 313). Uncertainty can be linked to a form of intuitive estimation in business situations that resist logical and statistical probability assessment. Uncertainty does not allow entrepreneurs to accurately map the relationship between inputs, outputs, and prices. For Knight (1921), the profit of the entrepreneur can be seen as compensation for bearing uncertainty. The expected returns to entrepreneurship tend to be low on average, but exhibit a high variance. Entrepreneurs exist in running businesses despite low absolute returns. They try to realize their business ideas from exercising judgment. With a subjective basis for processing, optimistic beliefs may help individuals to overcome uncertainty and may, therefore, influence their entrepreneurial activity. Entrepreneurship must consider uncertainty and obstacles inherent in the business creation process.

Mises ([1949] 1966) prepares the ground for understanding the intrinsic links between uncertainty and human action. Mises ([1949] 1966) constructs his entire economic theory from the basic axiom that human beings act. Mises ([1949] 1966) sees economic theory as a kind of body of certain knowledge that is strictly deduced from a small number of axioms. Humans are assumed to have desired ends, and to use the means at their disposal to pursue these ends. Mises' system involves interaction between heterogenous individuals as they may differ with regards to their motivations and in their entrepreneurial capacities in particular directions. In their entrepreneurial capacity, they form expectations of the future, and determine their use of current resources based on the anticipation of an expected fact. The decision to switch production techniques available might result from a change in entrepreneurial expectations. Then, "[i]n order to see his way in the unknown and uncertain future man has within his reach only two aids: experience of past events and his faculty of understanding" (Mises [1949] 1966, p. 337). According to Mises ([1949] 1966), uncertainty of the future is already implied in the very notion of action:

> The man acts and that the future is uncertain are by no means two independent matters. They are only two different modes of establishing one thing ... To acting man the future is hidden. If man knew the future, he would not have to choose and would not act. He would be like an automaton, reacting to stimuli without any will of his own.

> (Mises, [1949] 1966, p. 104)

For Mises, Austrian economics is a theory of human action, not of economic equilibrium; economic theory has its foundation in "praxeology" which deals with choice and action.[11] Praxeology is concerned with the pure logic of choice; it is independent of the particular psychological makeup of individuals. People live in a world characterized by uncertainty, in the sense that the likelihood of a particular course of action yielding a particular outcome "is not open to any kind of numerical evaluation" (Mises, [1949] 1966, p. 113). In such circumstances, people are unable to assign meaningful probabilities to the consequences of their actions. Dealing with uncertainty is crucial to the theory of human action. An individual is eager to reduce uneasiness arising from uncertainty. A science of human action is a science of the struggle of human beings to understand and cope with uncertainty. Mises' focus of the entrepreneurial function is on the market process.

It is recognized that entrepreneurship is fundamentally the driving force of the market. However, there is little consensus about how the entrepreneurial role should be incorporated into economic analysis. In standard neoclassical economic models, entrepreneurs are simply people with low risk aversion (Milgrom and Roberts, 1992). Entrepreneurs are, in the neoclassical economic approach, described as optimizers in that they act as profit maximizers or cost minimizers.[12] Entrepreneurship, like other factor inputs (land, labor, and capital) in the traditional production function, is a deployable scarce resource. Entrepreneurs are only assumed to act as residual profit claimants given their special risk-bearing appetite. Specifically, standard economic models define entrepreneurship as self-employment. The workforce is first divided into entrepreneurs and employees, and the available resources are then allocated among entrepreneurs. Kihlstrom and Laffont (1979) describe entrepreneurship as the occupational choice between employment and self-employment. In their model, the random utility derived from running a business is governed by a probability distribution with known parameters. Individuals choose the occupation that maximizes their expected utility with respect to this distribution. Kihlstrom and Laffont (1979) focus on risk aversion as the determinant which explains who becomes an entrepreneur and who works as a laborer. Risk-averse agents remaining in the firm are designated as employees or laborers. Unlike risk-averse laborers, entrepreneurs bear the risks associated with production, and contribute managerial skills. The entrepreneur functions in an equilibrium system incorporating firm formation. But entrepreneurship doesn't simply mean an occupation; it is an activity that considers the different aspects of a person.

---

11. Mises ([1949] 1966) starts by separating the problems of "truth" and "certainty," an epistemological problem, from the problem of probability, a problem that is, in his view, a praxeological problem. Then, two different types of probability are introduced: class probability (or frequency probability), and case probability (or the specific understanding of the sciences of human action). While class probability addresses an aggregate phenomenon, case probability addresses particular events or elements of a class.

12. Baumol (1968, p. 68) originally observe that neoclassical entrepreneurs are "automaton maximizers."

Individuals may differ not only in risk aversion, but also in their access to information. However, an entrepreneur, in neoclassical economic theory, has access to information required initially to perceive all alternative opportunities and all the possible consequences of acting on an opportunity. In fact, the exercise of judgment should involve private information available only to a few. Publicly available information is not sufficient for taking an important business decision. An entrepreneur takes a decision based on information that is not available to others. Profit opportunities are sometimes known to particular entrepreneurs, and not known to others due to information dispersion in markets. The entrepreneur may perceive the risk as much lower, because of the information in his or her possession.

Foss and Klein (2012) argue that entrepreneurial decision-making is about judgment in the face of Knightian uncertainty. Judgment cannot be traded or contracted on; the entrepreneur is required to start a firm in order to take economic actions based on his or her own judgment. Judgment refers to decision-making when the range of possible future outcomes is generally unknown. Entrepreneurial decision-making applies to situations in which there is no obvious rule that can be applied. Under conditions of uncertainty, judgment is the (largely tacit) ability to make decisions that turn out to be reasonable or successful *ex post* (Langlois, 2005). It can be viewed as the crucial role of the entrepreneur, and the fundamental reason for the existence of firms. Market participants make heterogeneous entrepreneurial judgments about future events. Knightian uncertainty is thus consistent with subjectivism of expectations.

Joseph Schumpeter is, without any doubt, the best-known contributor to the entrepreneurship field in economics. Schumpeter ([1912] 1934) considers the role of entrepreneurship in the economic growth process. According to Schumpeter ([1912] 1934, p. 66), entrepreneurial activities consist of the following five items: (1) the introduction of a new good or of a new quality of a good; (2) the introduction of a new method of production; (3) the opening of a new market; (4) the conquest of a new source of supply of raw materials or half-manufactured goods; and (5) the carrying out of the new organization of any industry, like the creation of a monopoly position or the breaking up of a monopoly position. For Schumpeter ([1912] 1934), an entrepreneur is a leader and "leads" the means of production into new channels. Entrepreneurship as a form of leadership is to be distinguished sharply from routine; willingness to recognize and exploit newly-perceived profit opportunities is the hallmark of entrepreneurial behavior. Schumpeter ([1912] 1934) emphasizes the entrepreneur's role as a major innovator who shifts the economy to new equilibrium states by upsetting the routine operation of the market process ("creative destruction" in Schumpeter's term). Experiential and localized knowledge explains firm heterogeneity and various aspects of competitive advantage, particularly differential innovation performance. The entrepreneur does something qualitatively new; he or she makes a new combination of already existing resources. The number of possible combinations is nearly infinite, and the entrepreneur intuitively selects a few that are possible. Thus, entrepreneurship tends to become an exceptional occurrence of massive importance. The entrepreneur's leadership qualities enable him or her to see the right way to act. Others will follow in his or her wake. Nelson and Winter (1982, p. 275), drawing from Schumpeterian insights, argue that "a central aspect of dynamic competition is that some firms deliberately strive to be leaders in technological innovations, while others attempt to keep up by imitating the success of the leaders." The leader-type entrepreneur often stimulates a cluster of innovating and imitating activity by other following entrepreneurs.

The Austrian tradition of viewing the market as a process helps us understand the functional role of entrepreneurship in the economy. Disequilibrium exists because economic agents are ignorant, and are often ignorant of their ignorance. Individuals pursuing entrepreneurship perceive profit opportunities more clearly than others. The individual entrepreneur must be a decision-maker who discovers the opportunities to enter a new market. Opportunity identification represents one of the important questions for the domain of entrepreneurship. An entrepreneur may be alert to a new superior production process before others respond. Following Hayek's (1968) thought, Kirzner (1973) argues that the existence of disequilibrium situations in the market implies profit opportunities. By contrast to Schumpeter's heroic innovator–entrepreneur aspect, Israel Kirzner focuses on the ordinary profit-seeking endeavors of market participants to discover opportunities for improvement. "For Schumpeter the essence of entrepreneurship is the ability to break away from routine, to destroy existing structures, to move the system away from the even, circular flow of equilibrium" (Kirzner, 1973, p. 127). Schumpeter emphasizes the ability of entrepreneurs to create new combinations beyond the current production possibility frontier. The Kirznerian entrepreneur is "routine-resisting" (Kirzner, 1997, p. 71) only at the stage of perceiving profit opportunities. Kirzner's "discovery" view of entrepreneurship can be distinguished from Schumpeter's "creation" perspective. According to Schumpeterian perspective, opportunities are created by the actions and imagination of entrepreneurs. That is, the "seeds" of opportunities to produce new products or services do not necessarily lie in previously existing industries or markets. A Kirznerian alert entrepreneur merely reacts to exogenously generated changes. In a situation where excess demand for some good exists, a "discrepancy" in the market is discovered by some market participants. It is the entrepreneurial element in human decision-making that identifies these market discrepancies as opportunities for profit. The Kirznerian entrepreneur might be described as a follower, in Schumpeter's sense, of the leading, originating innovators. Kirzner's focus on the entrepreneur is inspired by the objective of enabling us to see the "inside" workings of the capitalist system, while Schumpeter is concerned to enable us to see, "from the outside," what constitutes the essence of capitalism (Kirzner, 1999).

It is individuals who carry out entrepreneurial activities, no matter how they are defined. Their characteristics (personality, backgrounds, skills, etc.) matter. An entrepreneur can be defined according to a set of talents. Entrepreneurial talent includes both motivation and ability. Leibenstein (1987) describes as follows:

> Unquestionably there are entrepreneurs of unusual talents where modes of operation cannot be captured, described, and taught to others. But in my view, such people are at one end of the talent distribution found among entrepreneurs, which ranges from ordinary capacities, to the ability to slightly modify an existing firm, to the borderline of the great entrepreneurial innovators such as Henry Ford.
>
> (Leibenstein, 1987, p. 120)

According to Baumol (1990), entrepreneurial talent is assumed to be reasonably equally distributed across time and societies. Innovative entrepreneurship is incremental in nature. Baumol (1990) extends Schumpeter's five types of activities:

> To derive more substantive results from an analysis of the allocation of entrepreneurial resources, it is necessary to expand Schumpeter's list, whose main deficiency seems to be that it does not go far enough. ... Most important for the discussion here, Schumpeter's list of entrepreneurial activities can usefully be expanded to include such items as innovations in rent-seeking procedures, for example, discovery of a previously unused legal gambit that is effective in diverting rents to those who are first in exploring it.
>
> (Baumol, 1990, p. 897)

Abstracting from the social dimension of entrepreneurial behavior (which is a feature of the Schumpeterian entrepreneur), Kirzner focuses on the pure entrepreneurial function. For Kirzner, the role of the entrepreneur is to achieve some kind of adjustment necessary to move economic markets toward the equilibrium state.[13] For a profit opportunity to exist, the market must be in disequilibrium. When an entrepreneur identifies a profit opportunity, his or her action can be seen as moving the market toward equilibration. Alert entrepreneurs become deliberate coordinators of resources in markets exhibiting disequilibria. For Kirzner, "the changes the entrepreneur initiates are always toward the hypothetical state of equilibrium; they are changes brought about in response to the existing pattern of mistaken decisions, a pattern characterized by missed opportunities" (Kirzner, 1973, p. 73, original emphasis). In the equilibrium, economic agents have exploited all opportunities known to them. In the dynamic market economy, knowledge is neither complete nor perfect; markets are constantly in disequilibrium states. It is disequilibrium that gives scope to the entrepreneurial function. Kirzner

---

13. Kirzner defines the equilibrium state in neoclassical economics as "a state in which each decision correctly anticipates all other decisions" (1979, p. 110).

(1997) argues as follows: "[f]rom Mises the modern Austrians learned to see the market as an *entrepreneurially* driven *process*. From Hayek they learned to appreciate the role of *knowledge* and its enhancement through market interaction, for the equilibrative process" (Kirzner, 1997, p. 67, original emphasis). For Kirzner (1981, p. 55), "the innovative role assigned by Schumpeter to his entrepreneur finds its place naturally within the broader Misesian theory." There are three essential aspects of the broader Misesian theory of entrepreneurship, according to Kirzner (1981): (1) the recognition of the entrepreneurial element in each individual market participant; (2) the insight into entrepreneurship as the driving force behind the equilibrating tendencies within the price system; and (3) appreciation for the entrepreneurial basis for the social efficiency achieved by the market economy. Then, "action is the present choice between future alternatives that must, in the face of the foggy uncertainty of the future, now be identified in the very act of choice. It is this aspect of human action that renders it, for Mises, essentially entrepreneurial" (Kirzner, 1992, p. 128). What is missing in Schumpeter's account is an appreciation of the fact that the ordinary everyday profit-motivated endeavors of market participants to discover and exploit opportunities for improvement keep the market process working.

Kirzner emphasizes the quality of perception for recognizing a profit opportunity. The possession of idiosyncratic information allows people to see particular opportunities that others cannot see. The crucial role of entrepreneurs lies in "their alertness to hitherto unnoticed opportunities" (Kirzner, 1973, p. 39). Kirzner's concept of "alertness" is a singular property of individuals.[14] Alertness is "the ability to notice without search opportunities that have hitherto been overlooked" (Kirzner, 1979, p. 48). Profit opportunities tend to be discovered and grasped by routine-resisting entrepreneurial market participants. Entrepreneurial discovery takes place as entrepreneurs seek to seize the opportunities afforded by market frictions which are typically not known in advance. According to Kirzner (1997, p. 71), "[a]n opportunity for pure profit cannot, by its nature, be the object of systematic search. Systematic search can be undertaken for a piece of missing information, ..." Thus, Kirzner (1997) distinguishes discovery from search as follows:

> What distinguishes discovery *(relevant to hitherto unknown profit opportunities)* from successful search *(relevant to the deliberate production of information which one knew one had lacked)* is that the former *(unlike the latter) involves that* surprise *which accompanies the realization that one had overlooked something in fact readily available.*
>
> (Kirzner, 1997, p. 72, original emphasis)

---

14. For Lachmann, Kirzner's term "alertness" is considered as a behavioral prerequisite for capital formation which is founded on continuous "alertness to change and a willingness to make frequent changes, the switching 'on' and 'off' of various output streams as well as reshuffling capital combinations" (Lachmann, 1976, p. 148, original emphasis).

While searching involves some knowledge of what one is searching for, alert opportunity discovery is always accompanied by surprise. This implies that a better entrepreneur is one who possesses a superior perception of opportunity for profit. People have different information, because information is generated through their idiosyncratic life experiences. At any given time, only some people will know about particular market characteristics. Profits opportunities can only be defined *ex post*. Alert individuals, or opportunity finders, have a better grip on reality because they perceive it more accurately. The "surprise" element is at the heart of the life of an entrepreneur. A given external environment does not strictly determine decision-making alternatives and choices. There are substantial possibilities for the autonomy of individual choice. The entrepreneur has an extraordinary sense of "seeing" opportunities. As Kirzner (1992) suggests:

> The successful businessman—entrepreneur "sees" what other market participants have not yet seen. To see such opportunities will typically call for (a) superior imagination and vision (since the perceived opportunity to sell at the higher price is likely to exist only in the future) and (b) creativity (since such a profit opportunity is likely to take the form of selling what one buys in an innovatively different form, and/or different place, than was relevant at the time of purchase).
>
> (Kirzner, 1992, p. 129, original emphasis)

According to Kirzner (1979, p. 142), this sort of knowledge is largely "the result of learning experiences that occurred entirely without having been planned nor are they deliberately searched for." On the other hand, it is impossible for individuals to search for something that they do not know about. The term "search" implies entrepreneurs attempting to exploit opportunities that already exist. Then, searchers must already possess a sense of direction to guide them. Following Penrose's ([1959] 1995) arguments:

> The assumption that firms are "in search" of profits already implies some degree of enterprise, for it is only in the special case where the profitability of expansion in a given direction is obvious and the decisions to expand almost automatic that no particular quality of enterprise is required.
>
> (Penrose, [1959] 1995, p. 34, original emphasis)

Managerial mental models allow managers to perceive and interpret information. For Penrose ([1959] 1995, p. 24), the firm is "a collection of productive resources the disposal of which between different uses and over time is determined by administrative decision." Firms differ because of qualitative aspects of their managerial resources, such as creativity, level of ambition, ingenuity in fund-raising, and the ability to exercise their good judgment. The firm's productive opportunity set is itself a cognitive category (Penrose, [1959] 1995).

An opportunity exists only if it is perceived by the entrepreneur. The entrepreneur has to overcome mental and behavioral habits, and become liberated from the dictation of routine, in order to exploit opportunities successfully. Entrepreneurs do not know that a resource of alertness is at their disposal. If entrepreneurs are driven by alertness, their interpretations of market environments differ. In the alertness perspective on entrepreneurship, people are assumed to possess different knowledge and interpret the world differently. Entrepreneurs perceive external events, and formulate business plans according to their experience and knowledge. With new information and experience, they subsequently revise their plans in order to eliminate errors. Entrepreneurial alertness is associated with the interpretation framework. Knowledge shapes agents' interpretation of their environments. Experiences from everyday life are accumulated into a stock of knowledge that can be used to interpret incoming events. The interpretation framework helps an agent to identify problems and discover opportunities.

Herbert Simon did not discuss entrepreneurial behavior at great length. In his *Administrative Behavior*, Simon ([1947] 1997, p. 14) refers to the entrepreneurs "distinguished by the fact that their decisions ultimately control the activities of employees." For Simon (1958, p. 393), economic theory must address "situations where the alternatives of choice are not given in advance, but must be discovered; where the means–ends connection between choices and consequences are imperfectly known."

The following model considers two types of entrepreneurs: the Kirznerian (or Kirzner-type) entrepreneur and the Simonian (or Simon-type) entrepreneur. We present comparisons of how Kirznerian and Simonian entrepreneurs behave in the marketplace. The difference between these two entrepreneurs lies in the different decisions they make about their circumstances. For Kirzner, alertness is the essence of entrepreneurial activity. Alertness is distinct from search; alert individuals notice hitherto unexploited opportunities without search. They attempt to seize the noticed opportunities. For Simon, on the other hand, search is the essence of entrepreneurial activity. Non-alert individuals start to search for alternatives; the satisficers stop searching. Searching for more and better knowledge is often motivated by dissatisfaction with the information they have already obtained.

Thus, Kirznerian entrepreneurs are alert individuals; they discover hitherto unnoticed opportunities without search. On the other hand, Simonian entrepreneurs are non-alert individuals who search for alternatives that are good enough for them. For Kirzner, discovery replaces ignorance with newly-identified feasible states. Discovery is not simply stumbling on objective information that already exists out there, waiting to be found. Judgment about the existence of such opportunities may not be forthcoming at all. Kirznerian entrepreneurs perceive previously unnoticed profit opportunities created by exogenous changes. Alertness is activated to take advantage of exogenously generated changes. The Simonian entrepreneur, on the other hand, acts as a satisficer in the search process. Simonian entrepreneurs, with limited foresight, take actions to exploit already existing opportunities. Entrepreneurs can be either Kirzner-type or Simon-type in the model. The equilibrium number of entrepreneurs depends on the talent distribution. The concept of routine-based, boundedly rational decision-makers pervades Nelson and Winter's (1982) theory. Nelson and Winter (1982, p. 42) stress "the observed role of simple decision rules as immediate determinants of behavior and operation of satisficing principle in the search process for new rules."

Entrepreneurial activities can be directed towards the discovery of opportunities to earn profits. These activities are led by a group of people with a particular talent. Profit opportunities involve the possession of some entrepreneurial talent which is unavailable to potential competitors. Lucas (1978) analyzes a matching between firm size and entrepreneurial talent, given that entrepreneurial talent is unequally distributed; the most able entrepreneurs run the largest firms. However, it is unclear in Lucas' (1978) treatment why entrepreneurs would need firms at all. On the other hand, our model examines a matching between value-creating opportunity discovery and entrepreneurial talent, given that entrepreneurial talent is unequally distributed; the most able entrepreneurs are most likely to discover profit opportunities. In our model, opportunity discovery requires significant entrepreneurial talent. An entrepreneur is defined according to a set of talents. Entrepreneurial talent is considered here as the ability to detect opportunities. The ability to recognize an opportunity comes only from some specific talent. The issue is expectation and imagination asymmetry; the market cannot coordinate imaginations. Different individuals with unique, active, and free minds may construct heterogenous subjective expectations in the process. Each person in the model is endowed with a specific entrepreneurial talent level t, which varies across individuals. There is a distribution of talents in the population with the support of $[0, T]$. The entrepreneur can be viewed as being alert to the opportunities that exist already and are waiting to be noticed by entrepreneurial talent.

Opportunity discovery depends critically on the alertness of the entrepreneur. The model focuses on entrepreneurial talent, linking to alertness or "costless" discovery. An opportunity noticed by an alert entrepreneur uses no resources in the noticing of it. Kirznerian entrepreneurs need only be alert to profit opportunities. Alertness brings a kind of discovery that does not result from a conscious search process. Kirznerian entrepreneurs operate in the market conceived as a discovery process only. The discovery probability of payoff u is given by $p(t)$, where $p'(t) > 0$. With probability $p(t)$, an agent with talent t identifies the profit opportunity and becomes a Kirznerian entrepreneur who finds payoff u without search. A distribution of talent involves fundamental sources of new value that injects into the system. Individuals are heterogenous about their entrepreneurial talent. Talented individuals would be optimistic about their prospects, leading to entrepreneurial activity. Thus, our model describes a matching between discovering the profit opportunity and entrepreneurial talent, given that entrepreneurial talent is unequally distributed. The most talented entrepreneurs are most likely to discover the profit opportunity. The attainable payoff for an alert individual with talent t is $p(t)$ u.

With probability $1 - p(t)$, an agent with talent t becomes a Simonian entrepreneur, and starts to search for alternatives. The satisficing approach can be described as a decision-making strategy of the following form:

1. The Simonian entrepreneur sets an aspiration level before searching. The agent forms a payoff aspiration, denoted by v.
2. The Simonian entrepreneur begins to search for alternatives. Each individual has a constant search cost c. In the search process, the attainable payoff aspiration for a non-alert individual with talent t is given by $(1 - p(t))$ $v - c$.
3. The Simonian entrepreneur continues searching until the satisficing condition is obtained. The following condition is satisficing for a non-alert individual with talent t:

$$p(t)u \geq (1 - p(t))v - c. \tag{4.1}$$

The left-hand side of (4.1) is the attainable payoff for an alert individual with talent t, and the right-hand side is the attainable payoff aspiration for a non-alert individual with talent t. While Kirznerian entrepreneurs are described in terms of alertness or costless discovery, Simonian entrepreneurs are characterized by conscious search. Note that these individuals do not face any optimizing decision about their entrepreneurial activities. Fig. 4.1 displays these relationships. In this plane, the left-hand side of (4.1) is an increasing function of t, and the right-hand side is a decreasing function of t.

In Fig. 4.1, the two functions, $p(t)u$ and $(1 - p(t))v - c$, are measured vertically: $p(t)u$ slopes upward to t, and $(1 - p(t))v - c$ slopes downward to t. These two functions determine an equilibrium number of entrepreneurs t^*. People with more than talent t^* are Kirznerian entrepreneurs; t^* is the equilibrium cutoff. Thus, there is a cutoff t^*, such that people with more than talent t become Kirznerian entrepreneurs if, and only if, $t > t^*$. Fig. 4.2 displays the increase in the search cost (that is, $c_2 > c_1$). Then, as shown in Fig. 4.2, the cutoff t^* decreases in c (that is, $t^*_2 < t^*_1$). Thus, Simonian entrepreneurs decrease in the search cost.

4.5 SUMMARY

Economics formalizes the elements of complex decision problems so that a set of logical axioms can be used to analyze and compare alternatives. Represented by the expected utility of each alternative, the quantities for a set of alternatives can be used to rank a decision-maker's preferences for the alternatives. Rational decision-making would involve a comprehensive specification of all possible outcomes conditional on possible actions in order to

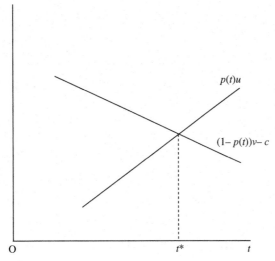

FIGURE 4.1   A cutoff t^*.

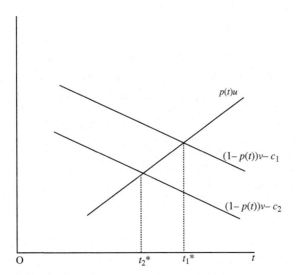

FIGURE 4.2   The increase in the search cost.

choose the single best action. However, such comprehensive calculation is not possible. Agents do not face a menu of pre-established alternatives from which to choose, but have to find them through a work of exploration. Herbert Simon examined the theory of global rationality underlying the traditional analysis of economics and management decisions, observing that actual decision-making behavior falls short of the ideal of global rationality in several ways. There are four principles about decision-making in Simon's arguments. (1) The principle of bounded rationality. Bounded rationality is an alternative conception of rationality that models the cognitive processes of decision-makers more realistically. The concept of bounded rationality implies that economic agents are, in practice, incapable of exercising global rationality. The capacity of the human mind for formulating and solving complex problems rationally is bounded. (2) The principle of satisficing. Optimizing is replaced by satisficing—the requirement that satisfactory levels of the criterion variables be attained. People simplify their decisions, primarily through the replacement of the goal of optimizing with the goal of satisficing, of finding a course of action that is good enough. An individual establishes his or her goal as an aspiration level. (3) The principle of search. Alternatives of action and consequences of action are discovered sequentially through search processes. An individual sequentially searches for alternatives, and selects one that meets the aspiration level. The search for alternatives is compatible with limited computational resources, and it terminates when an option is identified that exceeds the aspiration level. (4) The principle of adaptive behavior. An individual continually adjusts his or behavior to changing environments. Human rationality cannot be understood merely by considering the mental mechanisms that underlie human behavior. Instead, we should elucidate the relationship between the mental mechanisms and the environments in which they work.

Individuals pursuing entrepreneurship need to perceive opportunities more clearly than others. Opportunity identification represents one of the important questions for the domain of entrepreneurship. The entrepreneur endeavors to exploit this opportunity. Israel Kirzner focuses on the ordinary profit-seeking endeavors of market participants to discover and exploit opportunities for improvement. Kirzner's concept of alertness is a singular property of individuals. Alert individuals act on a previously overlooked profit opportunity. For Kirzner, the role of the entrepreneur is to achieve some kind of adjustment necessary to move economic markets toward the equilibrium state.

REFERENCES

Allais, M., 1952. Le comportement de l'homme rationnel devant le risque: critique des postulats et axioms de l'école Américaine. Econometrica 21, 503–546.

Baumol, W.J., 1968. Entrepreneurship in economic theory. Am. Econ. Rev. 58, 64–71.

Baumol, W.J., 1990. Entrepreneurship: productive, unproductive, and destructive. J. Polit. Econ. 98, 893–921.

Caplin, A., Dean, M., Martin, D., 2011. Search and satisficing. Am. Econ. Rev. 101, 2899–2922.

Foss, N.J., Klein, P.G., 2012. Organizing Entrepreneurial Judgment: A New Approach to the Firm. Cambridge University Press, Cambridge, MA.

Gigerenzer, G., Selten, R. (Eds.), 2001. Bounded Rationality: The Adaptive Tool-Box. The MIT Press, Cambridge, MA.

Gigerenzer, G., Todd, P.M., the ABC Research Group, 1999. Simple Heuristics That Make Us Smart. Oxford University Press, Oxford.

Gigerenzer, G., Hoffrage, U., Goldstein, D.G., 2008. Fast and frugal heuristics are plausible models of cognition: reply to Dougherty, Franco-Watkins, & Tomas (2008). Psychol. Rev. 115, 230–239.

Hamilton, W.D., 1964. The genetic evolution of social behavior. J. Theor. Biol. 7, 1–52.

Hayek, F.A., 1968. Competition as a discovery procedure. (M. S. Snow, trans.). Q. J. Austrian Econ. 5, 9–23.

Jones, B.D., 2002. Bounded rationality and public policy: Herbert A. Simon and the decisional foundation of collective choice. Policy Sci. 35, 269–284.

Kahneman, D., Tversky, A., 1973. On the psychology of prediction. Psychol. Rev. 80, 237–251.

Kihlstrom, R.E., Laffont, J.J., 1979. A general equilibrium entrepreneurial theory of firm formation based on risk aversion. J. Polit. Econ. 87, 719–748.

Kirzner, I.M., 1973. Competition and Entrepreneurship. University of Chicago Press, Chicago.

Kirzner, I.M., 1979. Perception, Opportunity, and Profit. University of Chicago Press, Chicago.

Kirzner, I.M., 1981. Mises on entrepreneurship. Wirtschaftspolitische Blätter 28, 51–57.

Kirzner, I.M., 1992. The Meaning of Market Process: Essays in the Development of Modern Austrian Economics. Routledge, London.

Kirzner, I.M., 1997. Entrepreneurial discovery and the competitive market process: an Austrian approach. J. Econ. Lit. 35, 60–85.

Kirzner, I.M., 1999. Creativity and/or alertness: a reconsideration of the Schumpeterian entrepreneur. Rev. Austrian Econ. 11, 5–17.

Knudsen, T., 2003. Simon's selection theory: why docility evolves to breed successful altruism. J. Econ. Psychol. 25, 97–123.

Knight, F.H., 1921. Risk, Uncertainty, and Profit. Houghton Mifflin, Boston.

Lachmann, L.M., 1976. On Austrian capital theory. In: Dolan, E.G. (Ed.), Foundations of Modern Austrian Economics. Sheed & Ward, Kansas City, MO, pp. 145–151.

Langlois, R.N., 2005. The entrepreneurial theory of the firm and the theory of the entrepreneurial firm. J. Manag. Stud. 44, 1107–1124.

Leibenstein, H., 1987. Inside the Firm: The Inefficiencies of Hierarchy. Harvard University Press, Cambridge, MA.

Lucas Jr., R.E., 1978. On the size distribution of business firms. Bell J. Econ. 9, 508–523.

March, J.G., 1978. Bounded rationality, ambiguity, and the engineering of choice. Bell J. Econ. 9, 587–608.

March, J.G., Simon, H.A., 1958. Organizations. John Wiley and Sons, Inc, New York.

Milgrom, P.R., Roberts, J., 1992. Economics of Organization and Management. Prentice-Hall, Englewood Cliffs, NJ.

Mises, L., 1949/1966. Human Action: A Treatise on Economics. Henry Regnery, Chicago, Original Work Published1949.

Nelson, R.R., Winter, S., 1982. An Evolutionary Theory of Economic Change. Harvard University Press, Cambridge, MA.

Newell, A., Simon, H.A., 1972. Human Problem Solving. Prentice Hall, New Jersey.

Penrose, E., 1959/1995. The Theory of the Growth of the Firm. Basil Blackwell, Oxford.

Rubinstein, A., 1998. Modeling Bounded Rationality. Cambridge, MA: The MIT Press.

Schumpeter, J.A., 1912/1934. The Theory of Economic Development. Harvard University Press, Cambridge, MA.

Simon, H.A., 1947/1997. Administrative Behavior: A Study of Decision-Making Processes in Administrative Organization. Free Press, New York.

Simon, H.A., 1955. A behavioral model of rational choice. Q. J. Econ. 69, 99–118.

Simon, H.A., 1956. Rational choice and the structure of the environment. Psychol. Rev. 63, 129–138.

Simon, H.A., 1957. Models of Man: Social and Rational. Wiley, New York.

Simon, H.A., 1958. The role of expectations in an adaptive or behavioristic model. In: Simon, H.A. (Ed.), Models of Bounded Rationality, vol. II. The MIT Press, Cambridge, MA, pp. 380–400.

Simon, H.A., 1959. Theories of decision-making in economics and behavioral science. Am. Econ. Rev. 49, 253–283.

Simon, H.A., 1969. The Sciences of the Artificial. The MIT Press, Cambridge, MA.

Simon, H.A., 1976. From substantive to procedural rationality. In: Latsis, S.J. (Ed.), Method and Appraisal in Economics. Cambridge University Press, Cambridge, MA, pp. 129–148.

Simon, H.A., 1978. Rationality as process and as product of thought. Am. Econ. Rev. 68, 1–16.

Simon, H.A., 1985. Human nature in politics: the dialogue of psychology with political science. Am. Polit. Sci. Rev. 79, 293–304.

Simon, H.A., 1989. The scientist as problem solver. In: Klahr, D., Kotovsky, K. (Eds.), Complex Information Processing: The Impact of Herbert A. Simon. Elbaum, Hillsdale, NJ, pp. 375–398.

Simon, H.A., 1990. Invariants of human behavior. Annual Review of Psychology 41, 1–19.

Simon, H.A., 1993. Altruism and economics. Am. Econ. Rev. 83, 156–161.

Simon, H.A., 1997. Models of Bounded Rationality, vol. III: Empirically Grounded Economic Reason. The MIT Press, Cambridge, MA.

Smith, V.L., 2008. Rationality in Economics: Constructivist and Ecological Forms. Cambridge University Press, Cambridge, MA.

Tversky, A., Kahneman, D., 1974. Judgment under uncertainty: heuristics and biases. Science 185, 1124–1131.

von Neumann, J., Morgenstern, O., 1944. Theory of Games and Economic Behavior. Princeton University Press, Princeton, NJ.

Williamson, O.E., 1985. The Economic Institutions of Capitalism. The Free Press, New York.

Williamson, O.E., 2009. Pragmatic methodology: a sketch, with applications to transaction cost economics. J. Econ. Methodol. 16, 145–157.

Chapter 5

Emergence of Prosocial Behavior

5.1 INTRODUCTION

Neuroscience research on mirror neurons suggests that humans have an innate capability to understand the mental states of others at a neural level.[1] Mirror neurons are located in our premotor cortex (Region F5), the part of our brain which controls motor activities like walking, grasping, turning, and pulling with our hands. A mirror neuron fires both when an agent acts and when an agent observes that action being performed by another. Rizzolatti and Craighero (2005) suggest that the mirror neuron mechanism resonates with the opening of Adam Smith's *The Theory of Moral Sentiments*: "[h]ow selfish soever man may be supposed, there are evidently some principles in his nature, which interest him in the fortune of others, and render their happiness necessary to him, though he derives nothing from it except the pleasure of seeing it" (Smith, [1759] 1981, p. 9). Rizzolatti and Craighero (2005) explain that this famous sentence by Adam Smith contains the two distinct ideas. First, individuals are endowed with a mechanism that makes them share the "fortunes" of others. By observing others, we enter into a "sympathetic" relationship with them. Second, because of our sympathy with others, we are compelled to desire their happiness. Others' unhappiness somehow intrudes into us; if others are unhappy, we are also unhappy. Humans have an innate interest in the fortunes of other people, and a desire for sympathy with others.

Adam Smith begins his *The Theory of Moral Sentiments* with a discussion of sympathy. Sympathy is the ability to imagine the situation where another agent stands, and to evaluate the agent's actions and feelings in that situation. Sympathy is more than pity or compassion. Smith ([1759] 1981, p. 10) defines sympathy broadly as "our fellow-feeling with any passion whatever." Sympathy forms the foundation of our moral judgment. Mirror

---

1. The pioneering research on monkeys identified mirror neurons in the premotor cortex firing when another monkey performed an action (Rizzolatti et al., 1996; Gallese et al., 1996). Since the discovery of mirror neurons in the monkey brain, much evidence from brain imaging has revealed the existence of a "mirror system" network in humans (Rizzolatti et al., 2001; Rizzolatti and Craighero, 2004).

The Cognitive Basis of Institutions. DOI: https://doi.org/10.1016/B978-0-12-812023-1.00005-3

neuron research suggests that we experience the same feelings ourselves as we observe in agents taking action. Smith ([1759] 1981) further identifies sympathy as a faculty not of the senses, but of the imagination. Smith's intuition—that individuals are endowed with an altruistic mechanism that makes them share the fortunes of others—is consistent with and supported by extensive neuroscience research on the likely role of the mirror system in sympathy.

The evolutionary biologist Robert Trivers showed how altruistic behavior in the animal world might be sustained by means of reciprocity (Trivers, 1971). The problem of sustaining altruism is, in his paper, conceptualized as a two-agent prisoner's dilemma game. The key aspects of the prisoner's dilemma are: (1) cooperation maximizes the total payoff to everyone involved in the interaction (i.e., mutual cooperation provides more benefits than mutual defection); however, (2) any player will receive a higher personal payoff by defecting, so a sizable temptation to cheat exists. If interactions occur just once in this situation, cooperation cannot be sustained, as both agents have reason to defect regardless of what the other agent chooses. The rational payoff maximization assumption predicts no cooperation in one-shot, anonymous encounters. There are gains for both players that can be obtained from mutual choices of cooperative strategies, but a non-cooperative choice is individually rational for each if they interact only once. However, if interactions are repeated, cooperation may be rational as predicted by the folk theorems. The players care about their future sufficiently (or do not discount too heavily), and have accurate information about the choices of the other players. Then, agents can make their cooperation contingent on cooperation from the other party, and make defection contingent on acts of defection from the other party. Reciprocators receive a large gain when interacting with other reciprocators. Furthermore, in Axelrod's (1984) study, the strategy "tit-for-tat," which embodies the fundamental idea behind reciprocity, plays an important role. The partner who plays tit-for-tat always cooperates with his or her partner in the first round of the repeated prisoner's dilemma game. Thereafter, he or she cooperates if the partner had cooperated in the previous round, and defects if the partner had defected in the previous round. In Axelrod's (1984) analysis, the strategy tit-for-tat can successfully invade simulated populations of partners engaging in prisoner's dilemma games, winning out over alternative strategies.

Altruism has been the research topic in many academic disciplines, including biology, psychology, philosophy, and economics. The term "altruism," however, has been used differently in order to fit the particular research contexts. In evolutionary biology, an organism is said to behave altruistically when its behavior benefits other organisms' reproductive fitness at a cost to its own fitness. The motivation for the behavior is not considered there. In economic analysis, however, the motivation of the player has to be explicitly considered in defining whether one's act is altruistic. There are scholars who explicitly assume that moral behavior is typically guided by repetitional concerns, suggesting that the motivation for behaving morally is self-interest. For example, Bateson et al. (2006, p. 413) suggest as follows: "[o]ur results therefore support the hypothesis that repetitional concerns may be extremely powerful in motivating cooperative behavior. If this interpretation is correct, then the self-interested motive of reputation maintenance may be sufficient to explain cooperation in the absence of direct return."

Khalil (2004) criticizes three major interactional theories of altruism: "egoistic," "egocentric," and "altercentric." The egoistic perspective maintains that altruistic assistance would be offered if one expects future benefit. This view is expressed by Trivers (1971) and Axelrod (1984), in which beneficence is modeled as a non-myopic, self-interested strategy to ensure future cooperation. Actually, the player who cooperates repeatedly is interested in maximizing his or her own expected utility. The egocentric view is affiliated with Becker (1976). The altruist in Becker's (1976) model helps the other because the utility function of the other is embedded in the altruist's utility function. Therefore, the altruist derives pleasure not because the other is assisted, but rather because the other's pleasure is already part of the altruist's utility. The altercentric view can account for resource sharing where the agent is built with a prosocial trait. Such a trait is modeled as springing from a "moral gene." The altercentric approach can be surmised from Frank (1988) and Simon (1993). They generally provide a Darwinian selection account of other-regarding, prosocial traits. In these models, prosocial agents proliferate because their conspecifics like them that way.

5.2   REPUTATION

Throughout their lifetime, people depend on frequent and varied cooperation with others. In the short-run, it would be advantageous to cheat. Cheating, however, may compromise one's reputation, and one's chances of being able to benefit from future cooperation. In the long-run, people are likely to cooperate in a mutually-beneficial manner. People may feel bad when they cheat their partners. This would make them better able to resist the temptation to cheat in the first round, and would enable them to generate a reputation for being cooperative.

The notion of altruism has been used in a debate, mostly within the fields of experimental economics and evolutionary anthropology (Gintis et al., 2005). The debate revolves around the question of whether ordinary people behave in the way predicted by a crude view of human beings, according to which humans have only self-regarding preferences. This view presents human beings as selfish utility maximizers. Opponents of this view maintain that human beings are norm-abiding agents, who have preferences for the well-being of others. Fehr and Fischbacher (2005) describe as follows:

> *Strongly reciprocal individuals reward and punish in anonymous one-shot interactions. Yet, they reward and punish more in repeated interactions or when their reputation is at stake. This suggests that they are motivated by a combination of altruistic and selfish concerns. Their altruistic motives induce them to cooperate and punish in one-shot interactions and their selfish motives induce them to increase rewarding and punishing in repeated interactions or when reputation building is possible.*

(Fehr and Fischbacher, 2005, p. 16)

The problem of identifying "purely" altruistic motivations in behavior is complicated by the fact that selfish agents might be induced to mimic altruistic behavior. Selfish agents may behave cooperatively or generously, in order to build up a useful reputation for altruism. Reputation is typically spread in societies. In societies where it is common to enter into relationships with people with whom one is indirectly (or not even indirectly) acquainted, reputation is often one's only source of information about them at the outset.

Social approval is considered as one of the important motives of individual behavior. In *The Theory of Moral Sentiment* ([1759] 1981), Adam Smith stressed the social approbation and disapprobation dimensions of human behavior:

> *We endeavour to examine our conduct as we imagine any other fair and impartial spectator would examine it. If, upon placing ourselves into the situation, we thoroughly enter into all the passions and motives which influenced it, we approve it, by sympathy with the approbation of this supposed equitable judge. If otherwise, we enter into his disapprobation, and condemn it.*

(Smith, [1759] 1981, p. 111)

We can view our own conduct from the externalized situations. The impartial spectator is a principle of moral self-reflection that develops over the course of our experience of exchanging places with others through the practice of imaginative sympathy. This capacity to reflect on ourselves derives from our capacity to see ourselves as others see us. We develop an otherwise impossible sense of propriety about our moral judgments.

Akerlof (1980) develops an economic model to show that disobedience to norms may involve a loss of reputation. In his model, a social custom, or a code of behavior, is introduced into an individual utility function as arguments. A reputation function depends on the dummy variable indicating whether the individual obeys or disobeys the code of behavior and the portion of the population who believes in the code.[2] It is specified in such a

---

2. In Bernheim's (1994) model, individuals care about both intrinsic utility and social status. When status is sufficiently important relative to intrinsic utility, many people conform to a rigid standard of behavior, despite heterogeneous intrinsic preferences.

way that the larger the number of code believers, the more reputation is lost by disobedience to the code. Certain groups of individuals can maintain a strong reputation over time. As in the literature on evolutionary game theory, Akerlof's (1980) model does not define a norm as a certain behavior followed by a number of people. Instead, it is defined as a moral expectation shared by a group of people. In order to change these taken-for-granted expectations and to establish a new code of behavior, it is not sufficient that one individual changes his or her behavior. A group of individuals must embrace and follow the new code of behavior to give it the status of a social norm.[3]

People want to achieve the reputation of being fair; they are fair because they care about their reputation. People may not be genuinely fair; they have to be rewarded for a good reputation. People often follow norms such as those of reciprocity and fairness, even when obedience is not in their immediate self-interest and there is no obvious sanction. Indeed, when individuals reciprocate each other, there exists for them an incentive to acquire a reputation by keeping to their promises and carrying out actions whose long-run benefits will outweigh the short-run costs. This mechanism allows individuals to establish mutually beneficial relationships. Individuals are influenced in their convictions by what they think others will do. Norms are constituted by expectations shared by members in a population, and are jointly recognized among them.

Social norms can be sustained if the pecuniary advantage from breaking norms is not sufficient to offset the forgone reputation effect. This is related to "indirect reciprocity." Indirect reciprocity is not based on repeated interactions between the same individuals (Boyd and Richerson, 1989; Nowak and Sigmund, 1998). Individuals have never met before, and the likelihood of meeting again is negligible. According to Alexander (1987), indirect reciprocity is arranged in the form of a chain; Person A helps Person B, Person B helps Person C, and so on. Person A is eventually helped by someone else who may not have been directly helped by him or her. Such chains of indirect reciprocity, according to Alexander (1987), serve to show how reciprocity can explain more generalized kinds of moral behavior. Indirect reciprocity is not limited to the same pair of individuals. Altruistic actions can be sustained if people who support others receive support in turn. To achieve such indirect reciprocity, building up a positive reputation is needed. For example, if individuals only cooperate with those of a reputation signaling that they have cooperated before, free riding is inhibited. Thus, reputation is a key element in indirect reciprocity.

---

3. For Brekke et al. (2003), sanctioning is influenced by a norm behavior. In their model, sanctioning takes the form of a lower or higher "self-image." Self-image depends not on one's perceived "type" as in Bernheim's (1994) model, but in the action taken itself. That is, the more one's behavior deviates from the norm behavior, the lower is one's self-image.

5.3 SOCIAL PREFERENCES

People are said to have entirely self-interested preferences if, and only if, they always choose in such a way as to maximize their own (expected) pecuniary payoffs. Traditional economic theory is built on the assumption that all people are entirely self-interested and do not care about the well-being of others. This assumption may be true for some, but it is certainly not true for all. People exhibit prosocial behavior when they do not always make choices that maximize their own pecuniary payoffs. Formal models of "social preferences" assume that people are self-interested, but are also concerned about the payoffs of others. That is, a player's utility function not only depends on his or her material payoff, but may also be a function of the allocation of resources within his or her reference group. People have social preferences if, and only if, they exhibit prosocial behavior and have relatively stable social preferences. More formally, given a group of N persons, let $x = (x_1, x_2, \ldots, x_N)$ denote an allocation of physical resources out of some set X of feasible allocations. The utility of individual i may be any function of the total allocation. The self-interested Person i's utility only depends on x_i. Individual i has social preferences if, for any given x_i, Person i's utility is affected by variations of x_j, $j \neq i$. A person is altruistic if the first partial derivatives of $u(x_1, x_2, \ldots, x_N)$ with respect to $x_1, x_2, \ldots, x_N$ are strictly positive.

In Charness and Rabin's (2002) paper, a two-person model of preferences assumes that a player's propensity to sacrifice for another player is characterized by parameters: the weight on the other's payoff when he or she is ahead and the weight on it when he or she is behind. Letting x_i and x_j be player i's and player j's material payoffs, consider the following simple formulation of player i's preferences as follows:

$$U_i(x_i, x_j) = (\rho r + \sigma s)x_j + (1 - \rho r - \sigma s)x_i, \quad i \neq j,$$

where

$$r = 1 \text{ if } x_j > x_i, \text{ and } r = 0 \text{ otherwise;}$$
$$s = 1 \text{ if } x_j < x_i, \text{ and } s = 0 \text{ otherwise.}$$

The weight i places on j's payoff may depend on whether j is getting a higher or lower payoff than i. The parameters ρ and σ allow for a range of different "distributional preferences." One form of distributional preferences is simple competitive preferences, which is represented by assuming that $\sigma < \rho \leq 0$. Then, players like their payoffs to be high relative to others' payoffs. Another form of distributional preferences is "inequality aversion," which corresponds to $\sigma < 0 < \rho < 1$. Then, player i likes material payoffs, and prefers that payoffs are equal. Players prefer to minimize disparities between their own payoffs and those of others.

Inequality aversion means that people resist outcomes that are perceived as inequitable. Then, people are willing to give up some material payoffs to move in the direction of more equitable outcomes. The model of Fehr and Schmidt (1999) intends to capture the idea that an individual may be uneasy, to a certain extent, about the presence of inequality, even though he or she benefits from the unequal distribution. The Fehr and Schmidt (1999) utility function, in addition to a standard neoclassical term, includes two new terms: positive deviations and negative deviations of one's own payoff, each weighted with its parameter. Given a group of N persons, the Fehr and Schmidt (1999) utility function of player i is given by:

$$U_i(x_i, \ldots, x_N) = x_i - \frac{\alpha_i}{N-1} \sum_{j \neq i} \max|x_j - x_i, 0| - \frac{\beta_i}{N-1} \sum_{j \neq i} \max|x_i - x_j, 0|,$$

where x_j denotes the material payoff Person j gets, α_i is a parameter that measures how much player i dislikes disadvantageous inequality (an "envy" weight), and β_i measures how much i dislikes advantageous inequality (a "guilt" weight). It is assumed that $0 \leq \beta_i \leq \alpha_i$ and $\beta_i < 1$. That is, an agent dislikes advantageous inequality less than disadvantageous inequality $(0 \leq \beta_i \leq \alpha_i)$, and an agent does not suffer terrible guilt when he or she is in a relatively good position $(\beta_i < 1)$.

In the two-player case, the utility function is simplified to:

$$U_i(x_i, x_j) = x_i - \alpha_i \max|x_j - x_i, 0| - \beta_i \max|x_i - x_j, 0|, \ i \neq j.$$

The second term in the utility function measures the utility loss from disadvantageous inequality, while the third term measures the loss from advantageous inequality. That is, the second term reflects how much player i dislikes disadvantageous inequality, discounted by an individual sensitivity parameter α_i, and the third term reflects how much player i dislikes advantageous inequality, again discounted by an individual sensitivity parameter more than β_i. The players do not like inequality for its own sake, but they dislike disadvantageous inequality more than they dislike advantageous inequality.

Player i's utility function is rewritten as:

$$U_i(x_i, x_j) = \begin{cases} x_i - \alpha_i(x_j - x_i), & \text{if } x_j > x_i, \\ x_i - \beta_i(x_i - x_j), & \text{if } x_j < x_i, \end{cases}$$

where $0 \leq \beta_i \leq \alpha_i$ and $\beta_i < 1$. For two players, the Charness and Rabin's (2002) distributional model has a similar piecewise linear form; their analysis emphasizes quasi-maximin preferences $(1 > \beta_i > -\alpha_i > 0)$, and can accommodate competitive preferences $(\beta_i < 0 < \alpha_i)$ as well as inequality-averse preferences $(\alpha_i > 0, \beta_i > 0)$. As a feature of the social utility function, players only care about final distributions of outcomes, not about how such distributions come about.

Bester and Güth (1998) show that, in a strategic interaction context, altruism may be favored by evolution. In their model, all members in a population interact with each other in pairs. Under the model specification, the altruist who encounters an egoist can earn a larger profit than an egoist who encounters an egoist. The consequence is that altruism will spread out in the population.

In Bester and Güth's (1998) model, all members are identical in the population. Two players, Player 1 and Player 2, play a symmetric game with "material payoffs" given by:

$$U_1(y_1, y_2) = y_1(ky_2 + m - y_1),$$

and

$$U_2(y_1, y_2) = y_2(ky_1 + m - y_2),$$

where $y_1 \geq 0$ and $y_2 \geq 0$ are the strategies of Player 1 and Player 2, respectively, and $-1 < k < 1$ and $m > 0$ are parameters. Each player's success depends not only on his or her own action, but also on the choice of the other player. If $k > 0$, the game exhibits "positive externalities" because a higher action by player i increases the success of player j. "Negative externalities" occur if $k < 0$. For example, in a production game with externalities, where y_1 and y_2 denote the each player's effort or input decision, Player 1's success can be defined as the difference between 1's output, $y_1 (ky_2 + m)$, and 1's (quadratic) effort cost, y_1^2.

Bester and Güth (1998) assume that the players do not necessarily maximize their material payoffs, but rather weighted sums of their own and their opponent's payoff. Player i's preferences are then described by:

$$V_i = \gamma U_i + (1 - \gamma)U_j, \quad j \neq i.$$

Here, the concern that player i expresses for player j is represented by the weight $1 - \gamma$. If the preference parameter γ is smaller than 1, that is, $\gamma < 1$, player i is said to be "altruistic." If $\gamma = 1$, player i is said to be "egoistic."

When an altruist invades a population of egoists and performs better than the opponent, then altruism will spread out and eliminate egoistic behavior in the process of evolution. In what follows, $R(\gamma, \gamma')$ denotes a player's success when his or her preference parameter is γ and the opponent's one is γ'. Then γ is restricted to the interval $[1/2, 1]$. By using the concept of evolutionarily stable strategies (ESS) (Maynard Smith, 1982), Bester and Güth (1998) define the evolutionary stability of a preference parameter γ^*, which lies in the interval $[1/2, 1]$, as follows.

A preference parameter γ^* is called "evolutionarily stable" if:

$$R(\gamma^*, \gamma^*) > R(\gamma, \gamma^*) \quad \text{for all } \gamma,$$

and

$$R(\gamma^*, \gamma) > R(\gamma, \gamma) \text{ whenever } R(\gamma^*, \gamma^*) = R(\gamma, \gamma^*).$$

According to the first requirement, an evolutionarily stable parameter γ^* is a best reply against itself. Then a population with parameter γ^* cannot be invaded by a small minority with deviant parameter γ. Furthermore, the second condition rules out that an alternative best reply, $\gamma \neq \gamma^*$, can spread out in the population if several parameters are equally successful. Bester and Güth (1998) provide the following results:

Let $k > 0$. Then $\gamma^* = (2 - k)/2$ is the unique evolutionarily stable preference parameter.
Let $k < 0$. Then $\gamma^* = 1$ is the unique evolutionarily stable preference parameter.

These results imply that the sign of the strategic interaction in the material function (the sign of k) determines the evolutionarily stable preference parameter γ^*. In Bester and Güth's (1998) model, the critical condition is whether the game exhibits positive externalities or negative ones. In particular, as $\gamma^* < 1$, the level of altruism is positively related to the positive parameter k. Altruism becomes more important when the strategic interdependence between the players is relatively high: $k \to 1$ implies $\gamma^* \to 1/2$. Complementarity (i.e., positive strategic interaction) thus leads to the altruistic preference parameter in the evolutionary process. The ESS concept is based on the idea that higher success reflects an advantage in reproduction. In the case of positive externalities, it is beneficial for the altruist to induce the opponent to increase the action level. Thus, the altruist will succeed in invading a society of egoists. Their analysis, however, restricts the set of preferences, not allowing broader preference parameter regions such as envy, as well as altruism. An envious person will value the material payoffs of relevant reference agents negatively. Human societies exhibit multiple stable equilibria with different mixes of altruistic and envious people, and different cultural rules enforcing behavioral norms.

5.4   ALTRUISM AND ENVY

The literature on socio-economic organizations has recognized the role of team spirit in motivating workers. An organization enjoys good performance as a result of high morale, and suffers bad performance as a result of low morale. The economic performance of an organization can vary because of the team atmosphere that prevails. However, in contract theory and principal—agent theory, which are based on material (or pecuniary) incentives, values and preferences are taken as given. Thus, the existing literature on incentives cannot offer a way of explaining the diversity of morale observed across organizations. There is something other than material incentives that matters in organizations.

The standard conception of the economic agent as a creature driven by material self-interest has provided powerful theoretical tools in the analysis of diverse problems. However, experimental support for this concept has been fragile. Results from experiments on public goods games, ultimatum games, trust games, and gift exchange games demonstrate that people in fact deviate from self-interest in systematic ways.[4] Aside from being concerned with their own payoffs, subjects appear to be concerned also with the payoffs of others. Preferences having this property are referred to as being socially interdependent, or other-regarding. The contents of self-interest are usually not specified by economic theory. Self-interest is not of the only human motivation, and it should be recognized as a special case. Departures from the standard self-interest assumption are potentially important for economics.[5]

In conventional economic theory, an individual is assumed to be calculating how to maximize his or her own exogenously given utility, independently of any psychological factor. In real life, however, people are affected by psychological and emotional factors. People feel loyalty or jealousy in society, which appears in economic literature only rarely.[6] Emotions generate behavior (Elster, 1989; Frank, 1988). An internal norm is a pattern of behavior enforced in part by internal sanctions, including shame, guilt, and loss of self-esteem, as opposed to purely external sanctions, such as material rewards and punishments (Gintis, 2003). Neglecting these psychological factors creates the serious risk that economists may not understand the diversity of norms and the changes in economic performances that are induced by changes in norms.

The purpose of this section is to go one step further. The analysis deviates from standard incentive theory because it focuses on the evolution of norms. Human beings are embedded in social structures. Individuals can acquire and internalize norms through socialization. Ben-Ner and Putterman (1998, 2000) suggest that the conjunction of (1) a postulated genetic basis for human behavioral predispositions; and (2) the demonstrable impact of environment on phenotypic variation in behavior opens up the possibility of a scientific research program for studying the influence of human environments on human preferences. Geneticists developed the distinction between a genotype, the sum of the genetic instructions provided to an organism, and a phenotype, the realized organism dependent on the interaction of those instructions with a particular environment. A similar distinction will be useful for the study of internal norms. I identify socialization as a socioeconomic action. Norms acquired in any given environment become

---

4. See, for example, Isaac and Walker (1988) and Fehr et al. (1993).
5. See Rabin (1993, 2002) and Fehr and Falk (2002) for studies on the psychological foundations of economics.
6. Examples of such exemptions include Akerlof (1980), in which the fear of a loss of reputation by acting differently from norms is discussed, and Mui (1995), in which the fear of inviting envy from others by doing well is introduced.

generalized reasons for behavior.[7] Values and preferences are transmitted through the internalization of norms. The human mind is an evolved information-processing mechanism, and through this mechanism, adaptation takes place (Tooby and Cosmides, 1992). Individuals behave in conformity with a particular social norm in the long-run.

Norms are acquired and internalized through an adaptation and imitation process. I introduce the probability of socialization, that is, a way that chance events affect population dynamics. Agents in a certain group are randomly paired to interact. Individuals try to socialize themselves to particular norms, and the actual formation of individual norms is influenced by the distribution of individuals with certain norms in the group. The socio-economic choices of agents with norms are self-enforcing, in the sense that the internalization of norms depends positively on the initial prevalence of agents with certain norms. The process by which norms evolve exhibits generalized increasing returns.[8] Norms generally take the form of conventions in the group. In the long-run, only one social norm persists. In this context, the social norm is an important element in the dynamics of economic performances, because morale is a norm-based motivation in organizations.

The model proposed in this section has the following features. There is a finite number of workers in an organization. The strategy of an individual is his or her effort level. An individual's utility function captures self-interest and other-regarding preferences. The individual's underlying motivation is modeled explicitly. Two possible types of other-regarding preferences, which exhibit "envy" and "altruism," are considered. Each psychological factor, which is denoted by a parameter, enforces a certain internal norm. Individuals are initially endowed with their norms, but they can acquire and internalize norms through socialization. In order to trace the evolution of norms through socialization, this section formulates a random-matching model in the group. While individuals act to maximize their utility, it is the prevalence of norms that determines the social norm they obey in the long-run. Individuals with different norms will typically take different actions. Chance events have large and persistent effects due to positive feedback. It is shown that different social norms generate different equilibrium effort levels within an organization. We have two steady states in the long-run: an equilibrium with high morale and an equilibrium with low morale. This multiplicity can explain why economic performance varies across organizations.

Some economic literature has analyzed the evolution of norms, considering other-regarding preferences or psychological factors. Eaton and Eswaran (2003) examine how preferences evolve by natural selection. They demonstrate that evolutionarily stable preferences can exhibit envy in a given

---

7. For a cultural explanation to social preferences, see Bowles (1998) and Bisin and Verdier (2001).
8. For the fundamental argument concerning increasing returns, see Arthur (1994).

environment. In preference maximization, players are not choosing effort levels to maximize relative fitness. According to Kandori (2003), the norm evolves over time, according to the actual effort levels taken by individuals. He shows that, owing to random shocks, the system moves from equilibria with higher effort to those with lower effort. In contrast to this section, however, they do not consider that norms and preferences are acquired and internalized through socialization. In the dynamics of the pattern of norms, the model presented in this section shows that effort levels can vary, depending on the prevailing norm within an organization.

5.4.1 The Model

We consider a group in which there are N individuals.[9] Each individual j chooses his or her effort level, e_j (≥ 0), $j \in \{1, \ldots, N\}$. Suppose that output is some function of each agent's effort, given by $f(\mathbf{e})$, where $\mathbf{e}$ is an N-dimensional vector of agents' effort levels. Consider a work situation in which each agent's compensation is determined as $f(\mathbf{e})/N$. It is painful to put forth effort, and the pain that an agent feels is given by $c(e_j)$ that is exogenous, where $c' > 0$ and $c'' > 0$.

Agent i's utility function is given by:

$$U_i(\mathbf{e}; \mu_i) = \frac{f(\mathbf{e})}{N} - c(e_i) + \mu_i \cdot (N-1) \cdot \frac{f(\mathbf{e})}{N}. \tag{5.1}$$

The first two terms of the right-hand side of (5.1) capture self-interest, and the last term represents other-regarding preferences. The parameter μ_i is the weight (or the "binding power" of the internal norm) that agent i's preferences put on all other agents' well-being. Also, $(N-1) \cdot f(\mathbf{e})/N$ are the aggregate compensations of all agents other than i in the N-member group under consideration. The last term is an attempt to formalize the discussion of internal norms, which are social and endogenous.

Suppose there are two possible types of norms that can be internalized by individuals. We say agent i's internal norm is enforced by "envy" if $\mu_i \equiv \mu^- < 0$.[10] With the binding power of this sort, i's utility is a decreasing function of the compensations of other agents. On the other hand, we say i's internal norm is enforced by "altruism" if $\mu_i \equiv \mu^+ > 0$. Then, i's utility is an increasing function of the compensations of others. Initially, each agent is endowed with a particular norm.

Each agent chooses his or her effort level, given the parameter μ_i. As the utility function U_i is strictly concave in e_i and the efforts are continuous, there would be a unique utility maximizer for U_i. High (or low) effort level corresponds to high (or low) morale in the group. We next present an

---

9. N is an even number in this analysis.
10. For the following analysis, we assume that $|\mu^-| \leq 1/(N-1)$.

equilibrium concept somewhat different from the standard Nash equilibrium of a game.[11]

Definition: Given the binding power of i's internal norm μ_i, an effort profile $\mathbf{e}^*$ is a morale equilibrium if:

$$U_i(\mathbf{e}^*; \mu_i) \geq U_i(e_i, \mathbf{e}^*_{-i}; \mu_i), \quad \forall i, \ \forall e_i.$$

A morale equilibrium holds so that each agent takes others' effort as given. It is necessary to compare morale equilibria, one with μ^+ and one with μ^-. We will show that equilibrium effort is higher with μ^+ (>0) than it would be with μ^- (<0).

For analytical simplicity, we assume that $f(\mathbf{e}) = \Sigma_j e_j$ and $c(e_i) = e_i^2/2$. Thus, (5.1) can be rewritten as:

$$U_i(\mathbf{e}; \mu_i) = \frac{\Sigma_j e_j}{N} - \frac{e_i^2}{2} + \mu_i(N-1)\frac{\Sigma_j e_j}{N}.$$

5.4.2   The Evolution of Norms

The model is constructed in discrete time, indexed by $t = 0, 1, 2, \ldots.$

We model the evolution of norms, namely, those that are acquired and internalized through socialization in the group. The model represents each individual's updating as a process of switching from one norm to another through learning from others. We introduce the socio-economic interaction between individuals in the group. At the beginning of time t, the number of agents with trait μ^+ (>0), that is, the binding power of the internal norm enforced by altruism, is denoted by $n(t)$. And let $\tau(t)$ be the fraction of individuals with trait μ^+ at the beginning of time t, that is, $\tau(t) = n(t)/N$. Thus, the fraction of individuals with trait μ^- (<0), that is, the binding power of the internal norm which is enforced by envy, is $1 - \tau(t) = (N - n(t))/N$ at the beginning of t.

We formulate a random-matching model in the group. The model introduces chance events that affect population dynamics through socialization. Individuals try to socialize themselves to particular norms at the beginning of each time. Socialization occurs as follows. With a certain probability of socialization, agent i is matched with an individual in the group, and then adopts the norm of that individual. The probability of socialization is allowed to depend on the fraction of individuals in the group. In this model, socialization acts as a type of complementarity.[12] That is, an agent with

---

11. This is also somewhat different from Kandori (2003), because we focus on the evolution of norms through socialization for the formation of preferences.
12. In Bisin and Verdier's (2001) model, the interior rest point under cultural complementarities is unstable while the other rest points are stable.

trait μ^- (or μ^+) has more incentives to socialize with an individual with trait μ^+ (or μ^-), as μ^+ (or μ^-) is the more widely dominant one in the group. Specifically, socialization to agent i with trait μ^- in t occurs with probability $q(\tau(t))$, which is a continuous, strictly increasing function in $\tau(t)$. On the other hand, socialization to agent i with trait μ^+ in t occurs with probability $q(1 - \tau(t))$, which is a continuous, strictly increasing function in $1 - \tau(t)$. With probability $1 - q(\tau(t))$ or $1 - q(1 - \tau(t))$, agent i is "naïve," and does not get randomly matched with somebody else in the group. Moreover, it is assumed that $q(0) = 0$ and $q(1) = 1$.

Let $P^{+-}(t)$ denote the transmission probability that an agent with trait μ^+ is socialized to trait μ^- at the end of time t. The socialization mechanism is then characterized by the following transition probability:

$$P^{+-}(t) = q(1 - \tau(t)) \cdot \left\{ \frac{N - n(t)}{N - 1} \right\}. \tag{5.2}$$

This represents the coefficient of norm transmission, where an agent with trait μ^+ (at the beginning of t) is matched with an individual with trait μ^- of the remaining individuals, and adopts and internalizes that trait at the end of t. The transition probability, $P^{++}(t)$, that an agent with μ^+ sticks to that trait in the socialization mechanism is given by:

$$P^{++}(t) = 1 - q(1 - \tau(t)) + q(1 - \tau(t)) \cdot \left\{ \frac{n(t) - 1}{N - 1} \right\}. \tag{5.3}$$

In the term $1 - q(1 - \tau(t))$ of the right-hand side of (5.3), an agent with μ^+ does not get matched with somebody else in the group, and sticks to that trait in t. In the last term, an agent with μ^+ at the beginning of t has a chance of socialization, but is matched with an individual with μ^+ of the remaining individuals in the group (the agent does not change his or her trait).

Similarly, for an agent with trait μ^- at the beginning of t, we get:

$$P^{-+}(t) = q(\tau(t)) \cdot \left\{ \frac{n(t)}{N - 1} \right\}, \tag{5.4}$$

$$P^{--}(t) = 1 - q(\tau(t)) + q(\tau(t)) \cdot \left\{ \frac{N - n(t) - 1}{N - 1} \right\}. \tag{5.5}$$

Given these transmission probabilities, the fraction $\tau(t + 1)$ of individuals with trait μ^+ at the beginning of time $t + 1$ is calculated to be:

$$\tau(t + 1) = \tau(t) + P^{-+}(t) \cdot (1 - \tau(t)) - P^{+-}(t) \cdot \tau(t),$$

Substituting (5.2) and (5.4), we obtain:

$$\tau(t + 1) - \tau(t) = \left\{ \frac{N - n(t)}{N - 1} \right\} \cdot \tau(t) \cdot \{q(\tau(t)) - q(1 - \tau(t))\}. \tag{5.6}$$

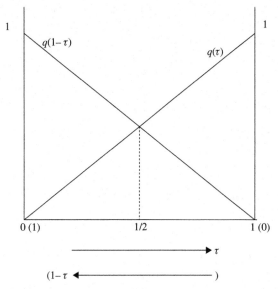

FIGURE 5.1 Population dynamics through socialization.

This is the equation for the population dynamics through socialization. It is easy to derive conditions that guarantee that the population dynamics of (5.6) converge to a "homogeneous" population in the limit. In Fig. 5.1, the probabilities of socialization, $q(\tau)$ and $q(1-\tau)$, are measured vertically: $q(\tau)$ slopes upward to τ and $q(1-\tau)$ slopes downward to τ. And $q(1-\tau)$ is a mirror image of $q(\tau)$. Assume that $\tau(0) > 1/2$ holds initially. Then, we have $\partial\{\tau(t+1) - \tau(t)\}/\partial\tau(t) > 0$ from (5.6). Thus, we have $\tau(t) \to 1$ for any $\tau(0) > 1/2$. Everyone adopts the same norm eventually. Similarly, the dynamics implies that $\tau(t) \to 0$ for any $\tau(0) < 1/2$.

To summarize:

Lemma 5.1: (Teraji, 2007).

For the fraction of individuals having the internal norm which is enforced by altruism, $\tau(t)$, in the group at the beginning of time t,

1. *$\tau(t)$ converges to 1 if $\tau(0) > 1/2$ holds initially,*
    and
2. *$\tau(t)$ converges to 0 if $\tau(0) < 1/2$ holds initially.*

Thus, the process by which norms evolve through socialization exhibits generalized increasing returns. Social norms take the form of conventions, that is, the modes of behavior to which a majority of agents subscribe. People follow a certain norm through socialization. In the presence of generalized increasing returns, chance events have large and persistent effects due to positive feedbacks. In the self-enforcing process that depends on the initial prevalence of norms, only one social norm emerges. The social norm, which is enforced by altruism or envy, persists in the long-run if a majority of agents adopt that norm initially. In the socialization mechanism, individuals behave in conformity with a particular social norm eventually.

Agent i's utility function, whose binding power of the internal norm is μ^+ or μ^- at the beginning of time t, is given by:

$$U_i(\mathbf{e}(t); \mu^+) = \frac{\Sigma_j e_j(t)}{N} - \frac{e_i^2(t)}{2} + (P^{++}(t) \cdot \mu^+ + P^{+-}(t) \cdot \mu^-) \cdot (N-1) \cdot \frac{\Sigma_j e_j(t)}{N},$$

$$U_i(\mathbf{e}(t); \mu^-) = \frac{\Sigma_j e_j(t)}{N} - \frac{e_i^2(t)}{2} + (P^{-+}(t) \cdot \mu^+ + P^{--}(t) \cdot \mu^-) \cdot (N-1) \cdot \frac{\Sigma_j e_j(t)}{N},$$

at the end of t. Agent i chooses i's effort level at the end of each time. Then, there is a unique effort level maximizing the utility function. The utility-maximizing effort levels for agent i in t, $e_i^*(t; \mu^-)$ and $e_i^*(t; \mu^+)$, are given by:

$$e_i^*(t; \mu^+) = \text{Arg Max}_{e_i \geq 0} U_i(\mathbf{e}(t); \mu^+),$$

$$e_i^*(t; \mu^-) = \text{Arg Max}_{e_i \geq 0} U_i(\mathbf{e}(t); \mu^-).$$

The first-order conditions are:

$$e_i^*(t; \mu^+) = \frac{1}{N} + (P^{-+}(t) \cdot \mu^+ + P^{+-}(t) \cdot \mu^-) \cdot \frac{(N-1)}{N}, \qquad (5.7)$$

$$e_i^*(t; \mu^-) = \frac{1}{N} + (P^{-+}(t) \cdot \mu^+ + P^{--}(t) \cdot \mu^-) \cdot \frac{(N-1)}{N}. \qquad (5.8)$$

Suppose that $\tau(0) > 1/2$ holds initially. From Lemma 5.1, $\tau(t) \to 1$ in the limit. Then, we have $P^{++}(t) \to 1$, and $P^{+-}(t) \to 0$ in (5.2) and (5.3). Thus, it holds that $e_i^*(t; \mu^+) \to \{1 + \mu^+ \cdot (N-1)\}/N$ in (5.7). Furthermore, in (5.4) and (5.5), $n(t)/(N-1) \to 1$ and $(N - n(t) - 1)/(N-1) \to 0$ for the remaining individuals other than agent i. Then, we have $P^{-+}(t) \to 1$ and $P^{--}(t) \to 0$. Thus, $e_i^*(t; \mu^-) \to \{1 + \mu^+ \cdot (N-1)\}/N$ in (5.8). Similarly, if $\tau(0) < 1/2$ holds initially, we have $e_i^*(t; \mu^+) \to \{1 + \mu^- \cdot (N-1)\}/N$ and $e_i^*(t; \mu^-) \to \{1 + \mu^- \cdot (N-1)\}/N$ in the limit.

Morale is a norm-based motivation. The social norm generates the equilibrium effort level in the long-run. Everyone chooses the same effort level in equilibrium because of a particular social norm that prevails. When there are multiple equilibria, these equilibria can be ranked. Morale is higher in the high-effort equilibrium than in the low-effort equilibrium. Thus, we have two morale equilibria: in one equilibrium with high morale, every agent's effort level is $\{1 + \mu^+ \cdot (N-1)\}/N$, and in the other equilibrium with low morale, every agent's effort level is $\{1 + \mu^- \cdot (N-1)\}/N$, where $\mu^+ > 0 > \mu^-$.

Thus, we have the following:

Proposition 5.1: (Teraji, 2007).

1. *Suppose that $\tau(0) > 1/2$ holds initially. Then, we have a morale equilibrium $e = (e, \ldots, e)$, where $e = \{1 + \mu^+ \cdot (N - 1)\}/N$.*
2. *Suppose that $\tau(0) < 1/2$ holds initially. Then, we have a morale equilibrium $e = (e, \ldots, e)$, where $e = \{1 + \mu^- \cdot (N - 1)\}/N$.*

The evolution of norms allows us to study long-run equilibria with different effort levels. Which norm is persistent determines the equilibrium effort level within an organization. If the norm is enforced by altruism, every agent chooses the high equilibrium effort level in the long-run. In this morale equilibrium, every agent is able to maintain high morale. Thus, norms inculcated through socialization can lead members to create incentives within an organization. On the other hand, if the norm, which is enforced by envy, persists, the low equilibrium effort level can be sustained. In this morale equilibrium, everyone shirks. Then, we have the decay of economic performance in the long-run. Thus, the economic performance can vary depending on the norm that prevails within an organization.

A key finding of path dependence is a property of "lock-in" by historical events (Arthur, 1994). In a world of increasing returns to scale (positive feedbacks), initial and trivial circumstances can have important and irreversible influences on the ultimate market allocation of resources. Path-dependent economics is a theory of equilibrium selection with positive feedbacks, and a path-dependent outcome is associated with characteristics of persistence and uniformity. The economy has a multiplicity of possible equilibrium solutions, where the dominant solution can be the suboptimal one.

In this section, effort depends on the prevailing norm within an organization. In a world of effort discretion, the economic performance can vary across organizations. Furthermore, the process by which norms evolve through socialization exhibits generalized increasing returns. One social norm emerges in the process that depends on the initial prevalence of norms. The social norm, enforced by altruism or envy, persists in the long-run. Individuals try to socialize themselves to one particular norm. Which norm is persistent determines the equilibrium effort level. It is possible for low-effort organizations to survive in the long-run. Then, the norm, enforced by envy, can be a cause of economic inefficiency. If the norm is enforced by envy, the private incentives do not exist for individuals to choose the high-effort equilibrium. In a world where norms are determinants to effort levels, the path-dependent low-effort equilibrium can exist.

These discussions connect with what Harvey Leibenstein referred to as X-inefficiency. A large number of X-inefficiency studies exist, dating from Leibenstein (1966). He referred to the difference between maximal effectiveness of utilization and actual utilization as the degree of X-inefficiency. According to him, individuals supply different amounts of effort, where effort is a multidimensional variable, under different organizational and environmental circumstances. Individual motivations and interactions between individuals are important. He presented a reasonable vision of human behavior within the organizational context, where X-inefficiency can persist stubbornly. Given effort discretion, the quantity and quality of effort can vary. Labor productivity is affected by the quantity and quality of effort input into the process of production. The introduction of effort variability allows for the existence of path-dependent high and low productivity firms in the long-run.

Individuals may make different decisions, depending on social contexts. Differences in context-dependent behavior can be interpreted as being generated psychologically, ethically, and sociologically. This section has provided theoretical insights into the diversity of economic performances within organizations. The approach taken here can shed light on issues on endogenous norms in society. This section has argued that diverse morale and the evolution of norms interact in nontrivial ways. Agents can acquire and internalize norms through socialization. The process by which norms evolve through socialization is self-enforcing, and depends on the initial distribution of individuals with certain norms in the group. In this formulation, different norms typically exhibit different morale equilibria in the long-run. The resulting dominance of a certain norm implies that a certain social structure appears more often than otherwise. The model has proposed two morale equilibria with different effort levels. If the norm, enforced by altruism, persists in the limit, a "good" morale equilibrium emerges, in which everyone chooses a high effort level and high morale is sustainable within an organization. Norms inculcated through socialization can lead individuals to create incentives. On the other hand, if the norm, enforced by envy, persists in the limit, a "bad" morale equilibrium emerges, in which everyone shirks. Morale can be thus a norm-based motivation. Given effort discretion, it is possible to expect a multiplicity of possible equilibrium solutions, where the dominant solution is suboptimal. The economic performance can vary, because of the norm that prevails in organizations.

5.5 ALTRUISTIC PUNISHMENT

The free-rider problem is one of collective action problems. Where a group of self-interested individuals may produce a shared good that will be available to every group member, they will seek to free ride on others' efforts. The dominant strategy is zero contribution at all times for all participants in standard public goods games. Thus, the group fails to produce the good.

A social norm can be seen as a behavioral regularity associated with a feeling of obligation. People come to hold normative attitudes about a state of affairs, believing that others should do what they are expected to do, especially when unexpected behavior causes harm. Normative incentives are frequently negative; costs are imposed on those who fail to conform to a behavioral regularity. People tend to follow prevailing norms conditional on observing others' compliance. Therefore, the potential of norms to guide behavior can break down if norm violations are not sanctioned. Strong reciprocity is often denoted as "altruistic punishment" (Boyd et al. 2003; Gintis et al. 2008). Individuals are strong reciprocators if they spend their own resources to punish others who show adverse behavior. Person A punishes Person B for defecting in a game (with Person A) with the consequence that, in a later game, Person B behaves more cooperatively toward Person C than he or she would otherwise have done. The punishment is altruistic, at least in its consequences, since Person A's infliction of punishment on Person B is both costly to Person A and beneficial to Person C. Individuals with altruistic preferences, armed with punishment mechanisms, are capable of enforcing widespread cooperation. As shown by Fehr and Fischbacher (2003):

> *If cooperators have the opportunity to target their punishment directly towards those who defect they impose string sanctions on the defectors. Thus, in the presence of targeted punishment opportunities, strong reciprocators are capable of enforcing widespread cooperation by deterring potential non-cooperators.*
>
> (Fehr and Fischbacher, 2003, p. 787)

Social norms emerge within groups to solve collective action problems. Norms are then prosocial: norms mandate individually costly behavior that is beneficial to others, or prohibit individually gratifying behavior that is harmful to others.

Hayek was well aware of the fact that individuals refrain from free riding because they fear their fellow's retaliation. That is, the solution to the collective action problems is to be found in punishment strategies implemented by other members of the group. Punishment strategies are invoked as remedies for the maintenance of social norms:

> *If deviant behavior results in non-acceptance by the other members of the group, and observance of the rules is a condition of successful cooperation with them, an effective pressure for the preservation of an established set of rules will be maintained. Exclusion from the group is probably the earliest and most effective sanction or 'punishment' which secures conformity, first by mere actual elimination from the group of the individuals who do not conform while later, in higher states of intellectual development, the fear of expulsion may act as a deterrent.*
>
> (Hayek, 1967, p. 78)

People comply with social norms because the threat of punishment makes it in their interest to do so. The importance of decentralized punishments (i.e., punishments carried out by individuals without the intervention of a central authority) is documented in experimental studies. Ostrom et al.

(1992) show the existence of such punishment opportunities in a common-pool resource use game. Systems without punishment mechanisms face severe rule-breaking, and they break down in the end. The fear of punishment has a positive effect on cooperation. Successful institutions are able to minimize expenditures on punishment by adapting their rules to local circumstances. Norms consist of voluntary efforts by group members to regulate their peers' behavior. Defections will be detected by the person concerned, and will subsequently be punished by the community.

How can cooperation among group members be developed and sustained? The cooperative behavior of humans has been addressed in many ways during the last decades. According to theoretical work (e.g., Fudenberg and Maskin, 1986), voluntary cooperation can be sustained even when the costs outweigh the benefits at single points in time. Game theoretic conclusions are deductive ones, based on players consistently and objectively processing information. In the laboratory, the most frequently used experimental design has been public goods games (Chaudhuri, 2011). A typical public goods game consists of a number of rounds.[13] The design nicely captures the social dilemma involved, without being too complex. The participants are told about the total number of rounds, and they are paid their winnings in real money at the end of the session. In each round, each participant is grouped with several other subjects under conditions of strict anonymity. Each participant is given a certain number of "points," redeemable at the end of the experimental session for real money. Each participant secretly chooses how many of their private tokens to put into the public pot. Experimenters multiply the number of tokens in the pot before it is distributed to encourage contributions. Each participant keeps the tokens that he or she does not contribute, plus an even split of the tokens in the pot. A self-interested player will contribute nothing to the common account, and will benefit from the public goods. In public goods experiments, subjects begin by contributing on average about half of their endowments to the public account. However, the level of contribution decreases over the course of multiple rounds.[14] When costly punishment is permitted, cooperation does not deteriorate. In a repeated voluntary contribution mechanism, contributors may choose to punish members of their own group because they believe that punishment will increase in the future (and, thereby, their own future payoffs will increase) or, in the case of altruists, because they simply want to benefit other contributors. Similarly, as bystanders who are not directly affected, the

---

13. A public good has two essential attributes: non-excludability and non-rivalry in consumption. A common-pool resource, however, is non-excludable but rival. The possibility of non-rival consumption is the major feature distinguishing public goods from common-pool resources.

14. Andreoni (1995a) and Sonnemans et al. (1998) investigate the effects of positive versus negative framing in a public goods setting. In their studies, cooperation is lower when the incentive structure is framed as a negative externality, rather than the usual positive externality.

contributors in one group may punish the free riders when engaged in indirect reciprocity (Alexander, 1987). When punishment is not permitted, on the other hand, the same subjects experience the deterioration of cooperation found in previous public goods games.

Fehr and Gächter (2000) indicate that many individuals are willing to punish unfair behavior at a personal cost in public goods games. Without punishment, mean contributions deteriorated over rounds to approach zero; with punishment, mean contributions remained significantly greater than zero. Free riders are often sanctioned. Enforcement often requires the intervention of bystanders who are not directly affected. Potential punishers are not themselves the victims, but have merely witnessed unfair behavior. This is called altruistic punishment, as individuals sacrifice for no direct benefits. It suggests that cooperation has evolved through the sacrifice of altruistic punishers who are ready to incur some costs to prevent unfair behavior. The existence of such altruistic punishers constitutes a threat that acts as a deterrent against norm violations. A norm is regarded as a rule governing individual behavior. Norms are enforced due to the expectations that norm violations will be punished. Recent research demonstrates that people in a diverse group of societies around the world engage in costly punishment of unequal behavior in third-party punishment experiments (Fehr et al., 2002). The third-party punishment game is played by three anonymous players. Player 1 and Player 2 are allocated a stake of money by the experimenter. Player 1 is given the opportunity to split the stake with Player 2. Player 2 simply receives whatever Player 1 allocates. Player 3 also receives a stake of money (about half what Player 1 and Player 2 split), and is informed of how Player 1 split the stake with Player 2. Player 3 has the option to punish Player 1 by paying some of his or her stake to have a proportionate amount reduced from Player 1's takings. Fehr and Fischbacher (2004) examine both the extent and possible causes of third-party punishment in one-shot dictator and prisoner's dilemma games. They find that a substantial number of third parties sanction violations of cooperative norms. In the experiments reported by Henrich et al. (2006), the amount of punishers varies considerably across societies, and is not explained by individual-level demographic variables such as age, gender, wealth, or income. What people punish in these experiments is violations of socially recognized, and perhaps culturally specific, norms of behavior. Cultural differences may account for variations in punishment behavior. Furthermore, Herrmann et al. (2008) provide evidence on punishment behavior in social dilemmas in 16 places around the world. They report that antisocial punishment is widespread and negatively correlated with norms of civic cooperation and the strength of the rule of law in a country.

Why does punishment occur? Individuals may feel aggressive emotions when others violate a norm. These emotions shape human preferences and, as a result, punishment of violators. Furthermore, we can see a relationship between gossip and social norms. Consider a situation in which Person A engages in norm-violating behavior towards Person B, and Person B tells Persons C, D, ... about the trespass. The diffusion of information about the trespass adds a group of potential third-party punishers (Persons C, D, ...) to the original second-party punisher (Person B). Person B may deliberately pass on information about Person A's trespass in order to increase the social pressure on Person A to reform his or her behavior. Knowing that, Person A may be deterred from defecting in the first place. Person B may gossip because of the direct benefits it provides, and gossiping may not be costly for Person B.

5.6 NORM COMPLIANCE

Societies have social norms or, for short, norms; members of the society are required to follow standards of behavior. Social norms are informal rules, as opposed to formal, legal rules promulgated by a court or a legislature. Social norms often direct individuals to undertake actions that are inconsistent with selfish actions. For example, in the dictator game, 50−50 division is generally viewed as norm-compliant (Andreoni and Bernheim, 2009).[15] People may deviate from such norms. In the case of legal compliance, individual incentives most often refer to deterrence (Becker, 1968). That is, individuals are deterred from criminal activities by a higher fine, and by a higher probability of conviction. Unlike legal rules, social norms are not supported by formal sanctions. Why do social norms not simply collapse from the violation? This section studies two distinct mechanisms on norm compliance. The incentive to comply with norms derives not only from the enforcement of costly punishment by others, but also from reputation building for oneself.

The importance of decentralized punishments (i.e., punishments carried out by individuals without the intervention of a central authority) is documented in experimental studies. Ostrom et al. (1992) show the existence of such punishment opportunities in a common-pool resource use game. In their experimental design, they kept subjects in constant groups, and did not reveal the number of rounds to be played. Since subjects knew that they would interact with their group members again, they could use punishment to discourage free riding, anticipating greater cooperation and greater payoffs in subsequent rounds. The fear of punishment has a positive effect on cooperation. In public goods experiments, subjects begin by contributing, on average, about half of their endowments to the public account. However, the level of contribution decays over the course of multiple rounds

---

15. The dictator game in theory gives rise to very inequitable distributions of resources. However, when the game is played for real, fair allocations figure prominently. Many game experiments offer abundant evidence that contradictthe hypothesis that all players are motivated only by their own material interest (see Camerer, 2003).

(Andreoni, 1995b). When costly punishment is permitted, cooperation does not deteriorate. Fehr and Gächter (2000, 2002) indicate that many individuals are willing to punish unfair behavior at a personal cost in public goods games. Costly punishment is administered by "third parties" (Fehr and Fischbacher, 2004). Potential punishers are not themselves the victims, but have merely witnessed unfair behavior. This is called "altruistic punishment," as individuals sacrifice for no direct benefits (Gintis et al., 2003). It suggests that cooperation has evolved through the sacrifice of altruistic punishers, who are ready to incur some costs to prevent unfair behavior. People are fair because they have a psychological motivation to restore fairness.[16] Then, punishment can be seen as a consequence of a sense of fairness.

Certain groups of individuals can maintain a strong reputation over time. Akerlof (1980) develops an economic model to show that social norms that involve pecuniary disadvantage to individuals may persist without erosion.[17] Disobedience to the norm may involve a loss of reputation. People want to achieve the reputation of being fair. People are fair because they care about their reputation. They may not be genuinely fair. They have to be rewarded for good reputation, and they have to be willing to comply with the norm. Individuals are influenced in their convictions by what they think others will do.[18] Conformity to the norm is conditional on expectations about other peoples' behavior. Norms are constituted by expectations shared by members in a population, and are jointly recognized among them.[19]

Thus, we have a set of solutions to the problem of norm compliance. The first solution is the punishment-based account. Following this account, people comply with norms because the threat of punishment makes it in their interest to do so. Altruistic punishment seems to have a solid foundation in human interaction. However, such costly punishment leads to a large increase in losses for altruistic punishers. The second solution is, on the other hand, the reputation-based account. Social norms can be sustained if the pecuniary advantage from breaking norms is not sufficient to offset the forgone reputation effect. This is related to indirect reciprocity. According to Alexander (1987), indirect reciprocity is arranged in the form of chain; a person is eventually helped by someone else, who may not have been directly helped by him or her.[20] Altruistic actions can be sustained if people who

---

16. For example, Rabin (1993) examines concerns for fairness.
17. Building on Akerlof's (1980) model, Naylor (1989) explains the logic of collective strike action.
18. Sliwka (2007) considers the notion of trust as a credible signal of a social norm.
19. Tirole (1996) considers the joint dynamics of individual and collective reputations.
20. See Nowak and Sigmund (1998) for a mathematical model of indirect reciprocity. Their model is based on image scoring; agents develop a positive reputation for cooperating, and only cooperate with others whose score is above a threshold (image score).

support others receive support in return. To achieve such indirect reciprocity, building up a positive reputation is needed.[21]

This section analyzes the interaction between the potential for costly punishment and building a personal reputation. The model considers two groups of agents in a society with one norm. Agents in one group (Group i) choose whether to comply with the norm by incurring some cost. They acquire utility from the reputation derived from complying with the norm. This utility depends positively on the proportion of motivated compliers. Individuals may differ in their motivation to comply with the norm. Punishment will be imposed on individuals who deviate from the norm. In the other group (Group j), there are agents who value compliance and potentially punish non-compliance (i.e., the sanctioning individuals). The model investigates individual punishment decisions. Agents choose to punish violators at some cost (decentralized punishment).

This section asks how individual values evolve endogenously over time and analyzes the long-run dynamics of norm formation. The present framework systematically investigates the different forces to account for the long-run stability of the norm. There are two scenarios as follows. In one scenario, there is some possibility that the erosion of the reputation effect induces individuals to break the norm. Then, the norm is enforced due to a higher level of punishment of noncompliance. Punishment would be used to enforce the norm if a substantial fraction of the people have little reputation-derived utility by obeying the norm. This section, however, suggests that altruistic punishment may play a limited role in sustaining the norm. In another scenario, everyone is motivated due to reputation formation, despite a lower level of punishment by others. For a lower level of punishment, effective reputation building provides a way to sustain the norm.

5.6.1 The Model

Consider a society populated by a continuum of agents at each period of time t. The population size is constant over time and normalized to 1. We assume that the population is composed of two groups, i and j, which differ in their characteristics. The first group i amounts to 1/2 of the whole population. Accordingly, the other half of the population belongs to Group j. Matching between the two groups takes place randomly. A member of Group i randomly matches with the opponent of Group j at each period. This may suggest a large society, in which one-shot encounters with unrelated strangers are common and information is rarely transparent.

There is one norm in the society. For simplicity, we assume that it is only possible to either follow the norm or not. An agent of Group i ("he") chooses

---

21. Engelman and Fischbacher (2009) assess the interplay of indirect reciprocity and strategic reputation building in an experimental helping game. When indirect reciprocity is not contaminated by incentives for strategic reputation building, they call this pure indirect reciprocity.

an action $x \in \{0, 1\}$ at each period. His action is represented by a discrete variable, one or zero. That is, $x = 1$ if the agent of Group i complies with the norm, and $x = 0$ if he violates it. Inertia is introduced with the assumption that every agent of Group i cannot switch actions at each point in time. He must make a commitment to a particular action in the short-run. Opportunities to switch actions arrive randomly; some fraction α, $0 < \alpha < 1$, of individuals is drawn randomly from Group i, and makes a new choice of either $x = 1$ or $x = 0$. Thus, we may interpret a norm as a prescription indicating how a person ought to behave in any situation at which he may be called to move.

An agent in Group i may deviate from the norm, but this deviation is costly. Punishment (such as ostracism) will be imposed on Group i members who deviate from the norm. If deviation from the norm is observed ($x = 0$), the opponent of Group j ("she") decides whether to punish the deviator, that is, chooses p on the closed interval $[0, 1]$. Here, with probability p, the agent of Group j punishes noncompliance, and with probability $1 - p$ she does not. The agent of Group j is only an outside party who happens to know that norm violation has occurred. Altruistic punishment is motivated to restore norm compliance, even though they are not expected to interact again in the future. It is costly for agents of Group j to punish norm violators. Thus, in the model, agents in Group i choose whether to comply with the norm, while agents in Group j value compliance and potentially punish violators. Punishment is then confined to interactions with others that share the same norm in the population.

Individuals in Group i are assumed to be heterogeneous with respect to their social concerns. We denote an agent's type of Group i as g. An agent of type $g = 0$ is not concerned with the social meaning of a certain action. A higher g implies higher social concerns. The distribution of g is assumed to be exogenous and uniform in the model. Let the uniform distribution of g be $F(g)$, with $g \in [0, G]$. Furthermore, let the density of the uniform distribution be $f = 1/G$.

In Group i, the short-run utility function of an agent of type g is given by:

$$U = x\{R(g, \mu) - D\} - (1 - x)p\, C. \tag{5.9}$$

Each individual in Group i is assumed to maximize the utility function (5.9), which constitutes the short-run equilibrium. The first term of the utility function (5.9) reflects the reputation value of a norm if the agent of Group i chooses to comply with the norm ($x = 1$). It depends positively on the agent's type g, and the fraction μ of individuals who comply with the norm in Group i. The agent internalizes the norm with the reputation value R. The strength of the norm is determined by the proportion of agents who follow it. Group i consists of μ compliers and $(1/2 - \mu)$ violators, where $0 < \mu < 1/2$. The fraction μ is given in the short-run, and is known by all members of Group i. A person who deviates from the norm forgoes reputation-derived utility. The second term reflects a fixed cost, $D > 0$, of compliance if the agent of Group i chooses $x = 1$. The final term reflects a fixed cost, $C > 0$, as the penalty for violation if the agent of Group i chooses not to comply with the norm ($x = 0$), and the opponent of Group j chooses to punish him with probability p. The punisher may disapprove of the violator, which reduces his level of satisfaction and his well-being.[22] To sum up, the agent of Group i has to pay a compliance cost, D, if $x = 1$, and the penalty for violation, C, if $(1 - x)\, p > 0$. It is assumed that $D > C$.

The reputation value R is assumed to take the following form:

$$R(g, \mu) = g\, \mu. \tag{5.10}$$

Absent of social concerns ($g = 0$), the agent of Group i would not follow the norm. An agent in Group i will consider following the norm when the value of his action is greater than the cost of violation. The higher g and the higher μ (the proportion of norm followers) are, the greater the utility from following the norm. The value of following it may take the form of an abstract reward, such as social approval from others. The reputation value can also be interpreted as the value of social image. People care about how others perceive them, and these concerns influence a wide range of decisions (Andreoni and Bernheim, 2009).

The model also encompasses individual punishers who are not direct victims for norm violation. They punish violators to restore norm compliance. For an agent of Group j, whether her opponent in Group i follows the norm or not is a random variable. Let the probability that her opponent in Group i chooses to comply with the norm be $\Pr(x = 1)$. The short-run utility function of an individual in Group j is given by:

$$V = \Pr(x = 1)B - \big\{1 - \Pr(x = 1)\big\}p\, T. \tag{5.11}$$

Each individual in Group j is assumed to maximize the utility function (5.11), which constitutes the short-run equilibrium. The first term of the utility function (5.11) reflects a psychological benefit, $B > 0$, received by the agent of Group j when her opponent from Group i chooses to follow the norm with probability $\Pr(x = 1)$. The second term is a cost T ($>B$) associated with punishment ($p > 0$) when her opponent from Group i violates the norm with probability $1 - \Pr(x = 1)$. Punishment can be psychologically costly. Punishers suffer a psychological cost if their opponents violate the norm. This psychological cost includes the punisher's anger (negative emotion) for violation of the norm.

The benefit–cost ratio b for the agent of Group j is given by:

$$b = \frac{B}{T}. \tag{5.12}$$

As the punisher's anger (T) increases, the benefit–cost ratio (b) decreases.

---

22. The cost C can be endogenized by assuming that individuals have beliefs about how they are judged for the norm deviation. This allows individuals to hold strong or weak beliefs, in the sense that the cost C can be high or low.

The reputation value of the norm affects individual short-run utility, as defined in (5.9). Individuals of Group i choose to comply with the norm if the utility from so doing is as great as the utility derived from not complying. The reputation value of the norm depends on the proportion of compliers in Group i. At each period, the fraction $(1/2)\alpha$ of individuals in Group i has a chance of making a new choice. In the aggregate, the proportion of compliers is $(1/2)\alpha$ $\Pr(x=1)$, and the proportion of violators is $(1/2)\alpha\{1 - \Pr(x=1)\}$. Here, α parameterizes the inertia of the adjustment process. In the adjustment process, it takes some time until individuals find out how many others follow the norm. The proportion μ will increase (or decrease) if $(1/2)\alpha\Pr(x=1)$ is larger (or smaller) than $(1/2)\alpha\{1 - \Pr(x=1)\}$ in Group i. Therefore, the long-run adjustment process is described by the following form:

$$\frac{d\mu}{dt} = \alpha\left[\frac{1}{2}\Pr(x=1) - \frac{1}{2}\{1 - \Pr(x=1)\}\right]$$

From the above equation, we have:

$$\frac{d\mu}{dt} = \alpha\left[\Pr(x=1) - \frac{1}{2}\right] \tag{5.13}$$

Thus, the speed of norm evolution is influenced by the rate α at which the group members acquire the knowledge necessary to appreciate the norm. If $\Pr(x=1)$ differs from 1/2, the proportion μ will increase or decrease at rate α until μ is equal to 0 or 1/2. The relationship between short- and long-run is summarized as follows. At each period t, the proportion of norm followers, μ, is given, and agents of Group i decide whether to follow the norm or not, given this proportion (short-run). The proportion μ is formed over a period of time, and it converges to a steady state (long-run).

5.6.2   Choice of "x"

An agent, who is drawn randomly from Group i, chooses $x \in \{0, 1\}$ to maximize his utility function:

$$U = x(g\,\mu - D) - (1 - x)p\,C,$$

rom (5.9) and (5.10). Maximization of this utility function constitutes the short-run equilibrium. If $g\,\mu - D > -p\,C$, he chooses to comply with the norm $(x=1)$; otherwise, he does not $(x=0)$. Thus, his best-response function is:

$$x = \begin{cases} 1 & \text{if } g\mu - D + pC > 0, \text{ or } g > \dfrac{D - pC}{\mu} \\[2ex] 0 & \text{if } g\mu - D + pC \leq 0, \text{ or } g \leq \dfrac{D - pC}{\mu} \end{cases}$$

It is useful to consider the probability of heterogeneous individuals choosing to comply with the norm in Group i, $\Pr(x = 1)$. Then:

$$\Pr(x = 1) = \Pr\left(\frac{D - pC}{\mu} < g\right)$$

$$= 1 - F\left(\frac{D - pC}{\mu}\right).$$

Using the density of the uniform distribution, $f = 1/G$, we have:

$$\Pr(x = 1) = 1 - \frac{D - pC}{\mu}f. \tag{5.14}$$

Thus, in the short-run, the probability to comply with the norm, $\Pr(x = 1)$, is a continuous increasing function in the probability of punishment, p. Only in the long-run, the incentive to comply with the norm adapts to the fraction of norm followers, μ. The strength of the norm is endogenously determined by the collective behavior of the group members.

5.6.3 Choice of "p"

An agent in Group j optimally chooses to punish the norm deviator in Group i with the probability of punishment, p. That is, at each period, the agent of Group j chooses $p \in [0, 1]$ to maximize her utility function:

$$V = \left(1 - \frac{D - pC}{\mu}f\right)B - \left(\frac{D - pC}{\mu}f\right)pT$$

from (5.11) and (5.14). This maximization also constitutes the short-run equilibrium. The first-order condition is:

$$\frac{\partial V}{\partial p} = \frac{C}{\mu}fB - \frac{D - pC}{\mu}fT + \frac{pC}{\mu}fT = 0.$$

which, for an interior solution, equates the marginal cost of punishment to the marginal benefit of punishment. Using the benefit−cost ratio b in (5.12), we have:

$$p = \min\left\{\frac{D - bC}{2C}, 1\right\} \tag{5.15}$$

For $(D - bC)/2C < 1$ in condition (5.15), it follows that:

$$\frac{\partial p}{\partial b} = -\frac{1}{2} < 0.$$

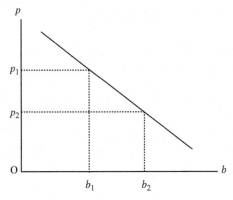

FIGURE 5.2 The relationship between b and p.

A lower benefit–cost ratio (b) implies a higher probability of punishment (p). In Fig. 5.2, if $b_1 > b_2$, then $p_1 < p_2$. As the punisher's anger at the norm deviation increases (or T increases), the ratio b decreases (see (5.12)), and the probability p increases. Once a norm deviation is observed, an agent of Group j feels anger towards the norm deviator, and attempts to reduce his payoff. Thus, anger plays an important role in punishment decisions.

5.6.4 Evolution of "μ"

The proportion of individuals who comply with the norm changes over time. From (5.13) and (5.14), the long-run adjustment process is described by:

$$\frac{d\mu}{dt} = \alpha \left[\frac{1}{2} - \frac{D - pC}{\mu} f \right]. \tag{5.16}$$

Using (5.16), a critical value of μ, μ^*, is defined as follows:

$$\mu^* = \min \left\{ 2(D - pC)f, \ \frac{1}{2} \right\}. \tag{5.17}$$

Condition (5.17) describes the critical value μ^* as a decreasing function of the probability of punishment, p, for $2(D - pC)f < 1/2$. Then, the value μ^* in condition (5.17) does not include the benefit–cost ratio b.

The adjustment process is not necessarily continuous. The proportion μ (> 0) is decreasing for $\mu < \mu^*$. Then, the adjustment process converges to $\mu = 0$ (everyone in Group i deviates from the norm). On the other hand, if μ^* is less than 1/2, μ is increasing for $\mu > \mu^*$. Then, the adjustment process converges to $\mu = 1/2$ (everyone in Group i follows the norm). Thus, there are two stable equilibrium points.

In a (μ^*, p) diagram shown in Fig. 5.3, there is one intersection between conditions (5.15) and (5.17). Condition (5.15) describes the probability of

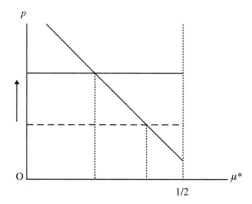

FIGURE 5.3 The increase in p.

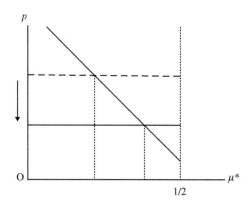

FIGURE 5.4 The decrease in p.

punishment (p) for a given benefit−cost ratio (b), which is a straight line; while condition (5.15) describes the critical value μ^* as a function of p, which is a declining line. Fig. 5.3 also describes the effect of a higher probability of punishment (p) due to a lower benefit−cost ratio (b). A lower b (or a higher T) implies a higher p in Fig. 5.2. In Fig. 5.3, a higher p implies that the intersection point between the two lines shifts to the left, which implies a smaller critical level of μ^* ($<1/2$). When μ is above the critical value μ^*, the adjustment process converges to $\mu = 1/2$. Anticipating a higher probability of punishment, more agents of Group i consider following the norm.

On the other hand, Fig. 5.4 describes the effect of a lower probability of punishment (p) due to a higher benefit−cost ratio (b). A higher b (or a lower T) implies a lower p in Fig. 5.2. In Fig. 5.4, a lower p implies that the intersection point between the two lines shifts to the right, which implies a larger critical level of μ^*. If the fraction of norm followers, which is above a larger

μ^*, is larger, the adjustment process converges to $\mu = 1/2$. For a higher reputation value, more agents of Group i consider following the norm. People conform to maintain their reputation. The strength of the norm is determined by the proportion of the members who follow it. Even though p is small, individuals choose to comply with the norm. If the norm is initially widespread, the equilibrium converges to one steady state, in which everyone complies with the norm. If instead the initial fraction of norm followers is small, the society ends up in another steady state with opposite features. Depending on the initial conditions, the society might converge to one or the other steady state.

We conclude the following:

Proposition 5.2: (Teraji, 2013).

a. *A higher p induces individuals to sustain the norm in the society with a smaller μ.*

b. *A larger μ induces individuals to sustain the norm in the society with a lower p.*

An agent does not automatically comply with a norm but deviates whenever the gains from deviation are sufficiently large. Proposition 5.2 (a) shows that the norm is enforced due to a higher level of punishment of noncompliance. The norm is enforced due to the expectation that the violation will be punished. There is some possibility that the erosion of the reputation effect induces individuals to break the norm. Punishment would be used to enforce the norm if a substantial fraction of people has little reputation-derived utility by obeying it. Punitive behavior is a way to restore norm compliance by penalizing violators. Proposition 5.2 (b), on the other hand, shows that everyone is motivated due to reputation formation, despite a lower level of punishment by others. For a lower level of punishment, effective reputation building provides a way to sustain the norm. Then, an agent must believe that a sufficient number of other agents will comply with the norm as well. Punishment and reputation are interacting mechanisms. Reputation mechanisms generate an environment where the execution of costly punishment is less frequent without taking away its deterrent force. The interaction of two mechanisms, punishment and reputation, not only comes closer to real life, but also provides a convenient way to norm compliance.

Even in societies that have strong governments, norms are a source of law, and often a cheap and effective substitute for law (Posner, 1997). The incentives for obeying laws are clear enough, but why do people obey norms? We develop a theoretical framework to consider the problem of norm compliance. Some norms often require people to sacrifice for the group. The existence of such norms presents an evolutionary puzzle. This section analyzes the interaction between the potential for costly punishment and building personal reputation. The model considers two groups, Group i and Group j, in a society with one norm. Agents in Group i choose whether to comply with the norm by incurring some cost. They acquire utility from the reputation derived from complying with the norm. This utility depends positively on the proportion of motivated compliers in Group i. Behaving in a manner that violates the norm is costly. On the other hand, in Group j, there are agents who value compliance and potentially punish noncompliance. They are altruistic punishers who are not direct victims for norm violation.

We have a set of solutions to the problem of norm compliance. The first solution is the punishment-based account. Following this account, people comply with the norm because the threat of punishment makes it in their interest to do so. Altruistic punishers are ready to incur some cost to prevent unfair behavior. They have a psychological motivation to restore norm compliance. Although altruistic punishment seems to have a solid foundation in human interaction, such a punishment leads to a large increase in losses for punishers. The second solution is, on the other hand, the reputation-based account. People want to achieve the reputation of being fair. People comply with the norm because they care about their reputation.

The present framework systematically investigates two scenarios to account for the long-run stability of the norm. In one scenario, there is some possibility that the erosion of the reputation effect induces individuals to break the norm. Individuals may have a selfish interest in violating it. Then, the norm is enforced due to a higher level of punishment for noncompliance. Punishment would be used to enforce the norm if a substantial fraction of people achieves little reputation-derived utility by obeying it. In another scenario, everyone is motivated due to reputation formation, despite a lower level of punishment by others. For a lower level of punishment, effective reputation building provides a way to sustain the norm. The two mechanisms, punishment and reputation, are interacting. The interaction not only comes closer to real life, but also provides a convenient way to norm compliance.

5.7 CORPORATE SOCIAL PERFORMANCE

Recent corporate scandals involving large companies have highlighted corporate social responsibility (CSR) as one of the principal issues confronting the free-market system. The term "CSR" has been defined in various ways.[23] According to one of the most frequently cited definitions, CSR is "a concept whereby companies integrate social and environmental concerns in their business operations and in their interaction with their stakeholders on a

---

23. Bowen (1953) first pointed out that corporate decision-making processes have to consider not only the economic dimension, but the social consequences deriving from their business behavior as well.

voluntary basis" (Commission of the European Communities, 2001, p. 6).[24] Companies have responsibilities to take ethical and moral issues into account while they make a profit. There is a growing interest in the relationship between corporate governance and the social performance of a company. Contemporary economic theory, however, has argued that so long as economic agents are rational and self-interested, they will be able to affect a better set of consequences than if their intentions are benevolent. In fact, according to Friedman (1970), the social responsibility of a business is to increase its profits. In Friedman's framework, firms are owned by shareholders who are principals, and business managers are agents with the duty to serve the interests of their principals. Managers are obliged by contract to shareholder value; it is their primary task to maximize profits. Following this view, CSR is a misuse of corporate resources that would be better spent on value-added internal projects or returned to shareholders.

The assumptions of neoclassical economic theory are subject to market forces. In the neoclassical analytical framework, the market system creates a decision-making situation that favors practices to ensure profit maximization at the expense of more morally preferable alternatives. However, the lens of neoclassical economic theory is distorting practical realities. A positive view of managers' support of CSR is articulated by Freeman (1984), who encourages consideration of external stakeholders, beyond direct profit maximization. Freeman's broad definition of stakeholders includes any group or individual who might affect the business objective, or might be affected by its realization. According to the stakeholder theory approach, it can be beneficial for the company to engage in certain CSR activities that stakeholders perceive to be important; unless stakeholders and their interests are dealt with, they might withdraw their support for the company. Freeman's (1984) work helps to re-conceptualize the nature of the company, to encourage consideration of new external stakeholders, legitimizing new forms of managerial understanding and action. CSR should be understood as a broad concept, since it takes in the whole set of philosophical and normative issues relating to the role of business in society. Stakeholder theory suggests a moral relationship between companies and stakeholders.

Corporate social performance (CSP) refers to voluntary activities. Moral management yields CSP. Supplying CSP is not just like managing traditional market-based activities. In fact, from empirical studies, the relationship

---

24. The Green Paper identifies four factors which lie behind the growing success of CSR concept (p. 4): (1) the new concerns and expectations of citizens, consumers, public authorities, and investors in the context of globalization and large-scale industrial change; (2) social criteria, which are increasingly influencing the investment decisions of individuals and institutions, both as consumers and as investors; (3) increased concern about the damage caused by economic activity to the environment; (4) transparency of business activities brought about by media and modern information and communication technologies.

between CSP and profits has not been adequately determined (Kolstad, 2007). CSP entails a company's recognition of broad responsibilities, and it should be concerned with more than just profit. Neoclassical economic theory, however, treats the firm as a black box, completely self-interested entrepreneur that arranges inputs so as to maximize material well-being. Friedman (1962) questioned the ability of business managers to pursue the social interest as follows:

> *If businessmen do have a social responsibility other than making maximum profits for stockholders, how are they to know what it is? Can self-selected private individuals decide what the social interest is? Can they decide how great a burden they are justified in placing on themselves or their stockholders to serve that social interest?*

(Friedman, 1962, pp. 133–134)

Instead, Friedman (1962) argues that a socially responsible firm is one that conducts business in accord with shareholders' desires, which generally makes as much money as possible, while conforming to the basic rules of the society. His position, based on free-market ideology, has come under increasing attack since the time of writing. According to the shareholder perspective, business is about economic and not social goals. In real life, however, economic decisions also have social consequences, and the boundaries between the social and economic world become blurred. Firms have responsibilities towards their environment that go beyond their legal and economic obligations. The stakeholder model reflects the modern understanding of companies as integrated in, rather than separated from, the rest of society. Within our economic system, there exists a growing cadre of companies where social, environmental, and ethical goals are on an equal footing with the profit motive.

On the contrary, the behavioral theory of the firm (Cyert and March, 1963; Leibenstein, 1987; Altman, 2001) can explain significant elements that are ignored in the neoclassical economic framework. A core idea in the behavioral theory of the firm is the importance of organizational slack, or X-inefficiency, in organizations. The firm may be operating inside of the production possibility frontier. Behavioral theory focuses on the role of discretionary resources in the corporate social strategy context. Ethical or moral issues may arise when companies engage in CSP and legitimize their activities to stakeholders. As a consequence, the ethical firm can be expected to be relatively more X-efficient, satisfying stakeholders' interests to some extent, rather than maximizing the material well-being. Thus, the behavioral theory of the firm has the potential to incorporate ethics or morality as a critical component of the underlying motivating structure. The behavioral theory of the firm fits closely with widespread conceptions of how social concerns are dealt with in business.

According to stakeholder theory, CSR includes various kinds of responsibilities or dimensions: each dimension of CSR can be examined in relation to the various stakeholders of the organization. To consider what it means for a company to be socially responsible, this section focuses on the "socially conscious" consumer. Managers do business in a way that maintains or improves the consumer's and society's well-being. Consumers may expect companies to protect the environment, and conduct business ethically. When consumers are given information about a company's level of social responsibility, they evaluate the company. Lack of awareness of the level of social responsibility is likely to be a major inhibitor of consumer responsiveness to CSR. The socially conscious consumer is a person who takes into account the public consequences of his or her private consumption or who attempts to use his or her purchasing power to bring about social change (Webster, 1975). This concept is based on the psychological construct of social involvement. The socially conscious consumer must be aware of social problems, be active in the community, and believe that he or she has the power to make a difference.

Consumers are a major stakeholder group particularly sensitive to CSP. Consumers' perceptions of a company as socially oriented are associated with a higher level of trust in that company and its products.[25] It is important for socially responsible companies to develop consumer trust and engage in CSR programs that are meaningful to their consumers. Trust is a fundamental asset in the company–consumer relationship. Commitment to a brand, defined as trust, is the consumer's desire to maintain the relationship. Socially oriented companies can use trust to improve their competitive performance (brand loyalty). Brand loyalty includes tolerance to pay a higher price for the product.[26] Ethical firms producing a particular product are not producing the same one as unethical firms. Even if the product is identical in all other characteristics, this same product can be differentiated by the extent to which it is produced in accordance with a specified set of ethical standards (Altman, 2005). Bhattacharya and Sen (2003) argue how non-product aspects of a company, such as CSR, can lead to consumer loyalty and other positive post-purchase outcomes. There are a number of studies in which consumers claim to be ready to pay higher prices for products from socially responsible companies, or to take the social responsibility profile of the producer into

---

25. Brown and Dacin (1997) support the idea that what consumers think about a company does influence their beliefs and attitudes toward the products of that company. They show that a high CSR grade leads to a higher evaluation of the company, and corporate evaluation is positively related to product evaluation. In their experiments, subjects were given a description of a fictitious firm, along with a CSR report card with various grades indicating above and below average community involvement. Subjects were asked to rate products made by the firm, as well as the firm itself.

26. In the "gift exchange game," Fehr et al. (1993) design competitive goods market experiments that allow for the emergence of reciprocal interactions. Buyers make a "gift" to the sellers by paying prices above the competitive level. Sellers in turn respond reciprocally by choosing quality levels above those dictated by their pecuniary interests.

consideration when comparing different brands (Creyert and Ross, 1997; Ellen et al., 2000; Mohr et al., 2001). From a demand-side perspective, a good reputation increases the value of the brand, which, in turn, increases the company's goodwill. Companies are faced with the challenge of becoming more attractive to socially conscious consumers. It is important to investigate how corporate decisions are received by consumers. Consumers are engaged in constructing the ethical identity of companies. Managers are encouraged to behave in an ethical manner, because information about a company's ethical behavior is thought to influence consumers' image of the company.

Corporate identity ("what the company is") emerges from the interaction between the company and its consumers. Identity theory at the individual level has been used to explain role-related behavior. Social structures affect the extent to which an individual commits to a particular role. Corporate identity denotes the characteristic way in which an organization goes about its business, or how it behaves and interfaces with the social environment. Corporate behaviors regarding ethical conformity, in light of normative expectations, represent a response mechanism to the social environment. Socialization, through which individuals learn the perspective necessary to perform roles, is expected to prime ethical identification. Corporate identity interplays with social concerns to influence the response to values, norms, and beliefs. CSP can be then regarded as a voluntary corporate commitment to exceed the explicit and implicit obligations imposed on a company by society's prevailing ideas and opinions. On the one hand, we have the interests of the economy arising from a largely unregulated market of companies pursuing their self-interest and, on the other hand, we have the interests of society arising from ethical or moral conformity that may suppress the effects of self-interest. Managers have to legitimate their activities within the field of tension between ethics and economic success. These aspects are strongly reflected in the dimensions of ethical or moral perceptions.

In this section, I identify CSP with the private provision of public goods (socially responsible or environmentally friendly activities), or the private redistribution of profits to social causes (Teraji, 2009). Social expenditures represent redistribution through community projects and support, philanthropy, training and educational programs, workers' rights initiatives, environmental abatement and protection, and alternatives to animal testing (Baron, 2008). What motivates a business to make a contribution to these public goods? The model provides a theoretical framework for analyzing corporate incentives to engage voluntarily in socially responsible activities. According to social exchange theory (Blau, 1964), individuals are motivated to take voluntary actions when they expect they could get something in exchange. That is based on the assumption that people base all their decisions on the calculation of costs and benefits in the pursuit of their own material well-being.[27] Some studies assume that private provision of public goods is motivated in part by a warm glow (Andreoni, 1989, 1990). If the warm-glow motive is strong enough, individuals may continue to make direct donations. CSP is modeled as a response to consumers' preferences over public goods (Besley and Ghatak, 2007). Bagnoli and Watts (2003) study the strategic use of CSP to appeal to "green" consumers with warm-glow preferences for public goods.[28] In this section, the manager does not engage in strategic CSP in an attempt to maximize only "pecuniary" profits. The manager may act to meet the expectations of socially conscious consumers, and thereby to receive the resulting "social satisfaction" from CSP. The "non-pecuniary" motivation of agents for being involved in activities that benefit others is an equilibrium outcome. The present model is related to Baron (2008) in the organizational forms for private provision of public goods. However, the model developed here provides an alternative explanation for understanding how different social preferences can be created in manager−consumer interactions. The model investigates how the consumer's taste for CSP is endogenously determined in the social environment. In response, the manager's taste for CSP is also endogenously determined in the environment.

5.7.1   Overview

Mainstream neoclassical theory assumes that, under conditions of perfect competition, a market economy will allocate scarce resources in an economically efficient way. Economic efficiency is what is called Pareto efficiency, a situation in which no improvement can be made to the material welfare of one person without making someone else worse off. Neoclassical economic theory has paid little attention to economic behavior that is inconsistent with maximizing the material welfare of the individual from a self-interested perspective. Commitment to general social interests is in breach of the conventional neoclassical postulate. In the pursuit of self-interest, perfect market structures and ethical behavior can be inimical to each other. If ethical behavior requires additional inputs into the process of production, the optimal market structures in Pareto-efficient resource allocation terms no longer exist. In order to overcome this problem, it must be understood that the design of

---

27. Many societies face the problem of how to provide public goods. Free riders are those selfish individuals who take advantage of the benefits provided by cooperators, without contributing to those benefits themselves. When people face strong material incentives to free ride, the self-interest model predicts that no one will contribute to the public good. However, if there are individual opportunities to punish others, those who cooperate may be willing to punish free-riding, even though this is costly for them, and even though they cannot expect future benefits from their punishment activities (Fehr and Gächter, 2000).
28. Eco-labeling is an example of what Baron (2001) calls strategic CSR: attempts to increase profits by attracting "green" consumers.

current institutions is the result of a reaction to social needs. In this section, instead of focusing on narrow self-interested behavior, I examine other configurations that provide the glue that links together individuals, helping realize potential improvements in society. In an alternative context, the firm intends to further public goods beyond its direct interests. The model provides an important foundation for CSP analysis by showing how one might evaluate the CSP of the firm from the ethical or moral perspective.

Contemporary economic theory has treated altruism in different ways.[29] For Becker (1976), self-interested behavior and altruism are not inconsistent for utility-maximizing economic agents. What counts in Becker's analytical framework is that individuals are rational, given their objectives and the constraints. For the theory of rational economic behavior, altruistic or ethical behavior can be incorporated into the utility function of the economic agent. The altruist derives pleasure not because the other is assisted, but rather because the other's pleasure is already part of the altruist's utility.[30] Thus, the donor would give if this donation increases his or her pleasure more by watching the pleasure of others. Altruistic behavior is caused by the fact that, as a factor in his or her utility function, the person also has the utility of the individuals whom he or she wants to benefit. However, a basic premise of the conventional neoclassical wisdom is that ethical or altruistic behavior cannot persist if it is inconsistent with competitive market forces. The firm has as its primary objective the maximization of its profits. Ethical behavior does not positively affect the firm's efficiency if the firm is operating at maximum from the start. In the perfectly competitive market, an ethical firm will lose market share to more efficient rivals, if it engages in inefficient activities. If managers want to work toward the betterment of society, they should do so as private individuals at their own expense. Yet, social and environmental problems have a direct impact to the company's welfare. If the company is not concerned with these problems, it may lose social acceptance and support. The company may be coerced into increasing its social output in order to survive. The "economic man" known to students of Walrasian economics acts on the basis of preferences that are self-regarding (excluding such intrinsic values as altruism, fairness, and vengeance), and are defined over a restricted range of outcomes (excluding honesty, as well as concerns about the process rather than simply the outcome of exchange per se) (Bowles and Gintis, 2000). The model of the economic man does not consider the process through which the premises from which man makes decisions are formed. The criteria for neoclassical rationality do not solve the question of what is good for us, or which action we should

---

29. Khalil (2004) critically examines major attempts by economists to account for altruism as reciprocity, as a source of vicarious satisfaction, or as an evolutionary trait respectively.
30. Altruism is frequently invoked by economists to explain relations between family members (Barro, 1974).

consider. Departures from the standard economic assumptions are potentially important in the social domain. A focus on processes is a reminder that the very idea of CSR suggests a need to consider the values, motives, and choices of those who are involved in formulating corporate policy. If government does not react to a change in social preferences, the company might be able to act decisively and quickly. Consequently, the company must be motivated to improve its social performance. The CSR programs can provide additional social benefits when such programs are central to the mission. Looking at social aspects is not a new phenomenon for companies. What is new, though, is the intensity and breadth of the efforts made by companies, as well as the increasingly strong societal demand for behaving more ethically and responsibly. Companies must learn how far they need to extend their responsibilities, what issues to take up, how to give meaning to those issues, and how to successfully combine economic, social, and environmental strategies.

Nowadays, society at large increasingly expects companies to behave as good corporate citizens.[31] The good citizenship metaphor emphasizes voluntary self-restraint and altruism concerning the realization of broad duties. Social responsibility derives from the moral legitimacy the corporation achieves in society. An exclusive focus on profit maximization is too limited, as it neglects the company's responsibilities towards socially conscious consumers. As consumer awareness of the need for CSP increases, managers increasingly recognize the responsibilities for implementing ethical programs to enhance social welfare. Through evaluation of the manager–consumer relationship, the manager chooses whether to integrate CSP within corporate strategy. CSP commitment increases the company's trust, making its operations more legitimate to consumers. Whether a company is being ethical or not is dependent on the manager awareness of creating trust in the manager–consumer relationship. Justifying the role of CSP as an extended scheme of corporate governance is an essential part of the requirements of a just society, understood as a joint venture for mutual benefit.

In this section, CSP is discussed as an outgrowth of "collaboration" between managers and consumers in the social environment. The term "collaboration," or "dialogue," describes the involvement of consumers in the manager's decision-making processes that concern social and environmental issues. The realization of the company's social responsibilities is shaped by the way the manager interacts with consumers. The model focuses on the views of both managers and consumers about CSP. The manager has a key role in the supply of CSP, using the company's resources to achieve its social goals. Active CSP is conceptualized as follows. The manager can anticipate the social preferences of consumers, and transform them into social

---

31. The term "corporate citizenship" is used to connect business activity to broader social accountability and service for mutual benefit, reinforcing the view that a corporation is an entity with status equivalent to a person (Waddell, 2000).

benefits. Social benefits then reflect the "price" a socially conscious consumer puts on the supply of CSP. If social benefits can be created in manager—consumer interactions, the consumers' ethical attitudes facilitate the corporate social responsiveness. The collaborative relations can reinforce positive responses with each other. The socially conscious consumers' attitudes to encourage CSP are influenced by the manager's view to practice CSP, and vice versa. Thus, both mangers and consumers are influenced by comprehensive perception toward CSP. This establishes the basis for understanding the central role of both ethical self-regulation and consumers' activism. Ethical self-regulation establishes a new context for manager—consumer interactions, and consumers' activism becomes effective in this context. This framework thus focuses on how CSP is understood and perceived by both managers and consumers. The model considers two key aspects of social preferences: the consumer's taste for CSP, and the manager's taste for CSP.

A public good is a voluntary contribution to society based on other-regarding attitudes. In this section, public goods can be provided by the firm through its CSP. Consumers may have social preferences for that public good. Ethical managers engage in impartial moral reflection on practicing self-restraint and altruism. CSP is also about managing change at the firm level, in a socially conscious manner. Then consumers become aware of a firm's level of social responsibility. The firm supplies both private and public goods by incorporating the social concerns of consumers. The motivation underlying CSP is modeled in terms of the resulting social satisfaction in manager—consumer interactions. The social satisfaction that the manager receives affects whether CSP is provided at the firm level. CSP will change our conception of what a for-profit organization is, and how its value is measured. Consumers experience the results of corporate behavior, so they influence the ethical standards by which corporate behavior is judged, and evaluate how well managers perform according to those standards. If the company succeeds in managing change in a socially responsible manner, this will have a positive impact for the entire environment.

This section offers an analytical framework to evaluate how the manager involves consumers in the decision-making processes. The collaboration is ultimately about exchanging opinions between them; in other words, it is about dialogue. A participatory dialogue allows consumers to translate their own judgments and voices into practice. In the model presented below, consumer engagement is important for the supply of CSP. The dialogue that needs to take place between the parties involved is guided and shaped by a confrontation of views and interests.[32] The dialogue has its added-value in

---

32. In research on public relations and the Internet, the possibility of interactivity between a company and its consumers is an issue of high relevance (Capriotti and Moreeno, 2007). Various forms of feedback are possible on corporate websites for visitors to ask questions, give opinions, or assess the CSR issues. These allow an assessment of, or opinions on, any of the issues on CSR.

coming to some kind of social consciousness. When consumers perceive that they can be increasingly effective in the supply of CSP, they will show more concern for CSP. If consumers are not free to participate in the collaboration, the level of perceived consumer effectiveness is limited. If consumers are insufficiently included in the dialogue, anticipated social benefits from the collaboration will be limited.

Furthermore, managerial perceptions of CSP are also important. CSP is part of the company's vision. A company's stance on CSP may strongly influence how ethically the company is perceived. The manager's "mental model" or "mind-set" has an important impact on the implementation of consumer dialogue. Mental models function as selective mechanisms or "filters" for dealing with experience. The manager thus has a filter that limits the resulting social satisfaction from the consumer dialogue. The filter makes it difficult for the manager to live up to the participatory ideals of the consumer dialogue. The filter depends on the manager's social and environmental awareness. The manager's decisions are thus based on the interplay between individual cognition and social institutions. Managerial perceptions are adapted to certain circumstances. Individuals follow an "ecological rationality," and rationality thus becomes context-based (Smith, 2003). Ecological rationality uses reason to discover the possible intelligence embodied in the rules, norms, and institutions of our heritage that are created from human interactions, but not by deliberate human design. People follow rules without being able to articulate them, but they can be discovered. Furthermore, norms are often considered to be the historical legacy of a traditional culture, supported and adhered to in a population. The term ethics refers to a set of moral norms that guide people's behavior. Norms define what actions are considered acceptable or unacceptable, according to shared understandings. In this way, this section focuses on moral "like-mindedness" among interacting individuals for CSP to succeed. Morality can be created and utilized in social interactions.

5.7.2 The Model

The firm in the model economy normally produces an identical (numeraire) private good which can be consumed. There are potential shareholders in the firm. The shareholders of the firm are assumed to be in contact with a manager. The manager operates the firm after accepting the contract. Let π be the (pecuniary) operating profit when the project is implemented. After the operating profit is specified, compensation is paid to the manager, and the remaining amount of profit is distributed to shareholders. Let $p\pi$, where $0 < p < 1$, denote the manager's compensation when the profit is specified. Similarly, $(1 - p)\pi$ is distributed to shareholders when the profit is specified.

Moral conduct refers to a pattern of behavior that goes beyond normal business management. The manager chooses units of social expenditures, g, which affect social welfare. By choosing g, the manager contributes to the private provision of public goods as CSP. The provision is then bounded. It is assumed that the manager picks up his or her own contribution g in some interval $[0, G]$, where $G > 1$ is given.

In this model, CSP is discussed as an outgrowth of "collaboration" between the manager of the firm and its consumers in the social environment. The model concentrates on two different dimensions of social preferences: the consumer's taste for CSP and the manager's taste for CSP. As a consequence, social preference formation interacts with economic decisions taken by others. The model analyzes the decision problems of the manager who receives the compensation alone and the manager who relies on social values.

Consumers choose whether they consume the private good, or become concerned about the CSP of the firm. Consumers have an initial endowment they can allocate between purchasing the private good and personal giving to social causes. They may care about moral management or CSP. The model considers the relative weight between the consumer's concern for the private good, and that for the amount of CSP supplied. Let the consumer's taste for CSP be described by a fraction α (≥ 0). The fraction α represents the value that a consumer receives from the amount of CSP supplied by the firm, g. It can be also considered as the "price" the consumer puts on the perceived benefits from the CSP of the firm. On the other hand, the consumption for the private good becomes $1 - \alpha$. Then the consumer benefits from both the private good and the amount of CSP supplied, g. More specifically, the consumer's utility function is given by:

$$U = \ln(1 - \alpha) + \alpha g. \tag{5.18}$$

The greater α is, the more utility a consumer gets from the CSP of the firm.

The manager allocates his or her time between producing the private good and providing CSP (the private provision of public goods). The manager provides CSP by allocating a portion of β (≥ 0) of his or her time to social causes. The fraction β is considered as the manager's taste for CSP. If the fraction β is positive, the manager chooses the amount of CSP supplied, g. On the other hand, the manager earns the payoff $\ln(1 - \beta)$ by allocating $1 - \beta$ of his or her time to producing the private good. Thus, the operating profit is specified as:

$$\pi = \ln(1 - \beta) - \beta g. \tag{5.19}$$

In (5.19), g reflects the firm's social expenditures, the cost of CSP. It may be also considered as the private redistribution of profits to social causes. The manager may view CSP as an expense rather than an investment.

Managerial perceptions of CSP are important. The manager may receive a non-economic or psychological benefit reflecting "social satisfaction" associated with the CSP of the firm. In the model, the manager perceives social satisfaction, h, from one unit of CSP provided by the firm. Thus, if the manager's taste for CSP is β, the manager gets βhg from providing the amount of CSP, g. Social satisfaction h reflects the manager's social and environmental awareness. From moral management or CSP, the firm has an ethical responsibility to society. The socially responsible firm seeks to meet the expectations of consumers, and to get a positive image from consumers in the economy. Thus, social satisfaction h depends on the consumer's taste for CSP. More specifically, let $h = h(\alpha)$ be a strictly increasing, strictly concave function of α. This implies that social satisfaction h increases when the value the consumer puts on the CSP of the firm increases. It is assumed that $h(0) = 0$. That is, if consumers are excluded from the collaboration, the value h is zero. When the profit is specified, the manager receives his or her compensation $p\pi$.

The manager's utility function is then given by:

$$V = p\pi + \beta h(\alpha)g. \qquad (5.20)$$

The manager struggles in deciding how to reconcile the benefit from CSP with the cost of CSP. The benefit from socially responsible programs is not guaranteed. To the extent that increased CSP results in improved benefit, the manger will be encouraged to become more socially responsible.

The consumer chooses α to maximize (5.18), subject to $\alpha \geq 0$. The first-order condition for the consumer's choice of α is:

$$-\frac{1}{1-\alpha} + g \leq 0. \qquad (5.21)$$

For $\alpha > 0$, (5.21) holds with strict equality. Then:

$$\alpha = \begin{cases} 0 & \text{if } g \leq 1 \\ \dfrac{g-1}{g} & \text{if } g > 1 \end{cases}. \qquad (5.22)$$

The optimal α depends on the amount of CSP supplied by the firm, g. From (5.22), we can say that the consumer is less concerned about the CSP of the firm if the social expenditures are not sufficient ($g \leq 1$). The consumer's benefits from CSP can increase with the amount of the public goods provided. Thus, the fraction α measures the consumer's awareness for CSP. Lack of awareness is likely to be a major inhibitor of consumer responsiveness to CSP.

For a positive α, we have:

$$d\frac{(g-1)/g}{dg} = \frac{1}{g^2} > 0.$$

Thus, α is a strictly increasing function of g if $g > 1$. The amount of CSP supplied by the firm has a significant impact on consumer responses.

The manager's choice problem is to choose g and β to maximize his or her utility in (5.20). Using (5.19), (5.20) can be rewritten as:

$$V = p\{\ln(1 - \beta) - \beta g\} + \beta h(\alpha)g. \qquad (5.20)'$$

In (5.20)', for a positive β, the first-order condition for the manager's choice of g over the interval $[0, G]$ is:

$$g = \begin{cases} 0 & \text{if } h(\alpha) \leq p \\ G & \text{if } h(\alpha) > p \end{cases}.$$

Thus, we can say that the manager is encouraged to provide CSP if his or her social satisfaction is larger than the "marginal" cost of CSP, or $h(\alpha) > p$. In the model, h increases in α. Then, the manager's choice of g is associated with the consumer's taste for CSP. The manager assesses the social demand for CSP, and then determines the optimal level of CSP to provide.

The manager chooses β to maximize (5.20)', subject to $\beta \geq 0$. The first-order condition for the manager's choice of β is:

$$-\frac{p}{1 - \beta} + (h(\alpha) - p)g \leq 0.$$

For $\beta > 0$, the above inequality holds with strict equality. Then, we have:

$$\beta = \begin{cases} 0 & \text{if } (h(\alpha) - p)g \leq p \\ \dfrac{(h(\alpha) - p)g - p}{(h(\alpha) - p)g} & \text{if } (h(\alpha) - p)g > p. \end{cases}$$

The optimal β depends on the manager's social satisfaction, $h(\alpha)$, the marginal cost of CSP, p, and the amount of CSP supplied, g.

5.7.3 Discussion

Vogel (2005), examining the links between ethics and profits, argues that:

> The emergence of 'companies with a conscience' represents a particular vivid expression of the contemporary reconciliation of social values and the business system. These are companies whose vision of social responsibility was integral to their business strategies from the outset. They were formed by individuals with strong personal social commitments who regarded their businesses both as vehicles to make money and as a means to improve society.
>
> (Vogel, 2005, p. 28, original emphasis)

Vogel (2005, p. 2) defines CSR, or business virtue, as "practices that improve the workplace and benefit society in ways that go above and beyond what companies are legally required to do." CSR belongs to strategy, and, in this respect, it is a matter of corporate policy. Companies choose to behave in a socially responsible way, in the same way they choose to spend more on marketing or production.

Standard economic analysis, built on the behavioral assumptions of neo-classical theory, pays little attention to ethics and morality beyond direct profit maximization. The self-interested economic agent has no ethical doubts. For instance, in the principal—agent literature (e.g., Grossman and Hart, 1983), the principal is the person whose material welfare should be maximized, and the agent should execute the orders of the principal by receiving adequate compensation. Efficiency, in the Pareto sense, is the main goal of the theory. The agency conception is based on the pecuniary conflict of interests of the principal (shareholder) and the agent (manager).

A model of business as an independent system has been represented by neoclassical economic theory. In a world of independent systems, the firm exists to maximize its profits. In the behavioral theory of the firm, on the other hand, organizational slack is a prerequisite for being able to afford corporate social strategy. Such strategies rely on individual managers making decisions based on their own values. Nowadays, society at large increasingly expects companies to behave as corporate citizens. Profits are not an end per se; they must be compatible with other social needs. To respond positively to moral responsibility, managers must find an alternative way of doing business. In reality, we recognize that the interactions among various factors of society are complex. From a behavioral theory perspective, this type of complexity is understandable as multiple and conflicting managerial goals. The individual is not separate from society, and thus is not separate from the corporation. Therefore, business is represented to be part of society, affecting it and being affected by other aspects of it. A relational view thus emerges, with corporations embedded in society. CSR is about the basic idea that businesses have to meet society's expectations in their practices. CSR involvement is fueled by various social demands, and enhances the company's access to various resources.

Consumer attention focuses on a company's decisions and actions. In particular, consumers may expect companies to behave ethically, and to be actively involved in helping society. Consumers then need to be aware of a company's level of social responsibility. Thus, companies are under pressure to behave in a socially responsible ways. Consumer expectations regarding the CSP of the firm are a motivating target, and must be considered carefully. Managers must grasp the nature of the values, attitudes, and behaviors of their consumers and accordingly, respond. The attitude would serve as a reference point in consumers' evaluations of a firm's involvement in CSP. CSP communicates that the company can be trusted to act as a partner that will respond to social needs. CSP is thus internalized in the consumer dialogue process. Managers have to develop communications that provide information about their social performance. Social interactions are driven towards motivating and sustaining mutual understanding through dialogue.

People construct the content in a subjective, meaning-creating process. Sense is a continuous process, oriented towards placing current experiences in a frame of reference. Social environments are characterized by a shared understanding among individuals about the nature of a particular issue and the available behavioral alternatives. The mental process in the case of CSP is directed at the creation of a common, context-bound view. Corporations are increasingly engaging as a form of social involvement in response to increased expectations for companies to contribute to CSP efforts. Companies bring their resources to rebuild trust in the social environment. By pursuing social and environmental objectives, socially responsible companies increase their trustworthiness, which in turn supports the process of value creation.

The manager needs to carefully examine the company's mission, vision, and value. Individuals voluntarily work together to create relationships in the pursuit of value creation. Value is a social phenomenon. It is created with the help of others, who value what they create. Working with others and for others can be a strong motivation to create new sources of value. Value is an important factor in a process of arriving at a decision about a course of action. Individuals commit to commonly shared values. They may incorporate their moral values into their economic decisions. Organizational ethics is a company's adoption of desired ethical standards. Ethical standards are formulated from a point of view that goes beyond the interests of a particular individual or group. Ethics sets out what kind of values, norms, and beliefs the good person should cherish. But ethics is only one of a number of dimensions of the decision processes. A dilemma arises in a situation where two (or more) conflicting standards appeal to the consciousness of an individual. An organization encouraging the view that profits are the only consideration will tend to be populated by individuals thinking and functioning on a lower ethical level.

Corporate activities are influenced by the level of moral conduct. The ethical component is silent when there is no moral issue associated with the decision. It comes into play when a moral issue is present. An advantage of behavioral modeling is that individuals can adapt appropriate behaviors to various situations. Wagner-Tsukamoto (2007) reconstructs an economic interpretation of moral agency by providing three levels of moral conduct. (1) Unintentional, passive moral agency. According to classical and neoclassical economic thought, the market economy itself comes with certain ethical ideals. The "invisible hand" of the market best serves consumers in getting a desirable product. Self-interested engagement in the market process itself thus reflects an ethical ideal. (2) Passive, intentional moral agency. Legal laws can be viewed as the enactment of moral minimum standards for all agents.[33] This reflects the enforced nature of moral agency. Findings

---

33. Friedman (1970) qualified his thesis on the social responsibility of firms by conforming to the basic rules of society, both those embodied in legal rules and ethical codes. Therefore, Points (1) and (2) position Friedman's view on business ethics, while Point (3) was not seen by Friedman.

regarding the enforcement of moral agency through legal laws are in line with findings that an agent's own enforcement of moral conduct is ineffective unless properly sanctioned in economic terms. (3) Active, intentional moral agency. Companies not only sell products on the basis of profitability, but also on certain ethical grounds. And such ethical grounds are pursued voluntarily by these companies. Companies engaged in moral agency exceed the minimum standards laid down by public ordering.

Personal belief systems and standards are often related to the social background of an individual. Institutional economics intervenes with institutions understood as incentive structures which order social interactions and resolve problems in organizational behavior ("governance structure" as Williamson, 1985). However, individuals always choose from a certain "bounded" situation that provides them with the premises they will use to make their decisions. The present framework focuses on "institutions" for voluntarily creating morality in social structures. The structures relate the company to social values. Managers stand in a particular relationship to the norms they must enforce. Ethical arguments can be derived from prevailing social norms. Social norms encompass an individual's perception of whether or not to engage in the behavior seen from others. If the individual perceives that others would encourage him or her to behave ethically, it is more likely that the individual would intend to behave ethically. The process of social identification and cohesion of social attitudes promotes moralization through an internalized focus on value and conformity. Simon (1993) argues that human beings do not behave optimally for their fitness. Decision-making that deviates from neoclassical rationality does not imply irrationality, or even errors. Simon coins the term "bounded rationality" to refer to rational choices given the physiological cognitive limitations of the individual decision-makers, in terms of acquiring and processing of information. Because of bounded rationality, individuals have a tendency to act on advice and respect norms. Then, "docility" contributes to the fitness of human beings in evolutionary competition. Here, docility means the tendency to depend on suggestions, recommendations, persuasion, and information obtained through social channels as a major basis for choice. Values, norms, and beliefs play a part in determining whether any individual does the right thing. Companies operating at a high level of moral conduct (active, intentional moral agency) take legitimate social interests into account. Morality is thus conceptualized by focusing on the behavioral manifestations of values, norms, and beliefs in social interactions.

Business and society are interwoven rather than distinct entities. A company's involvement in CSR programs involves answering the requirements of consumers, with particular focus on societal issues. This program not only enhances a company's ethical culture, but also its attention to CSR. A company develops its own meaning of CSR as a result of the interaction between the current confrontation with situations linked to CSR and company-specific capabilities. The decision problem in ethics is not only to choose among alternatives, but also to choose the alternatives. A proper balance is then created between economic and social aims. Value-driven companies develop their own interpretation of CSR on the basis of firmly embedded values, norms, and beliefs of society. Corporations are then social actors. If their economic actions are guided by social relations, their socially oriented actions are more likely to be shaped by their social relations. The corporate perspective on CSR and the resulting behavioral change are, therefore, products of social interactions.

Generally speaking, the idea of institutions is understood as systems of established and prevalent social rules that structure social interactions (North, 1990). Institutional variations, as codified in social structures, create differing environments for the resolution of public issues in which the role of corporations in society is under debate. Institutional variations can result from different perceptions or attitudes towards CSP. The acceptance of consumers into the corporate decision-making processes affects the way specific issues are evaluated, addressed, and resolved within the institutional context. Institutions can be conceptualized as internalized cognitive structures that reflect an individual's values, norms, and beliefs regarding social conduct. Such internalized structures are meant to dispose the individual towards socially desirable behavior. Cognitive differences among individuals thus result in institutional differences regarding the role of corporations in society. Morality depends on the like-mindedness of interacting individuals in order for CSP to prosper. Commitment encourages discretionary behaviors that result in positive goal outcomes. A commitment to good social conduct can be internalized by constructing and promoting moral like-mindedness among individuals, even though there is no apparent economic profit. Behavioral modeling can provide positive sets of ethical behaviors when the situation calls for an ethical response.

The framework presented here identifies CSP with the private provision of public goods, or the private redistribution of profits to social causes. In order to consider what motivates a business to make a contribution to public goods, the model focuses on manager–consumer interactions. The model analyzes two dimensions of social preferences: the consumer's taste for CSP and the manager's taste for CSP. As a consequence, social preference formation interacts with economic decisions taken by others.

The motivation underlying CSP is modeled in terms of social satisfaction that the manager perceives and receives in the interactions. Consumers experience the results of corporate social behavior, so they influence the ethical standards by which corporate behavior is judged, and evaluate how well the manager performs according to those standards. CSP communicates that the company can be trusted to act as a partner that will respond to social needs. The collaborative relations can reinforce positive responses with each other. Consumers' attitudes to encourage CSP are influenced by the manager's view to practice CSP, and vice versa.

CSP is considered as levels of economic, ethical, and discretionary activities of a business entity, as adapted to the values, norms, and beliefs of society. Managers operating at a low level of moral conduct do not take socially conscious consumers' interests into account, and their decisions are perceived to be immoral by consumers. No public good is then provided by firms. A behavioral view identifies an alternative source of inertia as the low level of moral conduct insensitive to the values, norms, and beliefs of society. On the contrary, managers operating at a high level of moral conduct focus on meeting the expectations of socially conscious consumers according to ethical requirements. Then the provision of public goods can be improved substantially. The corporate perspective on CSP, and the resulting behavioral change, are therefore products of social interactions. Managers who rely on social values facilitate the ethical development to be motivated beyond their self-interest. The primary force in determining the level of corporate moral conduct is the resulting social satisfaction in manager—consumer interactions. The social satisfaction that the manager receives is influenced by the consumers' expectations regarding CSP. More public goods can be provided when consumers are more likely to value CSP, and managers are attuned to the consumers' expectations about CSP.

5.8   CONSERVATION ON THE COMMONS

Environmental degradation and resource depletion have been globally pervasive concerns over the last few decades. The problem of the management of common-pool resources has received relatively increased attention in economics. A common-pool resource (e.g., forestry, water, fishery, pastures, etc.) is non-excludable but rival.[34] Joint users may harvest resources without gaining prior permission. Conserving common-pool resources is problematic, because many individuals rely on the benefits of extraction for their livelihoods. Excluding or limiting potential beneficiaries from using a common-pool resource is a nontrivial problem. The actions of one agent affect the resource stock, which in turn affects the well-being of another agent through decreased resource availability. Resources involving open access are much more vulnerable to over-harvesting than those with restricted access (Ostrom, 1990). Common-pool resource exploitation is popularly known as the Hardin's (1968) "tragedy of the commons." Hardin's (1968) metaphor is based on the assumption that human behavior is driven by selfish identity.[35] Individual selfish users are unwilling to pay the costs of conservation,

---

34. More precisely, according to Ostrom et al. (1999), common-pool resources are defined as natural or human-constructed resource systems in which (1) exclusion of beneficiaries through physical and institutional means is especially costly, and (2) exploitation by one user reduces resource availability for others.
35. The management of renewable resources has been traditionally studied within the neoclassical economic framework (Clark, 1976; Neher, 1990).

because the benefits of doing so are shared collectively. This section offers a novel explanation for the conservation of a common-pool resource, namely, economic agents who consider conservation to be their major concern are driven by pro-environmental (PE) identity.

In most microeconomics, economic agents are assumed to respond to financial incentives in a rational and consistent manner. The individual in standard economic theory has a stable and consistent preference ordering over actions. However, the definitional unity of the self is called into serious question by the empirical inconsistencies of economic choices, and by the need to construct multiple-self theories to explain them (Ainsle, 2001; Bénabou and Tirole, 2004). Our notion of identity follows in that it denotes a sense of self. The self-concept is an individual's belief about their personal qualities and abilities. Our self-concept or identity has a profound effect on the way we behave.[36] This section studies the identity of individuals as a source of PE behavior. Individuals have a sense of which type of self they are, i.e., selfish or PE. Individuals are inconsistent, and choose which identity to adopt with some probability. In this section, I consider identity switching in the conservation of a common-pool resource, arguing the interplay of selfish and PE identities.

Akerlof and Kranton (2000) emphasize the importance of identity for economic decision-making. They treat identity as an argument in a person's utility function, where identity is associated with different social categories and expected respective behavior. However, identity choice is very limited in their framework; their framework does not explain how an individual can change or remain with the same identity. People derive utility from thinking of themselves as valuable or productive according to social criteria, and their actions are shaped by the desire to maintain high levels of self-esteem. In Bénabou and Tirole's (2002) model, agents protect their self-esteem by engaging in self-deception, through selective memory and awareness. Bénabou and Tirole (2011) develop a general model of identity in which people care about "who they are," and infer their own values from past choices. In Bénabou and Tirole's (2011) setting, an agent who has built up enough of some asset continues to invest it, even when the marginal return no longer justifies it.

This section explores the idea of identity switching. In the model, there are two different types of identities: selfish and PE. Individual identity is composed of a weighted sum of these types. Beside material payoff, the PE identity obtains psychological payoff related to the conservation of a common-pool resource. It is interpreted that the PE identity derives "moral satisfaction" from the resource conservation. I analyze the implications for switching from a selfish to a PE identity in the conservation of a common-pool resource. A different probability of identity switching prescribes a

---

36. Self-confidence is examined as a means to enhance performance (Compte and Postlewaite, 2004).

different behavior, in which a different level of effort is selected. By increasing the probability of identity switching from selfish to PE, the agent is expected to provide a higher effort level to resource conservation. Then, resource sustainability can be attained.

5.8.1 The Model

A community is made up of a fixed number n of individuals. Agents have complete rights of access to a common-pool resource. They simultaneously decide the amount of resources to exploit from a common pool of renewable resources. The total stock of the resource in existence is denoted by F. Agents are assumed to be homogenous profit-maximizers. Each agent must choose a degree of labor or effort expended on the resource extraction e, where $0 \leq e \leq 1$. The resource extraction entails a (direct) cost $c(e)$. It is assumed that $c'(.) > 0$ and $c''(.) > 0$. In addition, to ensure an interior solution, it is assumed that $c'(0) = 0$, and $\lim_{e \to 1} c'(e) = \infty$. The agent has to decide whether to put in effort e at a cost $c(e)$ in the resource extraction, or put in effort $1 - e$ in the resource conservation.

There are two types of identities that an agent can take: selfish (S) and PE. Selfish and PE identities receive a material payoff related to the resource extraction e. Each identity receives a share of the total, equal to his share of total effort in the resource extraction. The material payoff $b(e)$ is then given by:

$$b(e) = \begin{cases} e & \text{if } ne \leq F \\ (1/n)F & \text{otherwise.} \end{cases} \qquad (5.23)$$

Besides material payoffs, the PE identity also obtains psychological payoff per unit of effort in the resource conservation defined by k, where $0 < k < 1$. It is interpreted as a degree of "moral satisfaction" that the PE identity derives from the resource conservation.

The selfish identity's payoff, U_S, and the PE payoff, U_{PE}, are given by:

$$U_S = b(e) - c(e) \qquad (5.24)$$

and

$$U_{PE} = b(e) + k(1 - e) - c(e). \qquad (5.25)$$

The timing of the situation at each point in time is the following:

Stage 0: The agent holds a selfish identity.
Stage 1: The agent keeps his identity or adopts a PE identity.
Stage 2: The agent chooses an effort e in the resource extraction.

Although an agent starts with a selfish identity at stage 0, he does not have to keep the same identity. At Stage 1, each agent gets an opportunity to change his identity. Let x be the agent's subjective belief about changing his identity, where $0 < x < 1$. The parameter x is the probability of identity switching from selfish to PE. With probability $1 - x$ the agent keeps the same identity, and with probability x he adopts a PE identity. Individuals are uncertain about their own preferences. With probability x, an individual has time-inconsistent preferences. Taking into account his sense of identity, each agent chooses his action optimally at Stage 3.

Using (5.24) and (5.25), the agent's expected utility at Stage 2 is written as:

$$(1 - x)U_S + xU_{PE} = (1 - x)\{b(e) - c(e)\} + x\{b(e) + k(1 - e) - c(e)\}$$
$$= b(e) + x\,k(1 - e) - c(e).$$

$$(5.26)$$

The agent chooses the optimal level of effort in the resource extraction to maximize his utility (5.26). The problem that he faces at Stage 2 is: $\max_e b$ $(e) + x\,k\,(1 - e) - c(e)$. The optimal level of effort equates to the marginal benefit of providing effort to its marginal cost.

5.8.2 Equilibrium and Behavior

First, I state the existence of an equilibrium as follows:

Proposition 5.3: *If $1 < (1/n)F$, there exists a unique equilibrium where an agent chooses a positive effort in the resource extraction. Otherwise, there exists no equilibrium in the interval $[(1/n)F, 1]$ and there may not exist an equilibrium in the interval $[0, (1/n)F]$.*

Proof: Suppose that $1 < (1/n)F$. Since $e \leq 1$, $e \leq (1/n)F$. Then, $b(e) = e$ from (5.23). Thus, from (5.26), the agent's payoff is given by $(1 - x)$ $U_S + x\,U_{PE} = e + x\,k\,(1 - e) - c(e)$. By differentiating this payoff with respect to e, the first-order condition for an interior solution is obtained as follows:

$$1 - xk = c'(e). \qquad (5.27)$$

Since $0 < x < 1$ and $0 < k < 1$, the left-hand side of (5.27) is a positive constant. Since $c'(e) > 0$ and $c''(e) > 0$ for $e > 0$, the right-hand side of (5.27) is strictly increasing; by the assumption ($\lim_{e \to 1} c'(e) = \infty$), $c'(e)$ goes through the positive real line $1 - xk$ as e goes from 0 to 1. Thus, there exists a unique value of e that satisfies (5.27). This unique equilibrium satisfies the second-order condition for a maximum.

Now suppose that $(1/n)F \leq 1$. Consider e in the interval $[(1/n)F, 1]$. Then, $b(e) = (1/n)F$, from (5.23). Thus, from (5.26), the agent's payoff is given by $(1 - x)\,U_S + x\,U_{PE} = (1/n)F + x\,k\,(1 - e) - c(e)$. Differentiating this payoff with respect to e yields $-xk = c'(e)$. Since the right-hand side is strictly increasing, there exists no equilibrium in the interval $[(1/n)F, 1]$.

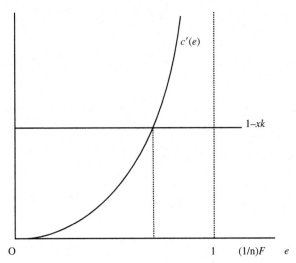

FIGURE 5.5 $1 < (1/n)F$.

Consider e in the interval $[0, (1/n)F]$, where $(1/n)F \leq 1$. Then, $b(e) = e$ from (5.23). The agent's payoff is given by $(1 - x)\ U_S + x\ U_{PE} = e + x\ k$ $(1 - e) - c(e)$, from (5.26). Differentiating this payoff with respect to e yields (5.27). Then, there may not exist an equilibrium in the interval $[0, (1/n)F]$. □

For $1 < (1/n)F$, the curve, $c'(e)$, intersects the real line, $1 - xk$ (Fig. 5.5). Thus, if the resource stock is sufficiently large compared with the number of individuals, there is a unique equilibrium in the interval $[0, 1]$. On the other hand, for $(1/n)F \leq 1$ (i.e., the resource stock is sufficiently small compared with the number of individuals), the curve, $c'(e)$, does not intersect the real line, $1 - xk$, in the interval $[0, (1/n)F]$ (Fig. 5.6).

Second, it is shown that identity switching from selfish to PE reduces the resource extraction. A significant departure from conventional models of the commons is that an amount of effort expended on the resource extraction depends on the agent's subjective belief about changing his identity. The subjective belief about identity switching affects resource extraction and conservation.

Proposition 5.4: *If $1 < (1/n)F$, an equilibrium effort in the resource extraction is strictly decreasing in the probability of identity switching from selfish to pro-environmental, x.*

Proof: If $1 < (1/n)F$, there exists a unique positive value of e that satisfies (5.27). Implicit differentiation of (5.27) with respect to x yields:

$$de/dx = -k\left(1/c''(e)\right). \tag{5.28}$$

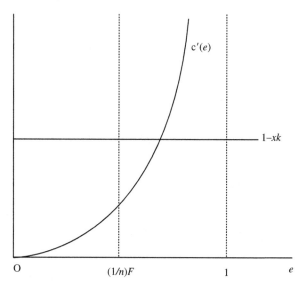

FIGURE 5.6 $(1/n)F \le 1$.

Since $0 < k < 1$ and $c''(e) > 0$ for $e > 0$, the sign of de/dx in (5.28) is negative. ☐

In Proposition 5.4, a different probability of identity switching prescribes a different behavior in which a different amount of effort is selected. By increasing the probability of identity switching from selfish to PE, the agent is expected to provide a higher effort level to resource conservation.

5.8.3 Identity and Resource Dynamics

The model in this section is set in discrete time, and time is denoted by $t = 0, 1, 2, \ldots$ Resource extraction and conservation occur at discrete points in time. This sub-section considers changes in the resource stock over time. If $F^{t+1} - F^t \ge 0$ for all $t \ge 0$, resource sustainability can be achieved. It is assumed that the stock of the resource F^0 is exogenously given at $t = 0$.

In the common-pool resource extraction problem, each agent would be better off if all would restrain their use, but it is not in the interest of the selfish identity to do so.[37] The PE identity can be deterred from overexploitation by moral satisfaction from the conservation.

---

37. Sethi and Somanathan (1996) study the evolution of three strategies, namely defecting, cooperating, and enforcing, among agents interacting in a common-pool resource game. The evolutionary process is described by a replicator dynamics equation, and differential survival is proportional to the relative performance value of each strategy.

Effort in the conservation renews the renewable resource, while effort in the extraction depletes the resource. For a fixed number n of individuals, the aggregate resource conservation is given by $n(1 - e^t)$ at t, while the aggregate resource extraction is given by ne^t at t. The interaction of these components governs the dynamics of the model. The next period's resource stock, F^{t+1}, is given by:

$$F^{t+1} = F^t + n(1 - e^t) - ne^t. \tag{5.29}$$

In the context here, F^t is given at the beginning of t. There exists no over-exploitation of common-pool resources over time if $F^{t+1} - F^t \geq 0$ for any t. Proposition 5.5 below shows that resource sustainability can be attained by increasing the probability of identity switching from selfish to PE.

Proposition 5.5: *Suppose that $1 < (1/n)F^t$ for $t \geq 0$. $F^{t+1} - F^t \geq 0$ if and only if $x \geq x^*$, where $1/2 = e(x^*)$.*

Proof: If $1 < (1/n)F^t$, there is a positive solution $e^t > 0$ to (5.27) for $t \geq 0$. From (5.29),

$$F^{t+1} - F^t = n(1 - e^t) - ne^t$$
$$= n(1 - 2e^t)$$

Consider x^* such that $1/2 = e(x^*)$. Since sign$\{de/dx\}$ in (5.28) is negative, $1/2 \geq e(x)$ for $x \geq x^*$. Therefore, for any t, $F^{t+1} - F^t \geq 0$ if, and only if, $x \geq x^*$. $\qquad\square$

Open access to resources can generate inefficiencies. In the common-pool resource extraction problem, the self-interested action is to engage in resource over-extraction. While all users can choose their own action, they do not communicate to craft management rules. The tragedy of the commons is called a social dilemma, because strategies that are individually rational can produce irrational results collectively. For efficient management, it is necessary that property rights over the resource be well-defined and enforced. However, the required institutions are absent in many settings, because of a lack of governance. Neither the state nor the market does a very good job in these situations, since they seek to impose external rules on the relevant agents. The law is conceptualized as a top-down mechanism to bring order through coercive enforcement. In the case of legal compliance, individual incentives most often refer to deterrence. People obey the law only when their expected compliance utility is greater than their expected violation utility. Rather, the agents can devise their own enforceable institutional arrangements to escape the self-interested action. Bottom-up processes are more appealing than authoritarian enforcement. Ostrom (1990) shows how local communities may succeed in overcoming the tragedy of the commons. Local problem solving does a better job of producing cooperation than the external imposition of rules. Local spheres must have some autonomy for systems to function properly; local actors resolve problems within their own sphere. The preservation of local autonomy improves outcomes for all the participants, allowing them to make locally informed decisions. For example, fishers using the same ocean grounds might come up with better and more durable solutions to the depletion of stocks than the set of rules imposed by state bureaucrats. Some agents have been able to avoid destroying their own resource base, while others have not, which raises the question of what differences might exist between these communities.[38]

It is not correct to assume that preferences and utility functions are fixed and exogenous, especially in relation to the environment. Welfare effects of changing preferences need to be addressed in environmental economics. People making voluntary contributions to public goods, such as blood donations and voluntary collection and recycling of waste, cannot be explained solely by pure self-interest. Many studies show that agents do not follow a purely self-interested strategy, but rather strike a balance between self and group interests (e.g., Charness and Rabin, 2002; Fehr and Gächter, 2000; Fehr and Schmidt, 1999). People care for equitable outcomes, and behave fairly and cooperatively in situations where the self-interest model would predict complete defection. Andreoni (1989, 1990) suggests that people derive utility from the act of giving, labeled the "warm glow effect." As important drivers of prosocial behavior, Bénabou and Tirole (2006) identify image-related rewards or punishment like concerns for social reputation and self-image.

A person's sense of self is, at least in part, derived from perceived membership in social groups. It is associated with different social categories, and how people should behave in these categories. Akerlof and Kranton's (2000) model shows that individuals have an incentive to choose particular actions such that their own individual characteristics fit with the socially determined self-image of the social categories to which they belong.[39] The utility function is an expression that characterizes what a person cares about. In Akerlof and Kranton's (2000) prototype model, identity depends on two social categories, Green and Red, and the correspondence of one's own and others' actions to behavioral prescriptions for their category. There are two possible activities: Activity 1 and Activity 2. Each person has a taste for either

---

38. For example, according to Cialdini et al. (1990), people are more likely to litter when the floor is already littered. That is, when the apparent social norm is to litter, people are more likely to conform to that apparent social norm by throwing their trash on the floor. When the apparent social norm is not to litter, people are more likely to conform to that norm by refraining from littering. This difference is magnified when the norm is made more salient.

39. Akerlof and Kranton (2002) construct a model of a student's utility where identity is salient. In a standard model of education, a student's utility will depend on his or her effort in school, and the pecuniary returns to this effort. In the school context, for example, social identities may include "jocks" or "nerds." Students also have ideas about how people in groups should behave.

activity. They consider an interaction between an individual with a taste for Activity 1 (Person 1), and an individual with a taste for Activity 2 (Person 2). If Person 1 (or 2) undertakes Activity 1 (or 2), he or she earns some positive utility. A person who chooses the activity that does not match his or her taste earns zero utility. As behavioral prescriptions, Greens should engage in Activity 1, and Reds should engage in Activity 2. For example, by choosing Activity 1, a person could affirm his or her identity as a Green. First, identity changes the payoffs from one's own actions. Person 1 who chooses Activity 2 would lose his or her Green identity. Second, identity changes the payoffs of others' actions (externality). If Person 1 and Person 2 are paired, Activity 2 on the part of Person 2 diminishes Person 1's Green identity. Third, the choice of different identities affects an individual's economic behavior. While Person 2 could choose between Green and Red, he or she could never be a "true" Green. Finally, the social categories and behavioral prescriptions can be changed, affecting identity-based preferences.

Group memberships might affect behavior in ways that cannot easily be reduced to material self-interest. The self-image then depends on the established social norms. Unlike legal rules, social norms are not supported by formal sanctions. Social norms are enforced privately in a decentralized way. Social norms reflect a historical trial-and-error process. Through this bottom-up process, the norm-based rules tend to reflect peoples' diverse situations. As the number of people complying with a norm increases over time, the expectation leads the norm to continue into the future.[40] A social norm is a behavior characterized by its being prevalent among group members. Social norms are "shared understandings about actions that are obligatory, permitted, or forbidden" (Ostrom, 2000, pp. 143–144). It is necessary that people believe a social norm exists, and know the class of situations to which the norm pertains.

Posner (2000) highlights the importance of norms as follows:

> *Why tax compliance? A widespread view among tax scholars holds that law enforcement does not explain why people pay taxes. The penalty for ordinary tax convictions is small; the probability of detection is trivial; so the expected sanction is small. Yet large numbers of Americans pay their taxes. This pattern contradicts the standard economic model of law enforcement, which holds that people violate a law if the benefit exceeds the expected sanction. Some scholars therefore conclude that the explanation for the tendency to pay taxes must be that people are obeying a norm—presumably a norm of tax payment or a more general norm of law-abiding behavior.*

(Posner, 2000, p. 1782)

---

40. Conventions describe what people generally do in a situation. Norms broadly encompass conventions. A specific convention is an actual behavioral regularity in a given group. In the theory of conventions, norms are considered as coordinating devices, and one of several alternatives that enable people to achieve their goals (Young, 1998).

Individuals may care about how their decision to comply with social norms is perceived by others, creating a role for social interaction. Elster (1989) argues that for a norm to be social, it must be shared by others, and should be at least partially maintained (or not maintained) based on their approval (or disapproval). According to Elster (1989), social norms are usually sustained through social emotions such as shame, guilt, or embarrassment.[41] Social norms often direct individuals to undertake actions that are inconsistent with selfish actions. Social norm activation can trigger identity switching from selfish to PE.

Throughout this section, agents are assumed to be homogenous. It would be more reasonable to describe agents who are endowed with different behavioral characteristics. Then one could express heterogeneity in strategies through different effort levels. Olson (1965) finds that certain types of inequality have a positive effect on collective action.[42] Heterogeneity has a double effect on the free rider problem (Baland and Platteau, 1999). As Olson (1965) argues, an increase in heterogeneity will make rich users internalize more the consequences of their actions, and they will invest less to use the collective good. However, on the other hand, this increase will diminish the propensity of poor users to take into account the impact on the resource that they may cause. According to Baland and Platteau (1997), the marginal cost of effort or cooperation is likely not constant across individuals, and instead it may depend on the opportunity cost of time, and the wealth of each individual. In a model of fishing, Dayton-Johnson and Bardhan (2002) present a U-shaped function of heterogeneity in the capacity to invest in fishing boats, where total income decreases, then increases with the degree of heterogeneity. An increase in heterogeneity in the ability to invest can then lead to an increase in the collective efficiency.

---

41. McAdams and Rasmusen (2007) define norms as behavioral regularities supported at least in part by normative attitudes. People suffer from guilt, disapproval, or shame from breaking the law. Guilt is disutility that violators feel about stealing, even though they believe they are certain not to be caught. Someone might lose utility from believing that other people disapprove of one, regardless of whether they take action that may materially affect one. Unlike disutility from guilt, disutility from disapproval arises only when one believes others have formed beliefs about one's behavior. Shame might arise when others discover the violation and think badly of the violator. The violator can imagine what others would think if they discover the violation. Unlike disapproval, the person feeling shame suffers disutility regardless of what others think.

42. Olson (1965) also predicts the effects of size on overall group performance. Olson (1965) argues that small groups are more likely to realize collective goals effectively. If the benefit involves rival consumption, then the per capita share of benefits falls as the collectivity increases. On the production side, the contributions of individuals may become less important to overall group performance as size increases.

5.9   SUMMARY

People are said to have entirely self-interested preferences if, and only if, they always choose in such a way as to maximize their own pecuniary payoffs. Self-interest is not of the only human motivation. Throughout their lifetime, people depend on frequent and varied cooperation with others. In the short-run, it would be advantageous to cheat. In the long-run, people are likely to cooperate in a mutually beneficial manner. This would make them better able to resist the temptation to cheat in the first place, and would enable them to generate a reputation for being cooperative.

People are, in real life, affected by psychological and emotional factors. Traditional economic theory is built on the assumption that all people are entirely self-interested, and do not care about the well-being of others. People exhibit prosocial behavior when they do not always make choices that maximize their own pecuniary payoffs. Formal models of social preferences assume that people are self-interested, but are also concerned about the payoffs of others. That is, a player's utility function not only depends on his or her material payoff, but may also be a function of the allocation of resources within his or her reference group. People have social preferences if, and only if, they exhibit prosocial behavior and have relatively stable social preferences.

REFERENCES

Ainsle, G., 2001. Breakdown of Will. Cambridge University Press, Cambridge, MA.

Akerlof, G.A., 1980. A theory of social custom, of which unemployment may be one consequence. Q. J. Econ. 94, 49−75.

Akerlof, G.A., Kranton, R.E., 2000. Economics and identity. Q. J. Econ. 115, 715−753.

Akerlof, G.A., Kranton, R.E., 2002. Identity and schooling: some lessons for the economics of education. J. Econ. Lit. 40, 1167−1201.

Alexander, R.D., 1987. The Biology of Moral Systems. Aldine De Gruyter, New York.

Altman, M., 2001. Worker Satisfaction and Economic Performance: Microfoundations of Success and Failure. M.E.Sharpe, New York.

Altman, M., 2005. The ethical economy and competitive markets: reconciling altruistic, moralistic, and ethical behavior with the rational economic agents and competitive markets. J. Econ. Psychol. 26, 732−757.

Andreoni, J., 1989. Giving with impure altruism: applications to charity and Ricardian equivalence. J. Polit. Econ. 97, 1447−1458.

Andreoni, J., 1990. Impure altruism and donations to public goods: a theory of warm glow giving. Econ. J. 100, 464−477.

Andreoni, J., 1995a. Warm-glow versus cold-prickle: the effects of positive and negative framing on cooperation in experiments. Q. J. Econ. 110, 1−21.

Andreoni, J., 1995b. Cooperation in public-goods experiments: kindness or confusion? Am. Econ. Rev. 85, 891−904.

Andreoni, J., Bernheim, B.D., 2009. Social image and the 50-50 norm: a theoretical and experimental analysis of audience effects. Econometrica 77, 1607−1636.

Arthur, W.B., 1994. Increasing Returns and Path Dependency in the Economy. University of Michigan Press, Ann Arbor, MI.

Axelrod, R., 1984. The Evolution of Cooperation. Basic Books, New York.

Bagnoli, M., Watts, S.G., 2003. Selling to socially responsible consumers: competition and the private provision of public goods. J. Econ. Manag. Strategy 12, 419–445.

Baland, J.M., Platteau, J.P., 1997. Wealth inequality and efficiency in the commons. Part I: the unregulated case. Oxf. Econ. Pap. 49, 451–482.

Baland, J.M., Platteau, J.P., 1999. The ambiguous impact of inequality on local resource management. World. Dev. 27, 773–788.

Baron, D.P., 2001. Private politics, corporate social responsibility, and integrated strategy. J. Econ. Manag. Strategy 10, 7–45.

Baron, D.P., 2008. Managerial contracting and corporate social responsibility. J. Public. Econ. 92, 268–288.

Barro, R., 1974. Are government bonds net wealth? J. Polit. Econ. 82, 1095–1117.

Bateson, M., Nettle, D., Roberts, G., 2006. Cues of being watched enhance cooperation in a real-world setting. Biol. Lett. 2, 412–414.

Becker, G.S., 1968. Crime and punishment: an economic approach. J. Polit. Econ. 76, 169–217.

Becker, G.S., 1976. Altruism, egoism, and genetic fitness: economics and sociobiology. J. Econ. Lit. 4, 817–826.

Ben-Ner, A., Putterman, L., 1998. Value and institutions in economic analysis. In: Ben-Ner, A., Putterman, L. (Eds.), Economics, Values, and Organization. Cambridge University Press, Cambridge, MA, pp. 3–19.

Ben-Ner, A., Putterman, L., 2000. On some implications of evolutionary psychology for the study of preferences and institutions. J. Econ. Behav. Organ. 43, 91–99.

Bénabou, R., Tirole, J., 2002. Self-confidence and personal motivation. Q. J. Econ. 117, 871–915.

Bénabou, R., Tirole, J., 2004. Willpower and personal rules. J. Polit. Econ. 112, 848–886.

Bénabou, R., Tirole, J., 2006. Incentives and prosocial behavior. Am. Econ. Rev. 96, 1652–1678.

Bénabou, R., Tirole, J., 2011. Identity, morals, and taboos: beliefs as assets. Q. J. Econ. 126, 805–855.

Bernheim, B.D., 1994. A theory of conformity. J. Polit. Econ. 102, 841–877.

Besley, T.J., Ghatak, M., 2007. Retailing public goods: the economics of corporate social responsibility. J. Public. Econ. 91, 1645–1663.

Bester, H., Güth, W., 1998. Is altruism evolutionarily stable? J. Econ. Behav. Organ. 34, 193–209.

Bhattacharya, C.B., Sen, S., 2003. Consumer–company identification: a framework for understanding consumers' relationships with companies. J. Mark. 67, 76–88.

Bisin, A., Verdier, T., 2001. The economics of cultural transmission and the dynamics of preferences. J. Econ. Theory. 97, 298–319.

Blau, P.M., 1964. Exchange and Power in Social Life. John Wiley & Sons, New York.

Bowen, H.R., 1953. Social Responsibilities of the Businessman. : Harper and Brothers, New York.

Bowles, S., 1998. Endogenous preferences: the cultural consequences of markets and other economic institutions. J. Econ. Lit. 36, 75–111.

Bowles, S., Gintis, H., 2000. Walrasian economics in retrospect. Q. J. Econ. 115, 1411–1439.

Boyd, R., Richerson, P.J., 1989. The evolution of indirect reciprocity. Soc. Networks. 11, 213–236.

Boyd, R., Gintis, H., Bowles, S., Richerson, P.J., 2003. The evolution of altruistic punishment. Proc. Natl. Acad. Sci. USA 100, 3531–3435.

Brekke, K.A., Kverndokk, S., Nyborg, K., 2003. An economic model of moral motivation. J. Public. Econ. 87, 1967–1983.

Brown, T.J., Dacin, P.A., 1997. The company and the product: corporate associations and consumer product responses. J. Mark. 61, 68–84.

Camerer, C.F., 2003. Behavioral Game Theory: Experiments in Strategic Interaction. Princeton University Press, Princeton, NJ.

Capriotti, P., Moreeno, Á., 2007. Corporate citizenship and public relations: the importance and interactivity of social responsibility issues on corporate websites. Public Relat. Rev. 33, 84–91.

Charness, G., Rabin, M., 2002. Understanding social preferences with simple tests. Q. J. Econ. 117, 817–869.

Chaudhuri, A., 2011. Sustaining cooperation in laboratory public goods experiments: a selective survey of the literature. Exp. Econ. 14, 47–83.

Cialdini, R.B., Reno, R.R., Kallgren, C.A., 1990. A focus theory of normative conduct: recycling the concept of norms to reduce littering in public places. J. Pers. Soc. Psychol. 58, 1015–1026.

Clark, C.W., 1976. Mathematical Bioeconomics: The Optimal Management of Renewable Resources. Wiley, New York.

Compte, O., Postlewaite, A., 2004. Confidence-enhanced performance. Am. Econ. Rev. 94, 1536–1557.

Commission of the European Communities, 2001. Green Paper: Promoting a European Framework for Corporate Social Responsibility. Commission of the European Communities, Brussels.Rabin, M., A perspective on psychology and economics. Eur. Econ. Rev. 46 (2002), 657–685.

Creyert, E.H., Ross, W.T., 1997. The influence of firm behavior on purchase intention: do consumers really care about business ethics? J. Consum. Mark. 54, 68–81.

Cyert, R.M., March, J.G., 1963. A Behavioral Theory of the Firm. Prentice-Hall, New Jersey.

Dayton-Johnson, J., Bardhan, P., 2002. Inequality and conservation on the local commons: a theoretical exercise. Econ. J. 112, 557–602.

Eaton, B.C., Eswaran, M., 2003. The evolution of preferences and competition: a rationalization of Veblen's theory of invidious comparisons. Can. J. Econ. 36, 832–859.

Ellen, P.S., Mohr, L.A., Webb, D.J., 2000. Charitable programs and the retailer: do they mix? J. Retailing 76, 393–406.

Elster, J., 1989. The Cement of Society: A Study of Social Order. Cambridge University Press, Cambridge, MA.

Engelman, D., Fischbacher, U., 2009. Indirect reciprocity and strategic reputation building in an experimental helping game. Games. Econ. Behav. 67, 399–407.

Fehr, E., Falk, A., 2002. Psychological foundations of incentives. Eur. Econ. Rev. 46, 687–724.

Fehr, E., Fischbacher, U., 2003. The nature of human altruism. Nature. 425, 785–791.

Fehr, E., Fischbacher, U., 2004. Third party punishment and social norms. Evol. Hum. Behav. 25, 63–87.

Fehr, E., Fischbacher, U., 2005. Human altruism—Proximate patterns and evolutionary origins. Analyse Kritik 27, 6–27.

Fehr, E., Gächter, S., 2000. Cooperation and punishment in public goods experiments. Am. Econ. Rev. 90, 980–994.

Fehr, E., Gächter, S., 2002. Altruistic punishment in humans. Nature. 415, 137−140.

Fehr, E., Schmidt, K.M., 1999. A theory of fairness, competition, and cooperation. Q. J. Econ. 114, 817−868.

Fehr, E., Kirchsteiger, G., Reidl, A., 1993. Does fairness prevent market clearing? An experimental investigation. Q. J. Econ. 108, 437−460.

Fehr, E., Fischbacher, U., Gächter, S., 2002. Strong reciprocity, human cooperations, and the enforcement of social norms. Hum. Nat. 13, 1−25.

Frank, R.H., 1988. Passions within Reason: The Strategic Role of the Emotions. W. W. Norton, New York.

Freeman, E., 1984. Strategic Management: A Stakeholder Approach. Pitman Publishing, Boston.

Friedman, M., 1962. Capitalism and Freedom. University of Chicago Press, Chicago.

Friedman, M., 1970. The social responsibility of business is to increase its profits. N. Y. Times Mag. 13 (32-33), 122−126.

Fudenberg, D., Maskin, E., 1986. The folk theorem in repeated games with discounting or with incomplete information. Econometrica. 54, 533−544.

Gallese, V., Fadiga, L., Fogassi, L., Rizzolatti, G., 1996. Action recognition in the premotor cortex. Brain 119, 593−609.

Gintis, H., 2003. The hitchhiker's guide to altruism: gene-culture coevolution, and the internalization of norms. J. Theor. Biol. 220, 407−418.

Gintis, H., Bowles, S., Boyd, R., Fehr, E., 2003. Explaining altruistic behavior in humans. Evol. Hum. Behav. 24, 153−172.

Gintis, H., Bowles, S., Boyd, R., Fehr, E. (Eds.), 2005. Moral Sentiments and Material Interests: The Foundations of Cooperation in Economic Life. The MIT Press, Cambridge, MA.

Gintis, H., Henrich, J., Bowles, S., Boyd, R., Fehr, E., 2008. Strong reciprocity and the roots of human morality. Soc. Justice. Res. 21, 241−253.

Grossman, S.J., Hart, O.D., 1983. An analysis of the principal-agent problem. Econometrica. 51, 7−45.

Hardin, G., 1968. The tragedy of the commons. Science 162, 1243−1248.

Hayek, F.A., 1967. Studies in Philosophy, Politics, and Economics. Routledge & Kagan Paul, London & Henley.

Henrich, J., McElreath, R., Barr, A., Ensminger, J., Barrett, C., Bolyanatz, A., et al., 2006. Costly punishment across human societies. Science 312, 1767−1770.

Herrmann, B., Thöni, C., Gächter, S., 2008. Anti-social punishment across societies. Science 319, 1362−1367.

Isaac, R.M., Walker, J.M., 1988. Group size effects in public goods provision: the voluntary contribution mechanism. Q. J. Econ. 103, 179−200.

Kandori, M., 2003. The erosion and sustainability of norms and morale. Jpn. Econ. Rev. 54, 29−48.

Khalil, E.L., 2004. What is altruism? J. Econ. Psychol. 25, 97−123.

Kolstad, I., 2007. Why firms should not always maximize profits. J. Bus. Ethics. 76, 137−145.

Leibenstein, H., 1966. Allocative efficiency vs. 'X-efficiency. Am. Econ. Rev. 56, 392−415.

Leibenstein, H., 1987. Inside the Firm: The Inefficiencies of Hierarchy. Harvard University Press, Cambridge, MA.

Maynard Smith, J., 1982. Evolution and the Theory of Games. Cambridge University Press, Cambridge, MA.

McAdams, R., Rasmusen, E.B., 2007. Norms in law and economics. In: Polinsky, M., Shavell, S. (Eds.), Handbook of Law and Economics, Vol. 2. Elsevier, Amsterdam, pp. 1573−1618.

Mohr, L.A., Webb, D.J., Harris, K.E., 2001. Do consumers expect companies to be socially responsible? The impact of corporate social responsibility on buying behavior. J. Consum. Aff. 35, 45−72.

Mui, V.-L., 1995. The economics of envy. J. Econ. Behav. Organ. 26, 311−336.

Naylor, R., 1989. Strikes, free riders, and social customs. Q. J. Econ. 104, 771−785.

Neher, P.A., 1990. Natural Resource Economics: Conservation and Exploitation. Cambridge University Press, Cambridge, MA.

North, D.C., 1990. Institutions, Institutional Change and Economic Performance. Cambridge University Press, Cambridge, MA.

Nowak, M.A., Sigmund, K., 1998. The dynamics of indirect reciprocity. J. Theor. Biol. 194, 561−574.

Olson, M., 1965. The Logic of Collective Action: Public Goods and the Theory of Groups. Harvard University Press, Cambridge, MA.

Ostrom, E., 1990. Governing the Commons: The Evolution of Institutions for Collective Action. Cambridge University Press, Cambridge, MA.

Ostrom, E., 2000. Collective action and the evolution of social norms. J. Econ. Perspect. 14, 137−158.

Ostrom, E., Walker, J., Gardner, R., 1992. Covenants with and without a sword: self-governance is possible. Am. Polit. Sci. Rev. 86, 404−417.

Ostrom, E., Burger, J., Field, C.B., Norgaard, R.B., Policansky, D., 1999. Revisiting the commons: local lessons, global challenges. Science 284, 278−282.

Posner, E., 2000. Law and social norms: the case of tax compliance. Va. Law. Rev. 86, 1781−2000.

Posner, R.A., 1997. Social norms and the law: an economic approach. Am. Econ. Rev. 87, 365−369.

Rabin, M., 1993. Incorporating fairness into game theory and economics. Am. Econ. Rev. 83, 1281−1302.

Rizzolatti, G., Fadiga, L., Gallese, V., Fogassi, L., 1996. Premotor cortex and the recognition of motor actions. Cogn. Brain Res. 3, 131−141.

Rizzolatti, G., Craighero, L., 2004. The mirror-neuron system. Annu. Rev. Neurosci. 27, 169−192.

Rizzolatti, G., Craighero, L., 2005. Mirror neurons: a neurological approach to empathy. In: Changeux, J.-P., Demasio, A.R., Singer, W., Christen, Y. (Eds.), Neurobiology of Human Values. Springer, Berlin, pp. 107−123.

Rizzolatti, G., Fodassi, L., Gallese, V., 2001. Neurophysiological mechanisms underlying the understanding and imitation of action. Nat. Rev. Neurosci. 2, 661−670.

Sethi, R., Somanathan, E., 1996. The evolution of social norms in common property resource use. Am. Econ. Rev. 86, 766−788.

Simon, H.A., 1993. Altruism and economics. Am. Econ. Rev. 83, 156−161.

Sliwka, D., 2007. Trust as a signal of a social norm and the hidden costs of incentive schemes. Am. Econ. Rev. 97, 999−1012.

Smith, A., 1759/1981. The Theory of Moral Sentiments. Liberty Fund, Indianapolis.

Smith, V.L., 2003. Constructivist and ecological rationality in economics. Am. Econ. Rev. 93, 465−508.

Sonnemans, J., Schram, A., Offerman, T., 1998. Public good provision and public bad prevention: the effect of framing. J. Econ. Behav. Org. 34, 143−161.

Teraji, S., 2007. Morale and the evolution of norms. J. Behav. Exp. Econ. 36, 48−57.

Teraji, S., 2009. A model of corporate social performance: social satisfaction and moral conduct. J. Behav. Exp. Econ. 38, 926–934.

Teraji, S., 2013. A theory of norm compliance: punishment and reputation. J. Behav. Exp. Econ. 44, 1–6.

Tirole, J., 1996. A theory of collective reputation (with applications to the persistence of corruption and to firm quality). Rev. Econ. Stud. 63, 1–22.

Trivers, R., 1971. The evolution of reciprocal altruism. Q. Rev. Biol. 46, 35–57.

Tooby, J., Cosmides, L., 1992. The psychological foundations of culture. In: Barkow, J., Cosmides, L., Tooby, J. (Eds.), The Adapted Mind. Oxford University Press, Oxford, pp. 19–136.

Vogel, D., 2005. The Market for Virtue: The Potential and Limits of Corporate Social Responsibility. Brookings Institution Press, Washington, DC.

Waddell, S., 2000. New institutions for the practice of corporate citizenships: historical, intersectoral, and developmental perspectives. Bus. Soc. Rev. 105, 107–217.

Wagner-Tsukamoto, S., 2007. Moral agency, profits and the firm: economic revisions to the Friedman theorem. J. Bus. Ethics 70, 209–220.

Webster Jr., F.E., 1975. Determining the characteristics of the socially conscious consumer. J. Consum. Res. 2, 188–196.

Williamson, O.E., 1985. The Economic Institutions of Capitalism. The Free Press, New York.

Young, H.P., 1998. Individual Strategy and Social Structure: An Evolutionary Theory of Institutions. Princeton University Press, Princeton, NJ.

Chapter 6

Cognition and Order

6.1 INTRODUCTION

Individual behavior is largely governed by cognitive rules. As an important development in the understanding of human cognition, we have the so-called "mental models" view (Johnson-Laird, 1983). People reason from their understanding of a situation, and their starting point is a set of mental models constructed from perceiving the real world. The mental model describes the kinds of mental representation individuals construct when they reason about the world. Mental models are organized knowledge structures that allow individuals to interact with their external environment. Mental models allow people to draw inferences, make predictions, understand phenomena, and decide which actions to take. Furthermore, the function of shared mental models is to allow group members to draw on their own well-structured knowledge, as a basis for selecting actions that are consistent and coordinated with other members.

Human behavior is depicted as being shaped by both environmental influences and internal dispositions. In the traditional theory of decision-making, an agent faces a problem, gathers relevant information from the environment, and processes the information in order to select the most appropriate response. However, the demarcation between the decision-makers and their environment is unidentifiable. Behavior, cognition, and environmental events operate as interacting determinants that influence each other. People are both products and producers of their environment.

The mind has some internal structure. Fodor (1983) proposes that mental phenomena arise from the operation of multiple distinct processes, rather than a single undifferentiated one. Perceptual processes such as vision are modular; they are organized into innate (i.e., hard-wired, genetically specialized) input systems. Each of these modules functions independently of the others, using its own dedicated processes, with each one responding to uniquely different inputs from the environment. In addition to Fodor's view that only perceptual systems are modular, evolutionary psychologists like Leda Cosmides and John Tooby propose massive modularity; many or most information-processing systems in the mind might be modular as well. Cosmides and Tooby (1992) argue that the mind as a whole, not just peripheral systems such as vision, can be regarded as a modular "Swiss Army knife" with a multitude of blades, each one dedicated to a specific task.

The Cognitive Basis of Institutions. DOI: https://doi.org/10.1016/B978-0-12-812023-1.00006-5

The mind is typically a collection of intellectual and conscious aspects, which are manifest in some combination of thought, perception, emotion, will, and imagination. Based on a scientific understanding of the brain, the mind is considered as a psychological and physiological phenomenon. The brain is one of the complex systems that we can observe and measure. The neuron is the basic signaling element of the brain. Most of our knowledge about the functional organization of neuron systems is based on the analysis of the firing patterns of individual neurons. The brain manifests distributed activation during cognition. According to Donald Hebb's *The Organization of Behavior* (1949), if groups of neurons converge on a single neuron, and if they have a tendency to fire together, they will be strengthened as a group. Information processing requires the functional cooperation of distributed neurons. The synapse is the transmission site where the signals are in spatio-temporal contingency. Such a synapse uses this locally available information to cause a local, input-specific synaptic modification. Since 1949, the Hebb's synapse concept has evolved to include several key features which form the basis of the contemporary understanding of its mechanism. The synaptic modification mechanism is now referred to as Hebb's rule. The consequence of such a learning procedure is a connectivity graph.

Independently from Hebb (1949), Friedrich Hayek's *The Sensory Order* (1952) proposes a brain theory essentially based on a neural network model.[1] Although *The Sensory Order* is in no way a direct economic or social analysis, it can be regarded as one of the most creative attempts in Hayek's writings.[2] Its publication links together the notions of evolution, spontaneous order, and the limits to explanation when dealing with complex phenomena. The brain must digest information in such a way as to simplify data and stimuli that would otherwise be infinitely complex and confusing. *The Sensory Order* distinguishes between the physical order existing in the external world and our phenomenal experience of it.[3] The essence of Hayek's attempt in theoretical psychology is to show how a structure can be formed which

---

1. Two books, Hebb (1949) and Hayek (1952), are categorized as "complementary rather than covering the same ground" (Hayek, 1952, p. viii).
2. Hayek's interest in human cognition goes back to the early 1920s, when he was still a student. In the winter of 1919−20 a fuel shortage and forced closure of the University presented Hayek with an opportunity to travel to Zurich, where he, as well as attending lectures in law and philosophy, "worked for a few weeks in the laboratory of the brain anatomist von Monakow, tracing fiber bundles through the different parts of the human brain" (Hayek, 1994, p. 64).
3. Hayek's theory of mind can be traced back to Ernst Mach's theory (Caldwell, 2004). Mach observed that perception consisted not just of sensations, but also of memories conjoined with them. Individual memory plays an essential role in classifying and retaining personal experience. While individual sensations themselves are never repeated, traces of some of them lingered on in memory. The difference between mental and physical phenomena arises from the different ways in which the elements are combined. For Mach, the world is seen as a continuum of connexions of elements in constant flux. The ego is a temporary continuum in the larger, more permanent continuum of the world.

discriminates between different physical stimuli and generates the sensory order that we actually experience.[4] The sensory order is a system of qualities that do not simply represent physical properties of the external world. The subjectivity of individual knowledge finds its foundation in the construction of the mind. The mind operates by assembling new sensory data into associations with our accumulated inventory of knowledge. Knowledge is forged by the connection of new sensory information to previous sensory experiences.

According to Hayek (1952), knowing the external world is a classification of sensory qualities by the mind. Hayek (1952) uses the term "sensory qualities" to refer to all the different attributes or dimensions with regard to which people differentiate their responses to different stimuli. What we know at any moment about the world is determined by the order of the apparatus of classification which has been built up by previous sensory linkages. The qualitative differences in perceptions that we experience depend on the specific pattern of neuron firings that a given stimulus produces within various neural networks. The experiences of individuals will differ, according to the pattern of neuron firings that each develops. Hayek's cognitive theory explains how different pieces of cognitive information cause different perceptions and, therefore, different actions. New linkages are established, depending on the pattern of ongoing neural activity. The structure of linkages governs our cognitive processes.

The mind is an order of relations. At the level of neurons in the brain, the classification of primary sensory impulses, and any further impulses they evoke, can take place on many successive levels. As a result of the multiple classification processes, the human mind produces a highly complex mental order.

Neuroeconomics is a recent research subject in economics, by virtue of its adoption of neuroscience as a basis for investigation of economic questions. Neuroeconomics is dedicated to an effort to provide hypotheses about how certain decision-making mechanisms work, about which areas of the brain are involved in which decision-making tasks, and about how variations in tasks produce different patterns of brain activation. Modularity is important for neuroeconomics. According to Camerer et al. (2004, p. 561), "the brain is like a large company—branch offices specialize in different functions, but also communicate to one another, and communicate more feverishly when an important decision is being made." Neuroeconomics studies the cerebral activity of economic decision-making. Modern neuroscientific techniques of functional neuroimaging, like functional magnetic resonance imaging and position emission tomography, are used to monitor certain brain responses to different tasks of economic decision-making. Camerer et al. (2004, 2005) think that neuroeconomics may help in improving economics,

---

4. Hayek's contribution to cognitive theory has received explicit recognition from neuroscientists (Edelman, 1989; Fuster, 2005).

by showing the deficiencies of existing economic models. Conventional economic models aim only to explain an individual's observable choices. In these models, decision-makers behave as if they had first constructed and stored a single list of all possible options ordered from best to worst, and then had selected the highest ordered of those available options. One of the most important insights of neuroscience is that the brain is not a homogeneous processor, but rather involves diverse specialized processes that are integrated in different ways when the brain faces different types of problems. According to Camerer et al. (2005), neuroscience is more about *where* things happen in the brain:

> *By tracking what parts of the brain are activated by different tasks, and especially by looking for overlap between diverse tasks, neuroscientists are gaining an understanding of what different parts of the brain do, how the parts interact in "circuitry," and, hence, how the brain solves different types of problems.*
>
> Camerer et al. (2005, p. 14, original emphasis)

Inspired by neuroeconomic studies, some economic models attempt to formalize the idea that judgment and behavior are the interaction between multiple, often conflicting, processes. Bernheim and Rangel (2004) emphasize the brain as operating in either a "cold" mode or a "hot" mode. Loewenstein and O'Donoghue (2004) emphasize the interaction between "deliberative" and "affective" systems. These models focus on dual processes that have primarily influenced psychology. The functioning of different brain areas, and the brain processes, derive human behavior.

Neuroeconomics can be conceived as an attempt to discover multi-level mechanisms to explain regularities on decision-making. Components and activities at one level can often themselves be decomposed into organized entities and activities at a lower level, which can themselves be decomposed into organized entities and activities at still lower levels. Neuroeconomics provides a substantial improvement of the mind-related economic approaches by adding and specifying important cerebrally-operated components of human decision-making. Behavior can be explained in terms of cognitive mechanisms; the mechanisms are in turn explained in terms of underlying interactions between brain regions. By recognizing different structural components of the mind, economics needs to build on the characterization of processes that contribute to decision-making.

6.2 SENSORY ORDER

In his *The Sensory Order* (1952), Hayek provided a theory of the process by which the mind perceives the world around it. Hayek's *The Sensory Order*, though it is in no way a direct economic or social analysis, links together the notions of evolution, spontaneous order, and the limits to explanation when dealing with complex phenomena.[5] In fact, Hayek stressed the importance of *The Sensory Order* as follows:

> *My colleagues in the social sciences generally find my study on* The Sensory Order: An Inquiry into the Foundations of Theoretical Psychology *(London and Chicago. 1952) uninteresting or indigestible. But the work in it has helped me greatly to clear my mind on much that is very relevant to social theory. My conception of evolution, of spontaneous order and of the methods and limits of our endeavours to explain complex phenomenon have been formed largely in the course of the work on that book.*
>
> Hayek (1979, fn. 26, p. 199)

Hayek's theory is underlined by various philosophical presuppositions.[6] Fleetwood (1995) distinguishes three phases within Hayek's work: the period up to 1936 or Hayek I; 1936–60 or Hayek II; and after 1960 or Hayek III.[7] As an exemplary member of the Austrian school, Hayek was deeply concerned with macroeconomic issues through the relationship between capital in its temporal existence and the operation of markets (Hayek I).[8] In Hayek II, the market is no longer a set of pipes channeling capital through roundabout channels, but rather an information processor, organizing and conveying the appropriate information to the relevant actors. "After 1936 Hayek develops an ontology and epistemology that permits not only events given in sense experience, but also conceptions or ideas held by agents. He now recognizes a further domain of reality, that is, a domain (metaphorically) deeper than, or beneath, the flux of events, namely the conceptions created by agents" (Fleetwood, 1995, p. 6). In Hayek III, he began to appeal to evolution to explain how a complex order could have come about. Hayek's conception of cultural and institutional evolution is a critique against the opposing viewpoint that society can reconstruct institutions to be more rational.

The brain must digest information in a way so as to simplify data and stimuli that would otherwise be infinitely complex and confusing. *The Sensory*

---

5. In the early 1950s, Hayek was led to investigate an ever-broadening range of fields, from biology and evolutionary theory, to systems theory, to cybernetics and theories of communication, all of which offered explanations of the principles underlying the complex phenomena with which they dealt (Caldwell, 2000).

6. Some researchers (Butos and Koppl, 1993; Caldwell, 2000, 2004; Horwitz, 2000; Rizzello, 1999; Streit, 1993; Teraji, 2014) argue that Hayek's cognitive theory spilled over to his later work on political and social theory.

7. According to Fleetwood (1995), prior to 1936 Hayek might be defined as a positivist; between 1936 and 1960 he adopted a synthesis of subjective idealist epistemology and empirical realist ontology. After 1960 he endorsed a position that Fleetwood (1995) calls quasi-critical realist or transcendental realist.

8. The Mises–Hayek, or the Austrian, business cycle theory focuses on the fact that production takes time, and examines the effects of a policy-induced change in interest rates on the allocation of resources in the process of production.

Order distinguishes between the physical order existing in the external world and our phenomenal experience of it. It aims to solve the problem of the relation between the physical order and the phenomenal order. The essence of Hayek's attempt in theoretical psychology is to show how a structure can be formed which discriminates between different physical stimuli, and generates the sensory order that we actually experience. The sensory order is a system of qualities that do not simply represent physical properties of the external world. The subjectivity of individual knowledge finds its foundation in the construction of the mind. The mind operates by assembling new sensory data into associations with our accumulated inventory of knowledge. Knowledge is forged by the connection of new sensory information to previous sensory experiences.

Conventional economics is about choices actually made, not about decision-making processes leading to the choices. In other words, conventional economic models include only variables that condition "what an agent chooses," and none that condition "how an agent chooses." This entails a "black box" view of the individual, meaning that it does not matter analytically how that behavior is actually generated (Gul and Pesendorfer, 2008). It is no doubt to exclude the need for psychological and physiological inquiry from conventional economics. In conventional economic models, any decision-maker can choose the highest ordered of all possible options in a mathematically consistent fashion. These models do not aim to explain the mechanisms by which choices are generated. However, the future is unknowable, though not unimaginable. Human action takes place under uncertainty; different individuals will typically hold divergent expectations about the same future event. As Lachmann (1994) argues:

> *Choice is an activity. A theory that refuses to concern itself with activity but nevertheless proposes to make use of its results must rest on the assumption that what happens during an activity does not matter to its results. It is therefore incompatible with any view which ascribes significance to states of mind and forms of actions.*
>
> Lachmann (1994, p. 224)

Following Hayek's theory of mind, we need to build on the characterization of processes that contribute to decision-making. Hayek's theory of mind sheds light on the process of choice; it describes the human mind as an adaptive classification system by which individual behavior is shaped. Furthermore, as Hayek (1988, p. 22) suggests, "[w]hat we call mind is not something that the individual is born with, as he is born with his brain, or something that the brain produces, but something that his genetic equipment ... helps him to acquire, as he grows up, from his family and adult fellows by absorbing the results of a tradition that is not genetically transmitted." Cross-fertilization between economics and neuropsychology is important in order to interpret the mind as the main source of knowledge. The brain and its cognitive mechanisms are the outcome of an evolutionary process. Human behavior is a result of a complex interaction between genetic forces and environmental forces.

People form expectations of future events by recalling what has followed in the past after the kind of events they now perceive around them. They learn through experiences that many events repeat themselves, according to more or less regular patterns. After the processes of interpreting their perceptions, individuals have to think of alternative courses of actions, and of probable consequences of each of the alternatives. Individuals then try to do the same in a given situation, by acting in a way that has proved to produce satisfying outcomes in similar situations in the past.

From its beginning, the Austrian school of economics has stressed the importance of uncertainty and ignorance in economic decision-making.[9] The Austrian theory is strongly individualistic, and consequentially psychological in nature. The theory is unique in its interest in implications of ideas, beliefs, and cognitive processes. As Boettke and Prychitko (1994, p. 228) argues, "Austrian theory is an analysis of acting minds, of individuals attempting to switch their present state of affairs for imagined better states. This invariably links methodological individualism with the concept of time and genuine uncertainty . . .." The Austrians take seriously the notion of human action in historical time. The Austrians fundamentally agree that economics should make the world intelligible in terms of human action (Vaughn, 1994). Economics must explain human action as the individual's responses to their subjective interpretations of their external environment.

The mind is a complex system, functioning as a decentralized spontaneous order. The set of connections between nerve fibers is a very complex network. These connections govern the organism's capacities for cognitive processes. The essence of Hayek's cognitive theory is the proposition that all of an organism's experience is stored in network-like systems of connections between the neurons of its cerebral cortex. Those connections have been formed by the temporal coincidence of inputs from various sources. Extensive and intricate systems of connections between distant cortical neurons are demonstrated in the brain. Hayek's cognitive theory is relational, in the sense that it is the structure of the connections between neurons that underpin the operation of the mind.

---

9. It is generally agreed that the 1871 publication of *Principles of Economics*, by Carl Menger, gave birth to the Austrian school of economics. Menger's primary contributions in economics include the subjective theory of value, the discovery of the law of marginal utility, the theory of the spontaneous emergence of institutions, and the conception of the production process as a series of successive temporal stages. Despite his contributions to marginal analysis and to value theory based on the concept of utility, Menger had a different methodological approach from the other economists who contributed to the marginal revolution. For instance, he rejected the use of mathematization on the grounds that economic theory is not studying "interdependencies," but "essences."

Following Butos and Koppl (1993), Horwitz (2000), McQuade and Butos (2005), and Wenzel (2010), Hayek's *The Sensory Order* (1952) can be summarized in the following six main points:

1. The overall theory:
Hayek's *The Sensory Order* is about how we perceive, acquire knowledge, and plan our actions. It provides a model of the mind of which working mechanisms are explained through the classification process. The sensory order is a representation of the physical order; it is a classification that takes place via a network of impulse connections. Hayek tried to show how the physiological impulses proceeding in the different parts of the central nervous system can be in such a manner differentiated from each other in their functional significance. The classification makes it possible to perceive and interpret external objects and events. The nervous system translates a set of stimuli in a series of impulses, which are transmitted through networks of connections. Thus:

> *The main aim of the theory presented is to show that the range of mental phenomena such as discrimination, equivalence of response to different stimuli, generalization, transfer, abstraction, and conceptual thought, may all be interpreted as different forms of the same process which we have called classification, and that such classifications can be effected by a network of connexions transmitting nervous impulses. From the fact that this classification is determined by the position of the individual pulse or group of impulses in a complex structure of connexions, extending through a hierarchy of levels, follow certain important conclusions concerning the effects which physiological or anatomical changes must be expected to have on mental functions.*
>
> Hayek (1952, p. 147)

For Hayek, "psychology must start from stimuli defined in physical terms and proceed to show why and how the senses classify similar physical stimuli sometimes as alike and sometimes as different, and why different physical stimuli will sometimes appear as similar and sometimes as different" (Hayek, 1952, pp. 7–8). Hayek recognized that the mind is a general classificatory device. The set of connections creates regularities in the behavior of the organism. The neural order is the connections between nerve fibers in the brain. The essence of Hayek's theory of mind is the proposition that all of an organism's experience is stored in network-like systems of connections between the neurons of its cerebral cortex.

2. Subjectivity:
Hayek's neural classification system is "subjective" rather than "objective." Individuals cannot observe phenomena externally; social reality cannot be explained by an objective process. A significant part of individual knowledge is idiosyncratic, and it is not transferable. That is:

> *If sensory perception must be regarded as an art of classification, what we perceive can never be unique properties of individual objects but always only*

properties which the objects have in common with other objects. Perception is thus always an interpretation, the placing of something into one or several classes of objects.

Hayek (1952, p. 142)

The subjectivity of individual knowledge finds its foundation in Hayek's explanation of the construction of the mind. Subjectivism is an important aspect of the Austrian approach. The subjectivism of knowledge renders the information relevant to economic activity inherently subjective. People are not only exposed to different facts, but also interpret these facts in different ways. What we know about the external world is an interpretation constructed by our cognitive classificatory apparatus. Individuals are fundamentally heterogeneous, and do not behave in the same manner. People process information differently, and they typically have different expectations with respect to many future contingencies. In the Austrian tradition, subjectivism implies that individuals hold different preferences, knowledge, and expectations, and more specifically, "the pre-supposition that the contents of the human mind, and hence decision making, are not rigidly determined by external events" (O'Driscoll and Rizzo, 1996, p. 1). Explanations must take these mental states as an ultimate starting point.

3. Incomplete and imperfect representation of the physical world:
The representation of the physical world is incomplete and imperfect. Therefore, it will give a distorted reproduction. Indeed, as the organism goes through the world and learns about it, the sensory order evolves into a gradual approximation of the physical order. That is:

We have seen that the classification of the stimuli performed by our senses will be based on a system of acquired connexions which reproduce, in a partial and imperfect manner, relations existing between the corresponding physical stimuli. The "model" of the physical world which is thus formed will give only a very distorted reproduction of the relationships existing in that world . . . ; and the classification of these events by our senses will often prove to be false, that is, give rise to expectations which will not be borne out by events.

Hayek (1952, p. 145, original emphasis)

The neural classification system is open to the external environment. The physical order is external to the brain, and it is different from the sensory order.[10] It is the neural order that produces the sensory order of phenomenal experience.

---

10. Simon (1959) also distinguishes the "objective" information which exists "out there" in the agent's environment from the information which "enters" his or her faculties and is processed by them. For Simon, people are information-processing organisms. The information which first enters an agent's faculties is selected from the total set of objective information. This selection process follows from the agent's "perception" and "attention." The information that first enters one's faculties is not necessarily representative of the objective information which is out there.

4. Learning and updating:
The acquisition of new knowledge changes the nervous system. This forms the basis for subsequent rounds of knowledge acquisition:

To acquire the capacity for new sensory discrimination is not merely to learn to do better what we have done before; it means doing something altogether new. It means not merely to discriminate better or more efficiently between two stimuli or groups of stimuli: it means discriminating between stimuli which before were not discriminated at all. If qualities are, as we have maintained, subjective, then, if new discriminations appear for the first time, this means the appearance of a new quality.

Hayek (1952, p. 156)

The updating mechanism entails continuous reclassification processes. The mind creates new classes by modifying or destroying the old ones every time the expectations deriving from a certain classification are disappointed by new experiences. That is:

The more this process leads us away from the immediately given sensory qualities, and the more the elements described in terms of these qualities are replaced by new elements defined in terms of consciously experienced relations, the greater becomes the part of our knowledge which it embodies in the definitions of the elements, and which therefore is necessarily true.

Hayek (1952, p. 170)

Hayek's *The Sensory Order* describes a system that engages in self-organizing activity that reconstructs the classes of objects and relations in response to external stimuli, and the correspondence of such activity with a changing external environment. This explains why we are likely to see the same thing differently on our second encounter with it, especially if this encounter is separated by a long interval from the first. One reason why is that our mind-set has been shaped or reshaped by experience during the interval.

5. Social embeddedness:
The organism's past experience will affect the structure of classification. As the sensory order is shaped by the environment, people interpret any new event in the environment in the light of experience. Cognition is embedded in the external environment. People perceive, interpret, and evaluate the world according to the mental frames that they have developed in interaction with their environment. Hayek's argument provides implications for the limits on what as individuals can know and explain. Individuals are, since their birth, endowed with a cognitive system, allowing them to perceive and give significance to external stimuli of any kind. These innate systems tend to assimilate the external stimuli according to their own classifying principles. An

agent is considered to be in a constant relation of tacit and explicit communication with the external world. Thus:

If the account of the determination of mental qualities which we have given is correct, it would mean that the apparatus by means of which we learn about the external world is itself the product of a kind of experience ... It is shaped by the conditions prevailing in the environment in which we live, and it represents a kind of generic reproduction of the relations between the elements of this environment which we have experienced in the past; and we interpret any new event in the environment in the light of that experience.

Hayek (1952, p. 165)

As Mises ([1949] 1966, p. 36) explains, "human knowledge is conditioned by the structure of the human mind." Therefore, an individual "does not himself create his ideas and standards of value; he borrows them from other people. His ideology is what his environment enjoys upon him" (Mises, [1949] 1966. p. 45). The mind directs and interprets sensory stimuli according to the neural connections which develop during an individual's life.

Since the construction of cognition takes place on the basis of interaction with the external environment, different minds think different things. Therefore, this connects with Hayek's view of localized, distributed knowledge and his view of competition as constituting a discovery process.

6. The map-model structure:

For Hayek (1952), the "map" provides a framework for evaluating impulses; it is an apparatus by means of which we learn about the external world. The map, shaped by experience, represents the individual's past. Thus, the map "is formed by connections capable of transmitting impulses from neuron to neuron" (Hayek, 1952, p. 115). On the other hand, operating within the map is a dynamic and changeable "model." The model is "the pattern of impulses which is traced at any moment within the given network of semi-permanent channels" (Hayek, 1952, p. 114). The mind forms a semi-permanent structure (the map) of understanding, while it forms an immediate, fleeting representation of the immediate environment (the model). The operation of a unified cognitive structure is composed of a mutable, but relatively stable map, and a more fluid model.

The map provides the framework for evaluating the stimuli affecting it at a particular time. The underlying map will gradually evolve to incorporate the information provided by every new experience. Therefore, a given combination of stimuli will seldom lead to the same course of action on any two occasions. We may find at any time that some of our knowledge is contradicted by experience. However, as our knowledge advances, it becomes more and more consistent with experience.

This semi-permanent structure provides the framework within which ... the impulses proceeding at any time are evaluated. It determines what further impulses any given constellation of impulses will set up, and represents the kinds of classes or "qualities" which the system can record, but not what particular events will be recorded at any moment. This structure itself in turn is liable to change as a result of the impulses proceeding in it, but relatively to the constantly changing pattern of impulses it can be recorded as semi-permanent.

Hayek (1952, p. 115, original emphasis)

The model of the physical world operates within the map. The model represents the present-time environment within which the organism is operating.

The pattern of nervous impulses which at any moment will be traced within the structure of connected fibres is, of course, a constantly changing pattern. The representations of the different part of the environment which the impulses produce will derive their significance exclusively from the fact that they tend to evoke certain other impulses. Each impulse representing an event in the environment will be the starting point of many chains of associative processes; in these the various further impulses set up will represent events which in the past have become connected for the individual with those which are represented by the impulses which evoke them.

Hayek (1952, p. 118)

The model represents both the actual state of the environment and its expected changes, in an adaptive process through which the organism can constantly check and correct expectations. As Hayek (1952) argues:

Even on a pre-conscious level the organism must live as much in a world of expectation as in a world of 'fact', and most responses to a given stimulus are probably determined only via fairly complex processes of 'trying out' on the model the effects to be expected form alternative courses of action. The reaction to a stimulus thus frequently implies an anticipation of the consequences to be expected from it.

Hayek (1952, p. 121, original emphasis)

The model predicts the effects of different courses of action and selects among the effects of alternative courses which one is apt to be desirable. Thus, "behavior will be guided by representations of the consequences to be expected from different kinds of behavior" (Hayek, 1952, p. 123).

Experience shapes the map, and the current environment is interpreted by the model. The map is something like a set of implications waiting to happen, and the model pulls out the implications relevant to the current environment from this set. The map—model framework thus corresponds to experience-based behavioral theory. A person must be able to take into account, in one's conscious representations, other experiences of which one is conscious: "'[m]emory' or 'recognition' means here no more than the reappearance in consciousness, in combination with circumstances with which it has become associated, of which has been consciously experienced before" (Hayek, 1952, p. 136, original emphasis). The map–model structure is subject to continuous changes, and can be updated. As a person learns, a new interpretation of reality takes over the old one through reclassification. Therefore, recognition consists in a process of clearing and improvement of one's mental classification apparatus. The human mind is an adaptive system interacting with its environment by performing a multilevel classification on the stimuli it receives. A conscious person is "guided to a large extent not only by his current perceptions but also by images and reproduction of circumstances which might be evoked by the existing situation" (Hayek, 1952, p. 135). The recognition of mental sets does not occur immediately through current experiences; it can only happen through a conscious reconstruction process. Several dimensions of the phenomenon can be distinguished. As a result, the mind produces a highly complex mental order, in which there are a countless number of different classes of events. Individuals have different experiences; these experiences determine their conceptions of the world. These conceptions are subjective, because they link to the history of an individual's confrontations with a changing environment.

6.3   MIND AND INSTITUTIONS

There is a clear analogy between Hayek's cognitive theory and his approach to social theory. As Horwitz (2000, p. 24) argues, "Hayek's theory of mind fits into his economic and social thought." The central topic of Austrian economics should be the epistemology of the economic agent over time. The methodological basis of the Austrian analysis of institutions resides in the subjective point of view. Individuals are heterogeneous, and do not behave in the same manner. The subjectivity of individual knowledge finds its foundation in Hayek's explanation of the construction of the mind. That is, "[w]e want to know the kind of process by which a given physical situation is transformed into a certain phenomenal picture" (Hayek, 1952, p. 7). The whole set of qualities is determined by the system of connections by means of which the impulses can be transmitted from one neuron to another. The perception of the world around us is conjectural, in the sense that it is informed by a set of classificatory dispositions which is itself the product of a kind of accumulated experience. The system of connections is acquired over time, by means of experience. The mind emerges from the complexity of the connections between the physical world and the sensory world.

Hayek's (1952) *The Sensory Order* deals with the interpretation and constraints of knowledge.[11] Constraints on what we can know, and hence on rationality itself, are principal themes in Hayek's work. Each individual only disposes of an incomplete and fragmented representation of the image of his or her own subjectivity. Knowing the world is a classification of sensory qualities by the mind. The human mind is not simply a mirror-like mechanism on which the objects of the external world are reflected.[12] The existence of the sensory order is explained by reference to the relationships between the impulses which are carried by neurons that occupy specific positions within the structure of the nervous system. In his paper "The primacy of the abstract," Hayek ([1969] 2014) argues:

> *I do not wish to deny that in our conscious experiences ... concrete particulars occupy the central place and the abstractions appear to be derived from them. But this subjective experience appears to me to be the source of the error with which I am concerned, the appearance which prevents us from recognizing that these concrete particulars are the product of abstractions which the mind must possess in order that it should be able to experience particular sensations, perceptions, or images. ... What I contend, in short, is that the mind must be capable of performing abstract operations in order to perceive particulars, and that this capacity appears long before we can speak of a conscious awareness of particulars.*

Hayek ([1969] 2014, pp. 315–316)

The order of mental qualities is the result of an act of classification of stimuli which is performed by the nervous system. Neurons tend to be connected depending on the pattern of ongoing neural activity. The process arises in each individual agent through the formation of neural maps, which provide the link between the physical order and the sensory order. The mind is only a classification structure. The nervous system acquires a structure, in which the position of any one neuron is defined by its connections to other nerve fibers. Without some degree of uniformity of minds, there can be no meaningful social interaction. However, one cannot state explicitly the classificatory rules which shape the mind's activity. One knows the external world as a result of some kind of tacit understanding of how the world works. Hayek (1952) argues:

> *A certain part at least of what we know at any moment about the external world is therefore not learnt by sensory experience, but is rather implicit in the means through which we can obtain such experience; it is determined by the*

---

11. From a contemporary Austrian perspective, "knowledge is a multifaceted, heterogeneous, disaggregated, often private or tacit and imperfect phenomenon" (Vaughn, 1994, p. 4).

12. According to Gray (1986), Hayek's theory of knowledge is Kantian. The human mind must possess *a priori* categories, or mental concepts, which allow us to make sense of the external world.

order of the apparatus of classification which has been built up by pre-sensory linkages. What we experience consciously as qualitative attributes of the external events is determined by relations of which we are not consciously aware but which are implicit in these qualitative distinctions, in the sense that they affect all that we do in response to these experiences.

Hayek (1952, p. 167)

There are three parallels between Hayek's cognitive and institutional theories (Wenzel, 2010). First, knowledge is limited. The sensory order is an imperfect representation of the physical order. Second, learning at the cognitive level and the generation of knowledge are important. The mind updates its understanding of the external environment. Third, knowledge has a social dimension; much of our knowledge is embedded in institutions. The mind learns from its environment, and much of our understanding of the world comes through the mind. According to Hayek (1948), the knowledge that a person can possess is dispersed, and each person can just possess a little piece of all the knowledge available in society. The dissemination of knowledge is crucial in society. In Hayek's theory of market economies, the market process is governed by rules that allow necessarily decentralized knowledge to be utilized effectively. Interactions of people in society make adaptation possible. At an individual level, actions are based on subjective perceptions of what exists. However, a correspondence between individual actions and an overall order is inherently problematic. People live in a world of expectations about interactions with others' actions. Institutions facilitate social interaction, since they restrict individual agents concerning their disposition to behave. Institutions reduce the context of action the individual has to behave in. The existence of institutions leads to individual representations of a given context of interaction that are to some degree intersubjectively compatible. In order to act in society, individuals must accept certain rules without consciously thinking about them. Such rules are themselves part of a spontaneous order that is not the product of conscious reason but, nonetheless, facilitates reasoned action. Institutions provide the framework which serves as the base of the formation of cognitive rules. Institutions lead to regularities in human behavior, and serve to coordinate the interaction between individuals. It is meaningful to discuss the social order only when almost all agents share the same perception of existing reality which includes others' actions.

The central challenge of organization is one of adaptation to changing or unforeseen circumstances. Well-performing economic systems are identified based on their particular adaptive qualities under a given set of conditions. When making important decisions, people rarely know either what options are available or their possible consequences. Intelligence, which is a response to uncertainty, allows people to create structures for interpretation and decision-making, through a process which is not reducible to deductive reasoning or information processing. There exists a causal link between the pattern or the regularity of behavior and the set of rules that gives rise to it.[13] Any institution has an observable component through the regularity of behavior. The nature of an institution consists in entities such as rules, beliefs, norms, and organizations. These entities are irreducible to the individual level. They necessarily involve social relations between individuals. Furthermore, institutions generate their own reproduction. That is, the individual's behavior reinforces the set of rules that produced it. Aggregating private knowledge and information, rules provide shared cognition. These rules provide the necessary clues for each individual to form his or her beliefs. It is crucial that individuals have a common understanding of the same situation or action, and they attribute to it the same meaning. Thus, rules provide the cognitive and informational resources necessary for the common understanding.

Institutions have been a central theme of investigation for Austrian economic theorists. Rizzo (2013, pp. 50−61) lists eight highly interrelated themes about Austrian economics: (1) the subjective (yet socially embedded) quality of human decision-making, (2) the individual's perception of the passage of time, (3) the radical uncertainty of expectations, (4) the decentralization of explicit and tacit knowledge in society, (5) the dynamic market processes generated by individual action, (6) the function of the price system in transmitting knowledge, (7) the supplementary role of cultural norms and other cultural products in conveying knowledge, and (8) the spontaneous evolution of social institutions.

The Austrian analysis of institutions resides in the subjective point of view.[14] Our ability to formulate and carry out plans depends on our capacity to discriminate mental events. The subjectivity of individual knowledge finds its foundation in Hayek's (1952) explanation of the construction of the mind. Knowing the external reality is a classification of sensory qualities by the mind. Individuals are fundamentally heterogeneous: they implement plans formulated on the basis of their local knowledge. Hayek (1948) explores the

---

13. Crawford and Ostrom (1995) develop the grammar of institutions, and propose that all institutional statements are coded using syntax. According to them, the general syntax of the grammar of institutions contains five components: "Attribute [A]," "Deontic [D]," "Aim [I]," "Conditions [C]," and "Or else [O]." Rules include all five components (ADICO), norms have four components (ADIC), and shared strategies have only three components (AIC).

14. According to Koppl (2006), certain currents in today's heterodox mainstream represent an "emerging new orthodoxy." The emerging new orthodoxy includes new institutional economics, old institutional economics, post Walrasian economics, constitutional political economy, complexity economics, and Austrian economics. There are five leading characteristics in the emerging new orthodoxy: bounded rationality, rule following, institutions, cognition, and evolution. People are assumed to have only bounded rationality. Hayek's *The Sensory Order* includes a theory of bounded rationality. People are rule followers. Institutions matter. Economists should construe the notion of cognition broadly. *The Sensory Order* should be of special interest to economics. Evolution does not always approximate an optimal solution.

concept of equilibrium from a subjectivist perspective. In equilibrium, different plans which individuals have made for action are mutually compatible. The analysis of coordination of individual actions is the main problem for economists supporting the Austrian tradition. Institutions are unexpected, composite, and organic phenomena of an "invisible-hand" character that originate in social interaction (Langlois, 1986). A market relationship can be seen as the non-intended outcome of repeated exchange. What is purposeful, but not intentional at the level of the individual, is spontaneous at the level of the social whole. For Lachmann (1976), the market process is characterized by inconsistency of individual plans. Inconsistency of individual plans is the direct consequence of the introduction of subjectivism to expectations. Furthermore, for Lachmann (1986), the market is best understood as an ongoing process, impelled by the diversity of aims and the divergence of expectations, in an ever-changing world of unexpected change. There is no *a priori* guarantee that the market process always generates large degrees of plan coordination. Lachmann (1986) contends:

> *To say that economic phenomena cannot be predicted in the sense we expect such an activity from a science is not to say that men are unable to form expectations about the future events in no way prevents us from making forecasts about the success of our actions, forecasts which may of course be falsified by later events. Indeed, the former compels us to undertake the latter. Making such forecasts is a human, not a scientific activity.*
>
> Lachmann (1986, p. 139)

Dealing with uncertainty is crucial to the theory of human action. The individuals' ability to coordinate their plans will deteriorate as the external environment becomes more volatile. The individual agent is always eager to reduce uneasiness arising from uncertainty. A science of human action is a science of the struggle of human beings to understand and cope with uncertainty. Institutions emerge as a result of individual agents attempting to reduce uncertainty. By making the behavior of others more predictable, institutions reduce the amount of information people need to behave effectively in society. Institutions embody knowledge about effective behavior. Institutions are largely not the intended product of human actions. Human beings are creating something they never know. Hayek (1979) argues as follows:

> *Many of the greatest things man has achieved are the result not of consciously directed thought, and still less the product of a deliberately coordinated effort of many individuals, but of a process in which the individual plays a part which he can never fully understand.*
>
> Hayek (1979, p. 150)

The set of formulating concepts used to deal with institutions basically relies on the notion of rule-following behavior. Institutions reflect behavior that is relatively predictable or nonrandomly patterned. Such patterns emerge as a result of following rules; they are systems of rules of conduct. Institutions are roughly regularities of behavior understandable in terms of rules. Institutions are means by which individuals are able to gather sufficient information in order to coordinate plans. Individual's beliefs, expectations, and plans are profoundly shaped by institutions. The convergence of expectations is required for plan coordination. Institutions play a crucial role in enabling individual agents to formulate accurate expectations for the successful coordination of plans. On the other hand, institutions are shared intersubjectively among different individual agents. Institutions are generally known systems of inter-personal rules which order repetitive interaction between individual agents. The individual agents have inter-personal rules, in order to interpret the world and to produce expectations about social interaction. These rules are influenced and shaped by institutions. Institutions thus make individual actions and expectations relatively compatible.

O'Driscoll and Rizzo (1996) indicate that:

> *Rules provide, as it were, safe bounds for behavior in a relatively unbounded world. Institutions are the social crystallization of rule-following behavior or, in other words, the overall pattern of many individuals following a similar rule .... Time and genuine uncertainty promote the following of rules and the development of institutions. The latter, in turn, serve to reduce, but not eliminate, the unboundedness of the economic system by providing the stable patterns of interaction.*
>
> O'Driscoll and Rizzo (1996, p. 6)

In social interaction, the parties potentially experience a range of different motivations. They have uncertainty which of these motivations is in fact experienced. For example, this is true of tipping in restaurants, where the customer's motivations and the waiter's ones are coordinated for mutual benefit. A tip is paid only after the service is delivered. Then, for the customer, one motivation is to reciprocate after satisfactory service, and another one is to free-ride on the waiter's effort and tips nothing. The former possibility will motivate a waiter to provide good service, and the latter will motive him or her to under-deliver. In such situations, a norm (such that one should, and usually does, leave a tip) enables the involved parties to coordinate on a specific mode of interaction (the waiter provides good service and the customer tips). However, norms are context-sensitive. In North America, tipping generally applies in restaurants. The fact that service is especially good in a restaurant may cue customers in North America to give a generous tip to the waiter. The same fact does not generally cue the same behavior, for example, in Japan. This makes it difficult for individuals to meet one another's expectations in all contexts.

North (2005) interprets institutions as closely connected to Hayek's theory of mind. Institutional studies generally emphasize both formal and

informal aspects of institutions; the former include laws, contracts, administrative rules, and procedures, while the latter include norms and routines.
From the 1990s, North linked together history, economics, sociology, and
cognitive science, and highlighted how belief systems constitute and regulate
the actions of individuals.[15] Here, the cognitive processes that underpin both
formal and informal aspects of institutions are explored. For Denzau and
North (1994) and North (2005), individuals construct mental models to interpret and produce expectations about their social environment. That is,
Hayek's mental "map" is the "mental model" as used by Denzau and North
(1994) and North (2005) to justify the importance of institutions. Mental
models are defined as the "subjective perceptions (models, theories) all people possess to explain the world around them" (North, 1990, p. 23).
Subjectivity in human interpretation of reality is a key factor in understanding individuals' actions. In order to avoid confusion, we use "mental models"
to refer to Hayek's "map." All experiences are organized from particular
points of view. Due to the variety of mental models, there are multiple possible framings of any given situation. There are different consequences
depending on the way people frame the situation. Mental models might be
hypothetical constructs of a certain set of experiences, through which individuals process information. People attempt to cope with over-complexity with
the help of mental models. The ability to construct such models develops
with the accumulation of knowledge.

Institutions are the sets of rules that allow a plurality of individuals to
coordinate their behavior, and to routinely solve typical problems that arise
in social interactions. Because of the existence of mutually-imposed constraints on individual actions, coordination implies a considerable degree of
order. The effects of institutions are largely connected to individual behavior
through language, tradition, custom, and so on. The fundamental institutions
like language, tradition, and custom are more constituents than products of
reason. Reason itself is traditional; the stock of traditions is "the result of a
process of winnowing and sifting, directed by the differential advantages
gained by groups from practices adopted for some unknown and perhaps
purely accidental reasons" (Hayek, 1979, p. 15). As Denzau and North
(1994) argue, "[t]he cultural heritage provides a means of reducing the divergence in the mental models that people in a society have and also constitutes
a means for the intergenerational transfer of unifying perceptions" (Denzau
and North, 1994, p. 15). Shared belief systems, or common interpretations,
emerge among individual agents in the population. Such belief systems are
considered as the result of individuals' attempts to cope with complexity,
with the help of simplifying mental models. Institutions facilitate social
interaction itself, since they restrict the individual agents concerning their

---

15. North's early work describes institutions as designed and efficient solutions to the coordination problems faced by individuals (North and Davis, 1971).

dispositions to behave. The constraints in the perception of alternatives can be expected to result in some similarities of individual choices. Each agent has little motivation to deviate from such similarities, as long as the consequences of similar behavior do not systemically diverge. The resulting belief systems constrain the repertoire of possible reactions to changes in the external environment.

North (2005) focuses on the evolution of belief systems that individual agents hold. Individual agents perceive "human landscape," interpret it, discover problems, and solve them. There is a link between institutions (macrolevel) and individual agents (micro-level). North (2005) summarizes this relationship as follows:

> There is an intimate relationship between belief systems and the institutional framework. Belief systems embody the internal representation of the human landscape. Institutions are structure that humans impose on that landscape in order to produce the desired outcome. Belief systems therefore are the internal representation and institutions the external manifestation of that representation.
>
> North (2005, p. 49)

Institutions structure individual agents. That is, ideas and ideologies shape the mental models that individuals use to interpret the world around them and make choices. In this sense, individual preferences are endogenous. How do institutions and belief systems coevolve? Regarding the nature of the coevolutionary mechanism, North (2005) suggests:

> This story of the Soviet Union is a story of perceived reality → beliefs → institutions → policies → altered perceived reality and so on. The keys to the story are the ways beliefs are altered by feedback from changed perceived reality as a consequence of the policies enacted, the adaptive efficiency of the institutional matrix—how responsive it is to alteration when outcomes deviate from intentions—and the limitations of changes in the formal rules as correctives to perceived failure.
>
> North (2005, p. 4)

Humans have imperfect understandings of their physical environment. Their beliefs are subjective and diverse. North (2005) argues that institutions reduce uncertainty. Here, institutions are the constraints that individual agents impose on their interaction, consisting of the formal rules of the game, informal norms, and the way they are enforced. Institutions are behavioral patterns or regularities. Institutions are constructed according to individuals' beliefs. Their beliefs are, on the other hand, changeable; individual agents confront novel experiences and resolve positively novel questions. For North (2005), the evolution of a society's institutions is, above all, a function of changes in the prevalent belief systems. That is, "the beliefs that humans hold determine the choices they make that, in turn, structure the changes in the human landscape" (North, 2005, p. 23). Institutions would stay stable, as long as the goal of coordination is achieved. However, institutions can change. Changes in individuals' beliefs account for institutional changes. Then, interactions between individual agents emerge into new behavioral patterns or regularities.

Cognitive science is considered a fundamental tool in order to understand decision-making processes and social relations. Learning makes possible the conversion of mental models. Individuals can confirm, modify, or completely abandon their mental models. This adjustment process takes place through learning. Collective learning arises from the interaction between individuals, which makes possible the sharing of mental models. Mental models are considered a crucial factor in institutional change.

6.4   COMPLEXITY AND SOCIETY

In his paper "The theory of complex phenomena," Hayek ([1964] 2014) considers the degree of complexity as follows:

> *The distinction between simplicity and complexity raises considerable philosophical difficulties when applied to statements. But there seems to exist a fairly easy and adequate way to measure the degree of complexity of different kinds of abstract patterns. The minimum number of elements of which an instance of the pattern must consist in order to exhibit all the characteristic attributes of the class of patterns in question appears to provide an unambiguous criterion.*
>
> Hayek ([1964] 2014, p. 260)

For Hayek, the emergence of patterns results from an increase in the number of elements between which simple relations exist. A particular order of events is something different from all the individual events taken separately; its properties emerge from the peculiar connection of its individual parts. We can consider any kind of phenomenon from the angle of the minimum number of distinct variables that a formula must possess in order to reproduce the characteristic patterns of structures. Hayek ([1964] 2014) argues:

> *It is, indeed, surprising how simple in these items, i.e., in terms of the number of distinct variables, appear all the laws of physics, and particularly of mechanics, when we look through a collection of formulas expressing them. On the other hand, even such relatively simple constituents of biological phenomena as feedback (or cybernetic) systems, in which a certain combination of physical structures produces an overall structure possessing distinct characteristic properties, require for their description something much more elaborate than anything describing the general laws of mechanics. In fact, when we ask ourselves by what criteria we single out certain phenomena as "mechanical"*

or "physical," we shall probably find that these laws are simple in the sense defined. Nonphysical phenomena are more complex because we call physical what can be described by relatively simple formulas.

Hayek ([1964] 2014, p. 261, original emphasis)

Kauffman (1993) considers a model of complexity based on two parameters: the number of components comprising a whole, N; and the degree of interdependence among these components, K. Each component contributes to the performance measure, or the "fitness," of the system, based not only on its own value, but on the values of K other components. Kauffman's NK model allows for a very general description of any system consisting of N components with K interactions between the components, in which there can be any number of states for each N. The NK model studies genomes with the property that the fitness of a genome depends on the interaction between the genes. In this setting, a genome is a system, and the gene positions are the components. The fitness of the resulting genome is a function of how the selected genes interact. In our context, complexity can be defined as the number of components of a certain piece of knowledge, and the degree of interdependence among them.[16] At each point in time, knowledge is scattered among a myriad of learning agents. By means of dynamic coordination, missing links between key complementary modules of knowledge can be built. Complexity provides a general context in which the generation of knowledge can be viewed as a collective process undertaken by a myriad of interacting agents. Systems where interacting individuals are better able to achieve dynamic coordination are likely to experience faster rates of enhancement of knowledge.

In Hayek's *The Sensory Order*, the human mind is a part of the natural and social environment. Then, social reality cannot be explained by an objective, intellectual process. Social institutions are implicitly open systems comprising a large number of interrelated elements. How do social scientists drive an understanding of society and social institutions? In his *Counter-Revolution of Science* ([1952] 1979), Hayek's answer is by means of a "compositive" method. The study of complex phenomena is distinguished from the study of simple systems. This distinction explains why the methods of physics cannot be used to study social sciences. Hayek ([1952] 1979) principally explores the pitfalls of scientific approaches in which methods suitable for the study of phenomena in physics are erroneously thought to also apply to social phenomena. Social sciences must discover "principles of structural coherence of the complex phenomena which had not been (and perhaps could not be) established by direct observation" (Hayek, [1952] 1979, p. 65). The concept of a scientific law that is valid for simple

---

16. There are two broad categories, namely, "computational complexity" and "dynamic complexity" (Rosser, 2012). While the former involves considering a system from the standpoint of computability, the latter is defined as occurring in dynamical systems that do not converge to a point, a limit cycle, or a smooth expansion or contraction endogenously.

phenomena, or a definite rule which links two events as cause and effect, is not applicable to complex phenomena. Social sciences study complex phenomena, while physics studies simple phenomena. The social scientific investigations focus on intentional human behavior and its unintentional consequences. The data of social sciences are the opinions of those who are involved in any action. That is, "[n]ot only man's action towards external objects but also all the relations between man and all the social institutions can be understood only by what men think about them" (Hayek, [1952] 1979, p. 57). Hayek ([1952] 1979) describes his attitude towards making social sciences irreducible to natural sciences or "scientism." Two arguments are advanced against scientism: the methods of physical sciences are not appropriate for the study of society as the facts of social sciences have a subjective nature, and social facts are too complex to be able to be measurable.

Mises ([1949] 1966) also points out the dangers of social sciences in imitating natural sciences too closely. That is:

Natural science ... leaves unpredictable two spheres: that of insufficiently known natural phenomena and that of human acts of choice. Our ignorance with regard to these two spheres taints all human actions with uncertainty .... When dealing with a social actor that chooses and acts and that we are at a loss to use the methods of the natural sciences for answering the question is why he acts this way and not otherwise.

Mises ([1949] 1966, p. 105)

For Mises ([1949] 1966), humans are not atoms, but acting beings. Taking this line of argument, Hayek ([1952] 1979) builds up his own ideas around the concept of complexity. The characteristic properties of complex phenomena can be exhibited only by models made up of a relatively large number of variables (about which we can get limited quantitative data). The advance of the physical sciences took place in fields where it proved that explanation and prediction could be based on laws which accounted for the observed phenomena as functions of comparatively few variables. However, when dealing with social phenomena, we have to build a theory of essentially complex phenomena. In his essay "The pretence of knowledge" ([1975] 2004), as Hayek describes, in the case of complex phenomena "our capacity to predict will be confined to such general characteristics of the events to be expected and not include the capacity of predicting particular individual events" (Hayek, [1975] 2014, p. 371). This recognition of the limits to our knowledge leaves the problem of action in future-oriented thinking open.

Hayek's (1952) *The Sensory Order* explains how physical states of the brain give rise to sensory perception. The collective action of individual neurons can carry out a highly complex hierarchical classification function. In the mental realm, what people know about the external world is an interpretation constructed by our cognitive classificatory apparatus. By recasting the problem of perceptual representation in terms of classification, Hayek (1952) suggests a specific framework of neural processing that accounts for human's subjective experience. *The Sensory Order* establishes a cognitive foundation for the methodological argument against scientism. Hayek (1952) argues:

> *The proposition which we shall attempt to establish is that any apparatus of classification must possess a structure of a higher degree of complexity than is possessed by the objects which it classifies; and that, therefore, the capacity of any explaining agent must be limited to objects with a structure possessing a degree of complexity lower than its own.*
>
> Hayek (1952, p. 185)

Thus, "[t]he qualities which we attribute to the experienced objects are strictly speaking not properties of that object at all, but a set of relations by which our brain classifies them" (Hayek, 1952, p. 143). These attributed qualities may therefore incorporate distortions which can lead to error. For Hayek, knowledge is an abstract representation of external reality, not a reflection of any intrinsic properties of that reality. Hayek (1952) understands the mind as a hierarchical order that produces knowledge. Knowledge is something produced by the mind's classificatory operation.

Concerning knowledge and rules, Hayek (1978) describes as follows: "[w]hat we call knowledge is primarily a system of rules of action assisted and modified by rules indicating equivalences and differences of various combinations of stimuli" (Hayek, 1978, p. 41). Individuals will not be aware of the resulting overall outcome of their actions. They are at most aware of the immediate effects of following a particular rule or deviating from it. Rules of conduct are adaptations to our ignorance of changing environments. Rule-following is a device for coping with the complexity of the environment in which we have to act on the one hand, and with the limits of our reason on the other.

Rules do not prescribe concrete actions; they are tacitly understood and unconsciously followed. Rules govern our perceptions and, particularly, our perceptions of others' actions. In *The Counter-Revolution of Science*, Hayek states that "[t]here is a great deal of knowledge which we never consciously know implicit in the knowledge of which we are aware, knowledge which yet constantly serves us in our actions, though we can hardly be said to 'possess' it" (Hayek, [1952] 1979, p. 217, original emphasis).

Hayek often refers to these rules of conduct as "abstract." The abstract character of the rules individuals follow is linked with the abstract nature of the resulting order. Hayek (1973) distinguishes between abstract orders and concrete ones. Regarding abstract orders:

> *We cannot see, or otherwise intuitively perceive, this order of meaningful actions, but are only able mentally to reconstruct it by tracing the relations that exist between the elements. We shall describe this feature by saying that it is an abstract and not a concrete order.*
>
> Hayek (1973, p. 38)

Thus, in his paper "The primacy of the abstract" ([1969] 2014), Hayek describes:

[W]e ought to regard what we call mind as a system of abstract rules of action (each "rule" defining a class of actions) which determines each action by a combination of several such rules; while every appearance of a new rule (or abstraction) constitutes a change in that system, something which its own operations cannot produce but which is brought about by extraneous factors.

Hayek ([1969] 2014, p. 322, original emphasis)

By contrast, regarding concrete orders:

Such orders are relatively simple or at least necessarily confined to such moderate degrees of complexity as the maker can still survey: they are usually concrete in the sense just mentioned that their existence can be intuitively perceived by inspection; and, finally, having been made deliberately, they invariably do (or at one time did) serve a purpose of the maker.

Hayek (1973, p. 38, original emphasis)

Individuals follow a rule whenever they behave responding to the established norms in their environment. Neural structures are reinforced according to how successful they are in promoting behavior that is well adapted to the environment. Rules of just conduct are generally prohibition of unjust conduct. Norms help mutual coordination between individual agents. In this context, behavioral patterns facilitate interaction by making the actions of any agent predictable to others. If, at the level of cognitive apparatus, the existing classification system generates disappointed expectations, the mind will rearrange sensory experiences into new configurations that allow better predictions to be made about the real world. All knowledge incorporated in rules is situational, and becomes obsolete in the course of time. The adaptation of individual knowledge to changing environments represents an interpretative process. Information is processed in the layers comprising the neural net. Learning consists of reinforcing some of the connections between the nodes and extinguishing others. A system is explained by both its constituent elements, and the connections by which they are related. A market is a system of social interaction characterized by a set of rules defining certain restrictions on the behavior of the market participants.

Simon (1957) introduces the notion of bounded rationality that questions the behavioral assumption of maximizing in economic theory. Because of their limited cognitive capacity, decision-makers do not seek the optimal alternative, but make a decision once they have found an alternative that satisfies their aspiration level. In Simon's (1957) notion of bounded rationality, the limitations to optimizing behavior derive from cognitive limitations that can, in principle, be overcome. Therefore, Simon's critique of economic rationality does not reject the notion of objective rationality itself: agents want to achieve a goal that optimizes their utility, but do not know the best ways to apply for realizing this goal. Under uncertainty, agents most often follow socially legitimated behavioral patterns, rather than the principle of optimizing behavior. Cognitive processes are not only quantitatively limited, but are socially influenced. By relying on rules, choices become informed by social context and the choice set of individuals is limited, thereby reducing uncertainty.

Individuals build up an understanding of the world based on their images of that world. The construction of personal knowledge depends both on the subjective perception of external stimuli, and on the interaction with the social environment. What the specific elements are, and how they are connected, is knowledge itself. Knowledge can be a system of connections that is also changing. A society can be viewed as a complex structure of rules that have evolved over a long period of time. Spontaneous orders are entirely based on abstract or general rules of conduct. The process by which new rules are adopted and diffused in society constitutes the driving force of economic evolution. Hayek's expression "twin conceptions of evolution and the spontaneous formation of an order" (Hayek, 1973, p. 23) refers to a close connection between an evolutionary process and a spontaneous order. If people follow certain rules of conduct, their separate actions will produce an overall order as an unintended result. Individuals themselves may not be able to articulate the rules they follow. The rules of conduct which contribute to the spontaneous formation of an order are themselves an unintended product of evolutionary processes. Using socially transmitted information, people can make predictions about the intent of others, preferentially assort with others who have similar or complementary intentions, and reap the advantages of coordinated activities. For Hayek, human cultural institutions are not rationally designed. Various cultural institutions are expected to embody the experience of generations; they are the "product of long experimentation in the past" (Hayek, 1978, p. 136).

6.5 SUMMARY

Friedrich A. Hayek (1952) provides a theory of the process by which the mind perceives the world around it. The essence of Hayek's theory of mind is the proposition that all of an organism's experience is stored in network-like systems (maps) of connections between the neurons of its cerebral cortex. Hayek's theory of mind shows how a structure discriminates between different physical stimuli, and generates the sensory order that we actually experience. What we know at any moment about the world is determined by the order of the apparatus of classification which has been built up by previous sensory linkages. Reason is an adaptation to the natural and social environments in which people live. Hayek's theory of mind explains how different pieces of cognitive information cause different perceptions and, therefore, different actions. The subjectivity of individual knowledge finds its foundation in the construction of the mind. The mind is an adaptive system, interacting with and adapting to its environment by performing a multi-level classification on the stimuli it receives from the environment. Different minds will map the world differently, such that their knowledge of the world is inevitably subjective and dispersed.

There are parallels between Hayek's mind and institutional theories. First, knowledge is limited. The sensory order is an imperfect representation of the physical order. Second, learning at the cognitive level and the generation of knowledge are important. The mind updates its understanding of the environment. Third, knowledge has a social dimension; much of our knowledge is embedded in institutions. The mind learns from its environment, and much of our understanding of the world comes through the mind.

Thus, subjectivity in human interpretation of reality is a key factor in understanding individuals' actions. All experiences are organized from particular points of view. There are different consequences depending on the way people frame the situation.

REFERENCES

Bernheim, B.D., Rangel, A., 2004. Addiction and cue-triggered decision processes. Am. Econ. Rev. 94, 1558–1590.

Boettke, P., Prychitko, D. (Eds.), 1994. The Market Process: Essays in Contemporary Austrian Economics. Edward Elgar, Aldershot.

Butos, W.N., Koppl, R.G., 1993. Hayekian expectations: theory and empirical applications. Const. Polit. Econ. 4, 303–329.

Caldwell, B., 2000. The emergence of Hayek's ideas on cultural evolution. Rev. Austrian Econ. 13, 5–22.

Caldwell, B., 2004. Hayek's Challenge: An Intellectual Biography of F. A. Hayek. University of Chicago Press, Chicago.

Camerer, C.F., Loewenstein, G., Prelec, D., 2004. Neuroeconomics: why economics needs brains. Scand. J. Econ. 106, 555–579.

Camerer, C.F., Loewenstein, G., Prelec, D., 2005. Neuroeonomics: how neuroscience can inform economics. J. Econ. Lit. 43, 9–64.

Cosmides, L., Tooby, J., 1992. Cognitive adaptations for social exchange. In: Barkow, J., Cosmides, L., Tooby, J. (Eds.), The Adapted Mind. Oxford University Press, Oxford, pp. 163–228.

Crawford, S., Ostrom, E., 1995. A grammar of institutions. Am. Polit. Sci. Rev. 89, 582–600.

Denzau, A.T., North, D.C., 1994. Shared mental models: ideologies and institutions. Kyklos 47, 3–31.

Edelman, G.M., 1989. The Remembered Present: A Biological Theory of Consciousness. Basic Books, New York.

Fleetwood, S., 1995. Hayek's Political Economy: The Socio-Economics of Order. Routledge, Lindon.

Fodor, J.A., 1983. The Modularity of Mind. Cambridge, MA: The MIT Press.

Fuster, J.M., 2005. Cortex and Mind: Unifying Cognition. Oxford University Press, Oxford.

Gray, J., 1986. Hayek on Liberty. Martin Robertson, Oxford.

Gul, F., Pesendorfer, W., 2008. The case for mindless economics. In: Caplin, A., Schotter, A. (Eds.), The Foundations of Positive and Normative Economics. Oxford University Press, Oxford, pp. 3–39.

Hayek, F.A., 1948. Individualism and Economic Order. University of Chicago Press, Chicago.

Hayek, F.A., 1952. The Sensory Order: An Inquiry Into the Foundations of Theoretical Psychology. University of Chicago Press, Chicago.

Hayek, F.A., [1952] 1979. The Counter-Revolution of Science, 2nd ed. Indianapolis, IN: Liberty Press.

Hayek, F.A., 1973. Law, Legislation, and Liberty, Vol. 1, Rules and Order. University of Chicago Press, Chicago.

Hayek, F.A., 1978. New Studies in Philosophy, Politics, Economics and the History of Ideas. Routledge, London.

Hayek, F.A., 1979. Law, Legislation, and Liberty, Vol. 3, The Political Order of a Free People. University of Chicago Press, Chicago.

Hayek, F.A., 1988. The Fatal Conceit: The Errors of Socialism. University of Chicago Press, Chicago.

Hayek, F.A., 1994. In: Kresge, S., Wenar, L. (Eds.), Hayek on Hayek: An Autobiographical Dialogue. University of Chicago Press, Chicago.

Hayek, F.A., 2014. In: Caldwell, B. (Ed.), The Collected Works of F. A. Hayek, Vol. 15, The Markets and Other Orders. University of Chicago Press, Chicago.

Hebb, D.O., 1949. The Organization of Behavior: A Neuropsychological Theory. Wiley, New York.

Horwitz, S., 2000. From The Sensory Order to the liberal order: Hayek's non-rationalist liberalism. Rev. Austrian Econ. 13, 23–40.

Johnson-Laird, P.N., 1983. Mental Models: Towards a Cognitive Science of Language, Inference, and Consciousness. Harvard University Press, Cambridge, MA.

Kauffman, S.A., 1993. The Origins of Order: Self-Organization and Selection in Evolution. Oxford University Press, Oxford.

Koppl, R., 2006. Austrian economics at the cutting edge. Rev. Austrian Econ. 19, 231–241.

Lachmann, L.M., 1976. From Mises to Shackle: an essay on Austrian economics and the Kaleidic society. J. Econ. Lit. 14, 54–62.

Lachmann, L.M., 1986. The Market as an Economic Process. Basil Blackwell, Oxford.

Lachmann, L.M., 1994. Expectations and the Meaning of Institutions: Essays in Economics. Routledge, London.

Langlois, R.N., 1986. Rationality, institutions, and explanation. In: Langlois, R.N. (Ed.), Economics as a Process: Essays in the New Institutional Economics. Cambridge University Press, Cambridge, MA.

Loewenstein, G., O'Donoghue, T., 2004. Animal Spirit: Affective and Deliberative Process in Economic Behavior. Working Paper, Carnegie Mellon University, Pittsburgh, PA.

McQuade, T., Butos, W., 2005. The sensory order and other adaptive classifying systems. J. Bioeconomics 7, 335–358.

Mises, L., [1949] 1966. Human Action: A Treatise on Economics. Chicago: Henry Regnery.

North, D.C., 1990. Institutions, Institutional Change and Economic Performance. Cambridge University Press, Cambridge, MA.

North, D.C., 2005. Understanding the Process of Economic Change. Princeton University Press, Princeton, NJ.

North, D.C., Davis, L., 1971. Institutional Change and American Economic Growth. Cambridge University Press, Cambridge, MA.

O'Driscoll Jr., G.P., Rizzo, M.J., 1996. The Economics of Time and Ignorance, 2nd ed. Routledge, London.

Rizzello, S., 1999. The Economics of the Mind. Edward Elgar, Cheltenham.

Rizzo, M.J., 2013. Foundations of The Economics of Time and Ignorance. Rev. Austrian Econ. 26, 45–52.

Rosser Jr., J.B., 2012. Emergence and complexity in Austrian economics. J. Econ. Behav. Org. 81, 122–128.

Simon, H.A., 1957. Models of Man: Social and Rational. Wiley, New York.

Simon, H.A., 1959. Theories of decision-making in economics and behavioral science. Am. Econ. Rev. 49, 253–283.

Streit, M.E., 1993. Cognition, competition, and catallaxy: in memory of Friedrich August von Hayek. Const. Polit. Econ. 4, 223–262.

Teraji, S., 2014. On cognition and cultural evolution. Mind Soc. 13, 167–182.

Vaughn, K.I., 1994. Austrian Economics in America: The Migration of a Tradition. Cambridge University Press, Cambridge, MA.

Wenzel, N., 2010. From contract to mental models: constitutional culture as a fact of the social science. Rev. Austrian Econ. 23, 55–78.

FURTHER READING

Hayek, F.A., 1967. Studies in Philosophy, Politics, and Economics. Routledge & Kagan Paul, London & Henley.

Hodgson, G.M., 1997. The ubiquity of habits and rules. Cambridge. J. Econ. 21, 663–684.

Chapter 7

Society and Knowledge

7.1 INTRODUCTION

Friedrich Hayek's thought largely rests on the concept of spontaneous order. Spontaneous orders in human affairs are patterns that arise as the unintended consequences of individual actions. Individuals must coordinate their actions with those of other persons. A mutual adjustment process of individual actions makes it possible to realize such an order.

According to Hayek (1973, p. 36), the term "order" refers to "a state of affairs in which a multiplicity of elements of various kinds are so related to each other that we may learn from our acquaintance with some spatial or temporal part of the whole to form correct expectations concerning the rest, or at least expectations which have a good chance of proving correct."

Many rules that govern our social interactions are emergent. A social system can be viewed as a massively complex structure of rules that have evolved over a long period of time. For a society to exist, rules must be habitually obeyed. For Hayek, rules are abstract principles that serve to "guide" individual behavior. Rules are behavioral patterns that individuals expect each other to follow. Rules coordinate and motivate interdependent behavior. Relying on rules is a device we have learned to use, because our reason is insufficient to master the details of complex reality. Some rules simplify and standardize mental models to operate in complex systems. Therefore, "what we refer to as knowledge is mainly a system for rules of action supported and modified by rules indicating similarities and differences between combinations of stimuli" (Hayek, 1978, p. 41). Hayek's concept of perception as classification has a counterpart in his concepts of rules and rule-following behavior. The mind itself is a particular order of a set of events taking place in some manner related to, but not identical with, the physical order of events in the external environment. For Hayek, rules make it possible for individuals to classify stimuli. The order of a group can be generated by the rules of conduct adhered to by its members. Cooperation can only proceed on the basis of common acceptance of rules that themselves have not been rationally constructed.

An equilibrium situation is considered as one in which individual action plans are fully coordinated. Individual knowledge, expectations, and the actions based on these, in the equilibrium situation, are mutually consistent.

The Cognitive Basis of Institutions. DOI: https://doi.org/10.1016/B978-0-12-812023-1.00007-7

Any change in the relevant knowledge leads individuals to alter their action plans, and it disrupts the equilibrium relations between their actions taken before and those taken after the change. Hayek (1948) argues a tendency toward equilibrium:

> In the light of our analysis of the meaning of a state of equilibrium it should be easy to say what is the real content of the assertion that a tendency toward equilibrium exists. ... In this form the assertion of the existence of a tendency toward equilibrium is clearly an empirical proposition, that is, an assertion about what happens in the real world ... The only trouble is that we are still pretty much in the dark about (a) the conditions under which this tendency is supposed to exist and (b) the nature of the process by which individual knowledge is changed.

(Hayek, 1948, pp. 44–45)

Events in society are the results of actions. Outcomes are understood in terms of a multitude of actions, related and unrelated. According to Hayek, our beliefs are themselves the product of a process of evolution. A system is explained by both its constituent elements and the connections by which they are related. In evolutionary dynamics, connections are continually changing and the recombinant process of connections may generate novelties.

Game theory provides an account of what behavior to expect in a given strategic interaction, that is, in an interaction where the player's resulting payoff, whatever that payoff is defined to include, is a function not only of one's own strategy, but of those of others as well. Provided that each player knows and understands the nature of the game and behaves rationally, the player's behavior is likely to conform to a Nash equilibrium of that game. That equilibrium outcome is a conjunction of strategies such that no player can improve his or her expected payoff by unilaterally switching to a different strategy. Using this equilibrium concept and the description of the nature of the underlying environment, game theory furnishes a prediction of what behavior and outcome to expect from that situation.[1]

Two approaches—epistemic and evolutionary—come together in coordination problems. An epistemic approach, grounded on individual beliefs and reasoning, explores a procedural individual rationality. An evolutionary approach, grounded on interaction networks and adaptation processes, explores self-organizational systems and emergent structures.

---

1. Quantal response equilibrium (QRE) extends the notion of Nash equilibrium to allow bounded rationality (McKelvey and Palfrey, 1995). It relaxes the assumption of best response and considers errors in choices, keeping the assumption of (statistically) accurate beliefs and equilibrium responses. According to QRE models, players are more likely to select better choices than worse choices, but they do not necessarily succeed in selecting the very best choice.

In an epistemic approach, Nash equilibrium is justified on the basis of pure introspection: deductive reasoning on the part of players produces the knowledge of equilibrium strategies that are identified and played. Refinements of the Nash equilibrium concept provide an attempt to identify a unique plausible outcome in cases of multiplicity. A common prior expectation about how the game should be played must be agreed on before play begins. In an evolutionary approach, on the other hand, the interaction between players is repeated over time. Instead of inferring players' behavior from an equilibrium notion, some plausible rules of behavior are postulated. For example, in Maynard Smith (1982), a symmetrical game for two players is played repeatedly by the members of some large population. Pairs of players are formed by independent random draws from the population. For each member of the population, the expected utility of playing a given strategy depends on the frequency with which the alternative strategies are being played in the population as a whole. It is assumed that, if one strategy yields a higher expected utility than another, the frequency with which the first strategy is played will tend to increase relative to that of the second. Evolutionary game theory formalizes the evolutionary arguments by assuming that more successful behavior tends to be more prevalent.

7.2   DISPERSED KNOWLEDGE

Knowledge varies by time and place. To the degree that people never occupy the same time and place, knowledge varies by subject. According to Hayek (1948), the knowledge that an individual can possess is dispersed, and each person can just possess a little piece of all the knowledge available in society.[2] Knowledge includes both scientific knowledge (of general rules) and particular local knowledge (of circumstances of time and place). Both production and exchange depend on particular local knowledge. Particular knowledge is held by other individuals as well as environmental circumstances. This is in contrast to the notion that the relevant kind of knowledge is scientific, and thus easily transmitted and used. Hayek's (1948) idea about the division of knowledge is articulated as follows:

> Today it is almost heresy to suggest that scientific knowledge is not the sum of all knowledge. But a little reflection will show that there is beyond question a body of very important but unorganized knowledge which cannot possibly be called scientific in the sense of knowledge of general rules: the knowledge of the particular circumstances of time and place.
>
> (Hayek, 1948, p. 80)

---

2. In 1936, Hayek became president of the London Economic Club, and delivered an address entitled "Economics and Knowledge" on November 10 of that year (Caldwell, 2014).

The "dispersed knowledge" problem presents a major challenge to society. The economic problem in society is, in fact, a problem of the utilization of knowledge not given to anyone in its totality. Relevant knowledge is never given or possessed in its totality by any one person. At any given time, the knowledge of any particular individual in society will be, at best, only partially consistent with both the knowledge held by other individuals and the facts of the external environment. Consumer preferences are continually changing, new modes of production are to be discovered, and tacit knowledge particular to time and place are dispersed among people. The economic problem is how to utilize the dispersed knowledge, while allowing for the meeting of mutual expectations of various heterogeneous agents. Economic agents, because of diverse experiences, may respond to the same objectively-defined stimulus in different ways. The problem of coordination in the economic process is associated with understanding how other individuals act.

How do we explain the development and coordination of the knowledge scattered among people? The problem is, in Kirzner's (1992, p. 147) words, "that of generating flows of information or of signals that might somehow stimulate the revision of initially uncoordinated decisions in the direction of greater mutual coordinatedness." A large part of the problem of coordination, in both production and exchange, involves making the best use of dispersed knowledge. Hayek (1948) argues that it is impossible to coordinate dispersed knowledge effectively on the basis of command. Hayek (1948) formulates this problem in the following way: "[t]he peculiar character of the problem of a rational economic order is determined precisely by the fact that the knowledge of the circumstances of which we must use never exists in concentrated or integrated form but solely as the dispersed bits of incomplete and frequently contradictory knowledge which all the separate individuals possess" (Hayek, 1948, p. 77). Knowledge is ever-changing, while information is something fixed (Boettke, 2002). Knowledge implies the capacity to process information, while information refers to measurable data. Knowledge is not considered as a static stock of information, but rather as the flow of new and ever expanding areas of the known. Human actions are based on connections between those actions and their effects. A high degree of coordination is observed in real life, which reflects the pivotal role of prices as both signals and incentives. The central question of economics as a social science is "how the spontaneous interaction of a number of people, each possessing only bits of knowledge, bring about a state of affairs in which prices correspond to costs, etc., and which could be brought about deliberately only by somebody who possessed the combined knowledge of all those individuals" (Hayek, 1948, pp. 50–51). Of course, some rules and accompanying governance institutions have been established through deliberate design, rather than evolving spontaneously. However, every society possesses an order without it having been deliberately created.

According to Mises ([1949] 1966):

The market phenomena are social phenomena. They are the resultant of each individual's active contribution.

(Mises, [1949] 1966, p. 315)

For Mises, "[a]ction is an attempt to substitute a more satisfactory state of affairs for a less satisfactory one. We call such a willfully induced alteration an exchange" (Mises, [1949] 1966, p. 37). The fact that each individual forgoes one good or service in exchange for another shows that the good or service acquired ranks higher on his or her value scale than the good or service foregone. Individuals are purposeful, as they use means to attain ends by way of exchange with others. Exchange is undertaken in the expectation of having something in return which renders two individuals tied to each other in a social relation. The study of purposeful human action is, for Mises, the key to understanding the market process. The state of equilibrium represents the perfect coordination of plans, as long as no external shock is introduced. However, the market can be grasped from a different angle that is less structural in character than exchange is. Mises ([1949] 1966, p. 245) argues the idea of a movement toward equilibrium: "[t]he final state of rest is an imaginary construction, not a description of reality. ... New disturbing factors will emerge before it will be realized. What makes it necessary to take recourse to this imaginary construction is the fact that the market at every instant is moving toward a final state of rest." As some "disequilibrium" is dissolved by an action, another one emerges. The dispersed knowledge, coordinated via the pursuit of individuals' intentions to act, guides the market process in the direction of a social order with relative coherence of prices and costs.[3] This is the argument of Hayek (1948): all economic problems are knowledge problems, arising out of uncertainty. Hayek supported the Mises' argument, and further elaborated on it, pointing out additional problems such as dispersed and tacit knowledge. The market mechanism functions to discover useful information, and distribute it widely.

What is striking in the process of market interaction is the interactive activity of a myriad of agents. Observed outcomes at the market level are often unintended consequences of the interaction of various actions that have been taken on the basis of subjectively held preferences, knowledge, and expectations. Agents, who individually possess only a tiny fraction of the total stock of knowledge available to society, generate an aggregate level of knowledge superior to that from any centrally devised planning mechanism.

---

3. During the 1930s, Hayek and Mises took part in what has become known as the "socialist calculation debates" over whether a nation's economy could be centrally planned without a market mechanism to generate accurate prices. According to them, a market system is superior to a planned economy, because the former can better discover, communicate, and use the fragmented and dispersed bits of information. Even if a planned economy could calculate, it would not be able to adapt with the speed and precision of the unplanned market economy. The burden of central administration makes a real, workable socialist economy strictly impossible.

We must take into account not only the mental states of the relevant individuals, but also the differences among these mental states. The market is understood as an ongoing process, impelled by the diversity of aims and resources, and the divergence of expectations. The market equilibrium can be viewed as the outcome of the interaction between several minds functioning independently from each other. Price movement signals changes to which individuals are motivated to adapt, even in the absence of any knowledge as to their causes. The signals sent to individuals in the market process lead to learning and the spread of information. There are as many subjective forms of knowledge as there are individual heterogeneous agents. When some individuals implement their plans formulated on the basis of their local knowledge, their actions generate changes in the relative prices which summarize the significance of the knowledge for the scarcity of various kinds of resources. And those price changes enable other individuals to adjust their own plans, without knowing anything about the details. Prices summarize existing market conditions, and, in that capacity, ensure that individuals do not have to engage in costly information collecting activities. Thus, the information provided by market prices enables people to form reasonably accurate expectations of others' plans. Through the market price, a coordinated utilization of resources based on the dispersed knowledge becomes possible.

A more fundamental problem with regard to knowledge relates to the creation of novelty. Novelty is fundamentally unpredictable. Since knowledge varies by time and place, an omniscient knowledge is impossible. For Hayek (1978), the term "competition" refers to a discovery procedure that approaches novelty by means of trial-and-error elimination.[4] The market system provides individuals with useful knowledge, and enables them to discover such knowledge by means of the competition process. The Austrian school of thought views the market as a process which is never really in equilibrium, but is always tending toward equilibrium. The market process tends toward equilibrium as coordination. Disequilibrium exists because economic agents are ignorant, and are often ignorant of their ignorance. The equilibrating tendencies reside within the profit incentives that originate in the prevalence of unexploited market opportunities.[5] The opportunities are found not just by

---

4. Targets are moving rather than fixed. While people are gaining additional knowledge by learning from earlier mistakes, some of their knowledge is, at the same time, becoming obsolete.
5. "Without economic calculation," Mises (1935, p. 105) describes, "there can be no economy. Hence, in a socialist state wherein the pursuit of economic calculation is impossible, there can be—in our sense of the term—no economy whatsoever." It is unclear how a socialist economy could be expected to achieve equilibrium. Movement towards equilibrium is the result of competition among rivals constantly adjusting their prices to exploit perceived gaps in the market. Public ownership would inevitably stifle this competitive process, and retard the movement towards equilibrium. Without such competition in a system of exchange, there is no way of knowing the opportunity costs of employing resources in particular uses. In the absence of a mechanism to reveal opportunity costs, there is no way of arriving at an allocation which puts resources to their most productive uses.

searching, which implies that the searcher knows what he or she is looking for, but more importantly by discovery. Kirzner (1973) argues that the existence of disequilibrium situations in the market implies profit opportunities. The role of entrepreneurs lies in "their alertness to hitherto unnoticed opportunities" (Kirzner, 1973, p. 39). The existence of ignorance creates a role for alertness. Local knowledge is an understanding of conditions unique to a particular time and place. Alertness implies that the economic agent possesses a superior perception of opportunity. While all agents possess some aspect of alertness, they cannot be alert to the same opportunities with the same level of proficiency in alertness. People with local knowledge will be alert to particular types of profit opportunities. As Kirzner (1979, p. 12) suggests, "[i]t would be good to know more about the institutional settings that are most conductive to opportunity discovery." This differential recognition of opportunities for profit raises important issues in the entrepreneurial market process. The market economy opens up arbitrage possibilities, because of the ignorance of individuals. Opportunities are conceptualized as pure arbitrage ones, which arises because the actors in the market do not have full information. Through their alertness, entrepreneurs can discover and exploit opportunities in which they can sell for high prices that which they can buy for low prices.

Individuals form subjective expectations about market data. As the market process unfolds, expectations change. Littlechild (1986) suggests the identification of ideal models of the market process based on "how the decision makers perceive of the world, how these perceptions change over time, how these additional information may be sought, and how the decision maker can limit his exposure to uncertainty" (Littlechild, 1986, p. 27). In Lachmann (1986), the evolution of the markets links to an economic phenomenon of price reduction generated by the standardization of products, which reduces the time necessary for price negotiation and excludes consumers from the production process. Due to this process, consumers become price-takers, while entrepreneurs become price-makers. It is possible to specify the markets according to their level of organization, their methods of negotiation, and the more or less flexible character of the prices.

For Kirzner (1997), "[m]ainstream microeconomics interprets the real world of markets as if observed phenomena represent the fulfillment of equilibrium conditions" (Kirzner, 1997, p. 63). In microeconomic models, individuals, who are ignorant about certain things, precisely know the extent of their ignorance. They can take steps through search activities and remedy this ignorance. However, individual agents are actually ignorant about what they are ignorant about. According to Kirzner (1979), ignorance of opportunities is associated with two types of knowledge: deliberated acquisition and non-deliberated acquisition. The former can be gained and learnt by deliberated search, while the latter can only be spontaneously absorbed from everyday life experiences. Experiences from everyday life are accumulated into a stock of knowledge that can be used to interpret incoming events. This sort of knowledge is largely "the result of learning experiences that occurred entirely without having been planned nor are they deliberately searched for" (Kirzner, 1979, p. 142).

Kirzner (1997) argues:

> *For Hayek the equilibrating process is thus one during which market participants acquire better mutual information concerning the plans being made by fellow market participants. For Mises this process is driven by the daring, imaginative, speculative actions of entrepreneurs who see opportunities for pure profit in the conditions of disequilibrium. What permits us to recognize that these two perspectives on the character of the market process are mutually reinforcing, is the place which each of these two writers assigns to competition in the market process. The Austrian approach includes a concept of competition which differs drastically from that encapsulated in the label "competition" as used in modern neoclassical theory.*

(Kirzner, 1997, p. 68, original emphasis)

The dissemination of knowledge is crucial in society. People differ in significant aspects, because of the uniqueness of their individual experiences. Interactions between people in society make adaptation possible. If equilibrium is to be useful, it must explain how people come to know enough to carry out mutually consistent plans. Much of Hayek's later work concentrated on how institutions other than the price system (such as the rules of conduct) furthered the societal tendency toward coordination of plans. Dispersed knowledge required for coordination can be facilitated not only by price signals, but also by a set of intersubjectively shared rules. In his article "The political ideal of the rule of law," Hayek ([1955] 2014, p. 160) argues that "if a multitude of individual elements obey certain general laws, this may of course produce a definite order of the whole mass without the interference of an outside force." A set of individuals possesses knowledge that no single one possesses (Foss and Foss, 2006). In economic models, we ought to keep carefully apart what the observing economist knows and what the agents whose actions are under examination are supposed to know. Much of economic importance depends on what people know. Each agent's information structure can be represented by an information partition. If an agent cannot distinguish one state from others, those states may be grouped into the same information set of his or her partition. Individuals' information partitions may be various and intersecting with others in intricate ways. Individuals learn the information that is dispersed in their environments. Social coordination requires institutional structures that encourage the use of dispersed knowledge. Human action is purposeful in the sense that individuals attempt to reach their goals. Mises ([1949] 1966) mentions that, due to the ever-presence of market uncertainty, there must be a substantial element of speculation in human action. At an individual level, actions are based on subjective perceptions of what exists.[6] However, a correspondence between

---

6. Lachmann (1976) fully embraces the notion of radical subjectivity, and argues that every economic actor will have a unique plan or a set of expectations about the future value of resources.

individual actions and an overall order is inherently problematic. We live in a world of expectations about interactions with others' actions. Uncertainty arising from others' actions generates economic problems. The success of a plan depends on the extent to which the plan is adapted to the actions of other agents. Because of their diverse experiences, individual agents will respond differently to the same stimulus. There is no guarantee that the process of plan revision will always result in greater overall plan coordination. That is, "different men in identical situations may act differently because of their different expectations of the future" (Lachmann, 1970, p. 36). A modern economy therefore requires the cooperation of a number of individual agents. Individuals are required to act in accordance with the same guidelines about how to act in various situations, which makes it possible for them to form accurate expectations of others' actions. Hayek's rejection of the possibility of socialist calculation and central planning is based on the inability of the collectivist economy to make the spontaneous adjustments that occur in a market economy. The coordination of individual efforts in society is not the product of deliberate planning, but has been brought about by means which nobody wanted or understood. It is meaningful to discuss social order only when all agents share the perception of existing reality which includes others' actions.

A Note on Common Knowledge

Common knowledge requires that each agent knows an event, each agent knows that the other agents know it, each agent knows that the other agents know that the other agents know it, and so on (Aumann, 1976). We consider a formal model of knowledge.[7] According to Aumann (1976), knowledge refers to events which are subsets of a set of states of the world. Let Ω be a set of possible states of the world. The set Ω is assumed to be finite. A state $\omega \in \Omega$ specifies every relevant aspect of the environment. One of these states is typically referred to as the "true state," while the other states correspond to possible alternative specifications of the environment. For each state $\omega \in \Omega$, there is a set of states $h_i(\omega)$ that agent i considers possible when the true state is ω. If $h_i(\omega)$ is a singleton, agent i knows the state of the world, but otherwise he or she is uncertain as to which of states in $h_i(\omega)$ is the true state. The knowledge of agent i is represented by an information partition H_i of the set Ω. Agent i does not know which is the true state, but knows only which element of an information partition H_i of Ω contains the true state.

Definition 7.1: An information partition H_i is a collection $\{h_i(\omega) \mid \omega \in \Omega\}$ of disjoint subsets of Ω such that:

(P1) $\omega \in h_i(\omega)$,
(P2) If $\omega' \in h_i(\omega)$, then $h_i(\omega') = h_i(\omega)$.

---

7. See, for example, Osborne and Rubinstein (1994).

Knowledge is defined through an exogenous information partition on Ω. We can think of $h_i(\omega)$ as the knowledge of agent i if the state is in fact ω. If the state of the world is ω, agent i's knowledge is given by the element $h_i(\omega)$ of the information partition H_i that contains ω. The finer is agent i's partition H_i, the more precise is his or her knowledge. Property P1 ensures that the true state ω is an element of agent i's information set (or knowledge). Property P2 says that if ω' is also deemed possible, then the set of states that would be deemed possible at ω' must be the same as those currently deemed possible at ω.

Example: Suppose $\Omega = \{\omega_1, \omega_2, \omega_3, \omega_4\}$ and that the information partition of agent i is $\{\{\omega_1, \omega_2\}, \{\omega_3\}, \{\omega_4\}\}$. Then $h_i(\omega_3) = \{\omega_3\}$, while $h_i(\omega_1) = \{\omega_1, \omega_2\}$.

We refer to a set of states $E \subset \Omega$ as an "event." Every state that an agent considers possible is constrained in E. We say that, at state ω, agent i knows event E if, and only if, $h_i(\omega) \subset E$. We define an agent's knowledge function K.

Definition 7.2: For any event $E \subset \Omega$, we have:

$$K(E) = \{\omega \in \Omega | h_i(\omega) \subset E\}.$$

The set $K(E)$ is the collection of all states in which agent i knows E. And K can be viewed as a function that associates, with any subset E of Ω, the set of states $K(E)$ in which the agent knows the event E. In the above example, suppose $E = \{\omega_3\}$. Then $K(E) = \{\omega_3\}$. Similarly, $K(\{\omega_3, \omega_4\}) = \{\omega_3, \omega_4\}$ and $K(\{\omega_1, \omega_3\}) = \{\omega_3\}$.

Because $K(E)$ is an event, the set $K(K(E))$ is also defined, and is interpreted as "the agent knows that he or she knows E." There are five axioms of knowledge. These describe what we know, and what we know about what we know.

First, for every state $\omega \in \Omega$, we have $h_i(\omega) \subset \Omega$. Therefore:

$$(K1) \quad K(\Omega) = \Omega.$$

This is called the axiom of awareness. That is, regardless of the actual state, the agent knows that he or she is in some state of the world. Equivalently, the agent can identify the set of possible states. The axiom of awareness excludes the existence of states of which we are not aware.

A second property of knowledge functions is that:

$$(K2) \quad K(E) \cap K(F) = K(E \cap F).$$

This property says that knowing that both event E and event F have happened is equivalent to knowing that the event "intersection of E and F" has occurred.

If $h_i (\omega)$ satisfies Property P1, the knowledge function also satisfies a third property:

$$(K3) \quad K(E) \subseteq E.$$

This property says that if the agent knows E, then E must have occurred. This is referred to as the axiom of knowledge.

Next, we cannot know something without knowing that we know it:

$$(K4) \quad K(E) \subseteq K(K(E)).$$

Property K4 says that if agent i knows E, then i knows that i knows E. This is called the axiom of transparency.

A fifth property states that we know what we do not know:

$$(K5) \quad \Omega \ K(E) \subseteq K(\Omega \ K(E)).$$

That is, if agent i does not know E, then i knows that i does not know E. This is referred to as the axiom of wisdom.

For simplicity, we consider two agents to define common knowledge.

Definition 7.3: Let K_1 and K_2 be the knowledge functions of agent 1 and agent 2, respectively. An event $E \subset \Omega$ is common knowledge between agent 1and agent 2 in the state $\omega \in \Omega$ if ω is a member of every set in the infinite sequence $K_1(E)$, $K_2 (E)$, $K_1(K_2 (E))$, $K_2 (K_1(E))$, and so on.

This definition implies that agent 1 and agent 2 know E, they know the other agent knows it, and so on.[8] A definition of common knowledge can be also stated in terms of information function. We say that an event $F \subset \Omega$ is "self-evident" between both agents if, for all $\omega \in F$, we have $h_i (\omega) \subset F$.

Definition 7.4: An event $E \subset \Omega$ is common knowledge between both agents in the state $\omega \in \Omega$ if there is a self-evident event F for which $\omega \in F$.

Theorem 7.1: *The two definitions of common knowledge, Definition 7.3 and Definition 7.4, are equivalent.*

Proof: Assume that E is common knowledge at ω according to Definition 7.3. Then, $E \supset K_i (E) \supset K_j (K_i (E)) \supset \ldots$ and ω is a member of each of these sets. Since Ω is finite, the infinite regression must eventually produce a set F such that $K_i (F) = F$. Therefore, F is self-evident. Next, assume that E is common knowledge at ω, according to Definition 7.4. Then, $F \subset E$,

---

8. On the other hand, mutual knowledge does not require that each agent knows that the other agents know it, and so on. There is mutual knowledge of a fact if each agent knows the fact. We say that an event $E \subset \Omega$ is mutual knowledge between both agents in the state $\omega \in \Omega$ if it is known to all agents, i.e., if $\omega \in K_1(E) \cap K_2 (E)$.

$K_i (F) = F \subset K_i (E)$. Iterating this argument, F is a member of every of the regressive subsets $K_i (K_j (\ldots E \ldots))$. Since $\omega \in F (\subset E)$, E is common knowledge by Definition 7.3. $\square$

Aumann (1976) uses the notion of common knowledge to prove that two individuals who share the same priors cannot "agree to disagree" in the following sense. If i and j start from common priors and update the probability of an event E (using Bayes' rule) on the basis of private information, then it cannot be common knowledge between them that i assigns probability η_i to E, and j assigns η_j to E with $\eta_i \neq \eta_j$. This is so even though they may base their posteriors on quite different information.

A probability measure p on Ω is the agent's prior belief. Agents i and j have the same priors; this is referred to as the assumption of common priors. For any state ω and event E, let $p(E| h_i (\omega))$ denote i's posterior belief. The belief of agent i, $p(E| h_i (\omega))$, is obtained by Bayes' rule. The event that "i assigns probability η_i to E" is $\{\omega \in \Omega \mid p(E| h_i (\omega)) = \eta_i\}$. Agents i and j may have different information, i.e., $h_i (\omega) \neq h_j (\omega)$.

Theorem 7.2: *Suppose that two agents, i and j, have the same prior beliefs over a finite set of states Ω. If it is common knowledge in some state $\omega \in \Omega$ that agent i assigns probability η_i to some event E, and agent j assigns probability η_j to E, then $\eta_i = \eta_j$.*

Proof: For two agents, there is some self-evident event F with $\omega \in F$ such that:

$$F \subset \{\omega \in \Omega \big| p(E\big|h_i(\omega)) = \eta_i\} \cap \{\omega \in \Omega \big| p(E\big|h_j(\omega)) = \eta_j\}.$$

Furthermore, F is a union of members of each information partition. Since Ω is finite, so is the number of sets in each union. Let $F = \cup_k A_k = \cup_k B_k$. For any nonempty disjoint sets C, D with $p(E| C) = \eta_i$, and $p(E \mid D) = \eta_i$, we have $p(E| C \cup D) = \eta_i$. Since for each k, $p(E| A_k) = \eta_i$, then $p(E \mid F) = \eta_i$. Similarly, $p(E| F) = p(E| B_k) = \eta_j$. $\square$

7.3 RULES OF CONDUCT

Hayek (1948) distinguishes "true" and "false" individualism. False individualism derives from Cartesian rationalism. Following false individualism, social processes must be subject to conscious control, if they are to serve human ends. Adopting the rationalist conception of individuals makes it difficult to understand spontaneous order. On the other hand, true individualism originates in British philosophy, particularly Scottish Enlightenment thought. Following true individualism, institutions are outcomes not of design, but of spontaneous conduct of individuals. Hayek's individualism stands in stark contrast to atomistic approaches which isolate an individual from society. We have Hayek's conception of the individual voluntarily submitting oneself to social rules. Certain economic phenomena arise from, but are irreducible to, individuals' actions. The spontaneous interplay of individual actions leads to the formation of institutions. As the social nature of persons is understood, the matter ultimately reduces to the structure of the social order itself. Hayek's discussions of spontaneous order in social life can be summarized as the following themes (Sugden, 1989):

1. Unintended consequences: Spontaneous orders are the unintended social consequences of individual actions, but not the execution of any human design.
2. Rule-following: Spontaneous orders are formed when individuals follow abstract rules of conduct. However, the individuals themselves may not be able to articulate the rules they follow. Agents follow rules unconsciously as if, in effect, programmed to do so. Rules of conduct typically "manifest themselves in a regularity of action which can be explicitly described, but this regularity of action is not the result of the acting persons being capable of thus stating them" (Hayek, 1973, p. 19). The desire to follow rules is part of a very basic fear of the unknown.
3. General predictability: The "'general" properties of a spontaneous order are predictable. The formed patterns have general characteristics that are highly predictable.
4. Specific unpredictability: The detailed features of a spontaneous order are unpredictable. Not knowing specific circumstances, individuals cannot predict exactly how other individuals will act. The problem is not merely the "quantity"' of knowledge that would be required in order to make accurate predictions, but also the "kind" of knowledge. Our knowledge of particular circumstances may come down to our unconsciously following rules that are adapted to the particularities of our own lives.
5. Division of knowledge: Each individual makes use of his or her specific knowledge in deciding how to act. Therefore, spontaneous orders embody a totality of knowledge that is not known to any single mind. A division of knowledge allows spontaneous orders to achieve feats of social organization that are beyond the capability of any planner.

Individuals make use of their acquired knowledge to draw up frameworks on how the diverse elements are connected in social systems. Frameworks, or mental models, enable individuals to anticipate the sequences and consequences of their actions in a context of uncertainty. Then, complexity provides a general context in which the generation of knowledge can be viewed as a collective process undertaken by a myriad of interacting individuals. In such a system of interactions, the outcome is unpredictable for each agent. The theory of spontaneous order rests on the assumption that individual knowledge is limited, incomplete, and imperfect in nature. Lack of knowledge as a feature of human decision-making renders it impossible for the agent to understand the contingencies and implications of one's own actions in entirety.

Hayek's general approach stresses the limited role of government and the importance of private institutions. For a liberal society to function, its members must be, in some measure, guided by common values or standards of judgment. Hayek viewed civilization as a direct product of certain habits, customs, and traditions. The power of foresight is always highly imperfect. In his book *The Constitution of Liberty*, Hayek points out the important role of ignorance in the advancement of civilization:

> *[T]he knowledge which any individual mind consciously manipulates is only a small part of the knowledge which at any one time contributes to the success of his action.*
>
> (Hayek, 1960, p. 24)

A necessary and sufficient condition for liberty to apply is the rule of law. Liberty means "[a] state in which a man is not subject to coercion by the arbitrary will of another or others" (Hayek, 1960, p. 11) or "that what we may do is not dependent on the approval of any person or authority and is limited only by the same abstract rules that apply equally to all" (Hayek, 1960, p. 155). For Hayek, choice within an impartial and impersonal framework provides the only means of sustaining individual freedom in society. The rule of law will be better upheld in an undesigned manner over time than through legislation. In this state, each person can use one's own knowledge for one's own purposes. In Hayek (1948), his argument is that "nobody can know *who* knows best, and that the only way by which we can find out is through a social process in which everybody is allowed to try and see what he can do" (Hayek, 1948, p. 15, original emphasis). Hayek (1960) defends individual freedom and the free market system on the basis of coping with ignorance. That is, "the case for individual freedom rests chiefly on the recognition of the inevitable ignorance of all of us concerning a great many of the factors on which the achievement of our ends and welfare depends ... It is because every individual knows so little and, in particular, because we rarely know which of us knows best that we trust the independent and competitive efforts of many to induce the emergence of what we shall want when we see it" (Hayek, 1960, p. 29).

Fleetwood (1995) describes the relationship between ignorance and coordination as follows:

> *Any notion of order that is more than a formal description of the conditions necessary for equilibrium must explain how agents initiate actions that are relatively spatio-temporally coordinated with one another under the really existing situation of incomplete knowledge ... Hayek reasons that actions may*

be coordinated if plans are coordinated, which depends upon the coordinations
of expectations, which in turn is based upon agents having access to knowledge
... of what others are doing or intend to do.

(Fleetwood, 1995, p. 94)

Rules, therefore, help the spontaneous creation of order through the coordination of plans and actions of individual agents. Hayek's concept of perception as classification has a counterpart in his concepts of rules and rule-following behavior. As Hayek ([1965] 2014, p. 49) notes, "[w]hat I have said about the need of abstract rules for the co-ordination of the successive actions of any man's life in ever new and unforeseen circumstances applies even more to the co-ordination of the actions of many different individuals in concrete circumstances which are known only partially to each individual and become known to him only as they arise." The mind is made up of abstract schemata that are a kind of "habitus," dispositions to think and to act in accordance with rules.[9] We can understand the actions of others who are equipped with similar, but not identical, systems for producing action patterns:

The question which is of central importance as much for social theory as for
social policy is thus what properties the rules must possess so that the separate
actions of the individuals will produce an overall order. Some such rules all
individuals of a society will obey because of the similar manner in which their
environment represents itself to their minds. Others they will follow spontane-
ously because they will be part of their common cultural tradition. But there will
be still others which they may have to be made to obey, since, although it would
be in the interest of each to disregard them, the overall order on which the suc-
cess of their actions depends will arise only if these rules are generally followed.

(Hayek, 1973, p. 45)

According to Hayek (1973), one of the main characteristics of human behavior consists of following rules of conduct. The word "rules" is sometimes used broadly to refer to any kind of directive for decision-making, or any behavioral regularity. Like prices, rules coordinate and motivate interdependent behavior. By emphasizing the importance of rules, as well as prices, Hayek's thought develops an interdisciplinary approach, as opposed to a narrow, economic one, to the explanation of coordination.[10] The rules of

---

9. The habitus of each agent is a product of individual history. For agents, the actions tend to be framed by their past experience and their current situation. Agents develop a practical sense of orientation that guides them in their actions. Therefore, each agent is able to follow rules without referring to them.

10. According to Caldwell (2014, p. 35), "he [F. A. Hayek] first encountered the idea while thinking about how a market system might work to coordinate human action in a world in which knowledge was both dispersed and subjectively held. But as he came to realize, such orders are ubiquitous: they can be found in nature, and they are integral for understanding the development of many of our most important social, cultural, and economic institutions."

conduct define the range of possible interactions in the various areas of human action, provide the individuals with the resources for understanding and communication, and, in the end, bring about social coordination. The rules of conduct thus constitute social practices. The rules of conduct which govern our actions are adaptations to our ignorance of the external environment in which we have to act. Following the rules of conduct helps people make decisions with some degree of certainty about which behavior is acceptable and which is not. In his essay "Kinds of rationalism," Hayek ([1965] 2014, p. 48) argues that "[t]he only manner in which we can in fact give our lives some order is to adopt certain abstract rules or principles for guidance, and then strictly adhere to the rules we have adopted in our dealing with the new situations as they arise." Rule-following behavior appears to be particularly pertinent for describing decision-making in uncertain and rapidly changing contexts. Rules are a device for coping with our ignorance of the effects of particular actions.

Rules of conduct are shared by individuals having a common cultural tradition. "This matching of the intentions and expectations that determine the actions of different individuals is the form in which order manifests itself in social life" (Hayek, 1973, p. 36). The rules one individual is expected to follow influence the choices made by other individuals. If people have widely divergent expectations, some of their actions will invariably fail and need to be revised. Culture limits the range of actions that people are likely to take in a particular situation, making their conduct more predictable, and thereby facilitating the formation of reliable expectations. Shared mental models can give rise to behavioral regularities, to the extent that they can be observed in the population. As a consequence, following the rules of conduct mutually reinforces sets of expectations to maintain a degree of social order. Behavioral patterns emerge endogenously, reflecting a socially constructed reality. Given the human need for rules, there is a tendency to repeat those patterns as a guideline for action in future instances of similar behavior.[11] Individuals do not appear as isolated entities; they live in a context of traditional rules, which constitutes the ground for common understanding and shared meaning. While specific predictions about complex phenomena cannot be made, a theory derived from observing broad patterns of behavior under certain conditions may enable us to expect those patterns to recur.

Abstraction is a device that allows our minds to cope with literally limitless amounts of information. The way one knows the reality is based on the

---

11. O'Driscoll and Rizzo (1996) propose the concept of "pattern coordination," based on the distinction between typical and unique aspects of future events. While typical events are the ones that an observer perceives as being repeated regularly, unique events are the ones that occur only once. According to pattern coordination analysis, if the market is able to coordinate typical events, it is no more the case when unique characteristics of human actions are taken into account.

mind's abstraction. Abstraction occurs whenever an individual responds in the same manner to circumstances that have only some features in common. We understand others when we share with them common abstract rules of understanding. As rules become more abstract, they allow for increased coordination and reduced conflict. Adopting rules that abstract from specifics and focus on a few relevant factors allows us to make predictions for entire classes. For those classes, we can make reasonably confident predictions about correct behavior. As Hayek (1973, p. 29) puts it, "abstract concepts are a means to cope with the complexity of the concrete which our mind is not fully capable of mastering." People deal with the complexity by filtering it, deeming some feats relevant, while ignoring others. Therefore:

> We never act, and could never act, in full consideration of all the facts of a particular situation, but always by singling out as relevant only some aspects of it; not by conscious choice or deliberate selection, but by a mechanism over which we do not exercise deliberate control.
>
> (Hayek, 1973, p. 30)

From human interactions in a given community, behavioral patterns emerge unintentionally. Although people need the rules of conduct, the rules are not necessarily used as a means towards an end. The distinction between the abstract rules in a spontaneous order and the specific rules in an organization is a fundamental one for Hayek. For the specific rules in an organization:

> What distinguishes the rules which will govern action within an organization is that they must be rules for the performance of assigned tasks. They presuppose that the place of each individual in a fixed structure is determined by command and that the rules each individual must obey depend on the place which he has been assigned and on the particular ends which have been indicated for him by the commanding authority.
>
> (Hayek, 1973, p. 49)

On the other hand, for the abstract rules in a spontaneous order:

> [T]he rules governing a spontaneous order must be independent of purpose and be the same, if not necessarily for all members, at least for whole classes of members not individually designated by name. They must ... be rules applicable to an unknown and indeterminable number of persons and instances. They will have to be applied by the individuals in the light of their respective knowledge and purposes; and their application will be independent of any common purpose, which the individual need not even know.
>
> (Hayek, 1973, p. 50)

For Hayek, the rules of organization differ significantly from the sort of rules that may lead to the formation of spontaneous orders. The imposition of rules that did not emerge spontaneously may very well reflect the domination of a specific set of interests enforced through the coercive capacity of organizations. However, spontaneous orders can emerge out of systems of deliberately constructed rules. That is, within the rules of organization set by the commanding authority, the interactions between its members might establish new rules for the coordination of their activities. These rules will be spontaneously evolved rules of organization. They can be viewed as routines, built up and improved on within the firm (Nelson and Winter, 1982). When discussing "the need for an evolutionary theory," Nelson and Winter (1982, p. 36) observe that their "... basic critique of orthodoxy is connected with the bounded rationality problem." Routines are the regular and predictable behavioral patterns of firms. Firm behavior is determined by simple decision rules. Routines are considered to be reflective of historically given decisions and behaviors that have come to govern the actions of a firm. Cognitively, these routines act as a mechanism to coordinate the skills of employees (routines as organizational memory). Addressing the interplay between changing external environments and changing routines, Nelson and Winter (1982) try to bring bounded rationality together with tacit knowledge in the notion of routine.

Human knowledge is at once fundamentally practical and fundamentally abstract. It is embodied in the abstract schemata that compose the mind and manifests itself through the rules that guide individual actions, quite often without realizing it. It is knowledge made up of "know-how," as opposed to a propositional knowledge, one which "know that." Tacit knowledge is the part of our knowing that we are unable to communicate to others. Tacit knowledge refers to a non-deductive and non-scientific kind of knowledge. Even the most highly formalized knowledge invariably follows from an intuition, which is simply a manifestation of tacit knowledge. The knowledge need not be articulable.[12] The reason is that it may take too much time to articulate. For example, people can speak and write grammatically correct English, without being able to articulate the rules of English grammar. That is, we can know more than we can tell.[13] Hayek (1973) asks whether inarticulate rules always guide our mental activity, relating tacit knowledge and practical learning of rules:

> *So long as individuals act in accordance with rules it is not necessary that they*
> *be consciously aware of the rules. It is enough they know how to act in*

---

12. This is illustrated by Hayek's own experience: "I have always regarded myself as a living refutation of the contention that all thinking takes place in word or generally in language. I am as certain as I can be that I have often been aware of having the answer to a problem—of "seeing" it before me, long before I could express it in words. Indeed, a sort of visual imagination, of symbolic abstract patterns rather than representational pictures, probably played a bigger role in my mental processes than words" (Hayek, 1994, pp. 134–135).
13. According to Polanyi (1969), "tacit knowing is the fundamental power of the mind, which creates explicit knowledge, lends meaning to it and controls its use" (Polanyi, 1969, p. 156).

accordance with the rules without knowing that the rules are such and such in
articulated terms.

(Hayek, 1973, p. 99)

Rules of conduct are largely tacit. Respecting these rules, to a great
extent, conditions the coordination of individuals' actions. However, the
individuals are not capable of explaining these rules. Hayek (1967) argues
that tacit rules must exist for regularity of the conduct of individuals to
exist:

[T]he term rule is used for a statement by which a regularity of the conduct
of individuals can be described, irrespective of whether such a rule is known
to the individuals in any other sense than that they normally act in accor-
dance with it.

(Hayek, 1967, p. 67)

Furthermore, Hayek (1967) points out that rules often limit the range of
possibilities within which the choice is made consciously:

By eliminating certain kinds of action altogether and providing certain routine
ways of achieving the object, they merely restrict the alternatives on which a
conscious choice is required. ... [T]he rules which guide an individual's
action are better seen as determining what he will not do rather than what he
will do.

(Hayek, 1967, p. 56)

Habit can be understood as a form of social embeddedness of economic agents.[14] By acting on the basis of habits, individuals make their
behavior predictable for other people. The limits of human cognitive
capacity can be seen as the source of uncertainty in the economic process. In the face of uncertainty, each agent has to make decisions.
Uncertainty brings the rules of social life that individuals rely on to make
decisions. The problem of uncertainty provides a starting point from
which it becomes possible to understand why economic behavior is rule-
driven.

Solving social problems is limited by insuperable complexity. The model
of cognitive limits in human beings is grounded on the theory of the abstractness of the human mind. Our cognitive processes follow abstract schemata.
The notion of the abstractness of the human mind is combined with that of
rule-governed behavior. Though people are able to use social rules, they are
not necessarily able to spell them out. Coordination can only proceed on the
basis of common acceptance of rules that themselves have not been

---

14. Granovetter (1985) pioneered the concept of embeddedness, emphasizing the potentially beneficial effects of group life on economic development. Communities are said to have such beneficial effects by mitigating asymmetries in information.

rationally constructed. Hayek (1978) explains the role of abstraction in the emergence of new behavior in the following terms:

> *It is the determination of particular actions by various combinations of abstract properties that makes it possible for a causally determined structure of actions to produce actions it has never produced before and, therefore, to produce altogether new behaviour not commonly expected from what it usually described as a "mechanism." Even a relatively limited repertory of abstract rules that can thus be combined into particular actions will be capable of "creating" an almost infinite variety of particular actions.*
>
> (Hayek, 1978, pp. 48–49, original emphasis)

The process of human association may not proceed entirely unconsciously, but largely so. The social consequences of human action are unintended. In his paper "The theory of complex phenomena," Hayek ([1964] 2014) describes systems in which a certain combination of elements produces an overall structure possessing distinct characteristic properties:

> *The "emergence" of "new" patterns as a result of the increase in the number of elements between which simple relations exist, means that this larger structure as a whole will possess certain general or abstract features which will recur independently of the particular values of the individual data, so long as the general structure (as described, e.g., by an algebraic equation) is preserved. Such "wholes," defined in terms of certain general properties of their structure, will constitute distinctive objects of explanation for a theory, even though such a theory may be merely a particular way of fitting together statements about the relations between individual elements.*
>
> (Hayek, [1964] 2014, pp. 261–262, original emphasis)

The concept of emergence refers to the possibility that, when certain elements or parts stand in particular relation to one another, the whole that is formed has properties that are not possessed by its constituent elements taken in isolation (Lewis, 2012, 2015). Emergence implies that the system as a whole is able to perform functions or solve problems that elements cannot achieve separately. Any whole is then said to be a "higher-level" entity. The higher-level entities possess emergent properties that arise as a result of interactions between their lower-level parts; they are distinct from the properties of the lower-level entities taken in isolation. For Hayek, the rule-governed, relationally-defined social wholes that structure individuals' interactions are explanatorily irreducible factors. Emergent properties are irreducible to the properties of the lower-level elements of which the higher-level whole is formed. Micro-level behaviors and their interactions generate emergent patterns at the macro-level. The connections between the components are variable, and the components themselves are changed by their interactions. The behavior of a complex system cannot be systematically characterized in advance. For example, according to Markose (2005), the von

Neumann models based on cellular automata have laid the ground rules of modern complex system theory regarding: [15]

1. the use of large ensembles of micro-level computational entities or automata following simple rules of local interaction and connectivity;
2. the capacity of these computational entities to self-reproduce, and also to produce automata of greater complexity than themselves; and
3. the use of the principles of computing machines to explain diverse system-side or global dynamics.

A large number of agents interact in a dynamic way. Complexity theory starts from the assumption that much of the observed complexity in the world can be explained by relatively simple interactions between components of the system of interest. Systems attain complex phenomena through local interactions of agents with local information. These agents respond to simple rules, but there is no centralized command and control. Hayek ([1955] 2014, pp. 160−161) suggests the following: "this mutual adjustment of the elements to their immediate neighbors creates an equilibrium of the whole, an order in which every element takes its appropriate part. We meet here what Michael Polanyi has described as the spontaneous formation of a polycentric order, an order which is not the result of all the factors being taken into account by a single centre, but which is produced by the responses of the individual elements to their respective surroundings."[16] A complex system consists of a set of elements which interact with each other. Their coordination is guided as by an "invisible hand" toward the joint discovery of a hidden system of things (Polanyi, 1951). Polanyi's "polycentric order" and his later work on "tacit knowledge" may have been especially helpful to Hayek as he attempted to formulate his own ideas about cultural evolution and complexity theory.

In his paper "The pretence of knowledge," Hayek ([1975] 2014) suggests that, in social sciences, we have to deal with phenomena of "organized complexity":

> *Organized complexity here means that the character of the structures showing it depends not only on the properties of the individual elements of which they*

---

15. The prototype of what is called a cellular automaton is informally described as follows. Consider a finite chessboard in which each square can be one of different colors at each moment in time. Assume that we have a rule that specifies what the color of each square should be as a function of its four neighboring squares, that is, north, south, east, and west. Now let an initial pattern of colored squares be given. We then turn the system on and let the rule-of-state transition operate, examining the pattern that emerges as time passes. We watch the squares taking different colors at each time step in accordance, and look for the steady state pattern (if any).
16. The concept of polycentricity was first envisaged in Polanyi (1951). Polycentricity emerges as a non-hierarchical, institutional, and cultural framework that makes possible the coexistence of multiple centers of decision-making with different objectives and values.

are composed, and the relative frequency with which they occur but also on the manner in which the individual elements are connected with each other.

(Hayek, [1975] 2014, p. 365)

Emergence is the major consequence. Something emerges when several agents follow simple rules.[17] It is often referred to as a holistic phenomenon, because the whole is more than the sum of the parts and is produced when agents interact and mutually affect each other. No single agent can know, comprehend, or predict actions and effects that are operating within the system as a whole. In his essay "Notes on the evolution of systems of rules of conduct," Hayek ([1967] 2014) considers the concept of "the overall order of actions," which is an emergent property:

The overall order of actions in a group is in two respects more than the totality of regularities observable in the actions of the individuals and cannot be wholly reduced to them. It is so not only in the trivial sense in which a whole is more than the mere sum of its parts but presupposes also that these elements are related to each other in a particular manner. It is more also because the existence of those relations which are essential for the existence of the whole cannot be accounted for wholly by the interaction of the parts but only by their interaction with an outside world both of the individual parts and the whole. If there exist recurrent and persistent structures of a certain type (i.e., showing a certain order), this is due to the elements responding to external influences which they are likely to encounter in a manner which brings about the preservation or restoration of this order; and on this, in turn, may be dependent the chances of the individuals to preserve themselves.

(Hayek, [1967] 2014, p. 282, original emphasis)

According to Foster (2006), the term "complex" related to a state of "ordered complicatedness." Foster (2006, p. 1076) argues: "[d]espite the apparent complicatedness of a complex system, the set of rules that it adopts and the associated processes can be simple to represent analytically." Each unit adjusts its behavior in accordance with the behavior of its neighbors. As each unit adjusts to its neighbors, the whole group eventually synchronizes the behavior.

The components of a complex adaptive system are commonly modeled as agents. The number of agents in the system is, in general, not fixed. They are implicitly assumed to be goal-oriented. The environmental conditions to which an agent reacts are affected by others' activities. Such interactions are

---

17. The most oft-quoted example of this principle is Craig Reynold's "boids" program. The simulated birds, or boids, on a computer screen can wheel in union and avoid obstacles as if conducting highly coordinated maneuvers. Each boid in the simulation obeys only three simple models: (1) maintain a minimum distance from other objects in the environment, (2) match velocities with other boids in the neighborhood, and (3) move toward the perceived center of mass of voids in the neighborhood.

initially local, but their consequences are often global. The processes in complex adaptive systems are often non-linear. Self-organization can be defined as the spontaneous emergence of global structure out of local interactions. Self-organization can be described in terms of information. From an information perspective, self-organization occurs when information is aggregated and becomes more than the sum of the parts. New information is created locally, but some information is destroyed somewhere else. As Hodgson and Knudsen (2010) point out, "[s]elf-organization means that complex structures can emerge without design, but these structures themselves are subject to evolutionary selection" (Hodgson and Knudsen, 2010, p. 56).

7.4 SALIENCE

Schelling's (1960) focal point idea is important as an explanation of how players coordinate. Each member of the population expects every other member to behave in accordance with the relevant regularity. A shared vision of what should be obvious to each player leads to the emergence of a focal point. Focal points do not depend entirely on *a priori* reasoning. Schelling (1960) contends that coordination is inherently dependent on empirical evidence. Individuals often coordinate at points that, in some sense, "stand out" from the others. There is sometimes a logic to decide what solution has this unique property. However, this uniqueness often depends on the experiences and contingent associations of the individuals involved.

Schelling (1960) observes that any feature of a coordination equilibrium that draws attention to itself, making it stand out among the equilibria, will tend to produce self-fulfilling expectations that this salient equilibrium will result. In explaining the concept of salience, Schelling (1960) describes players as searching for a "key," "signal," or "suggestion" that is hidden in their decision problem:

> *Finding the key, or rather finding a key—any key that is mutually recognized as the key becomes the key—may depend on imagination more than on logic; it may depend on analogy, precedent, accidental arrangement, symmetry, aesthetic or geometric configuration, casuistic reasoning, and who the parties are and what they know about each other.*
>
> (Schelling, 1960, p. 57, original emphasis)

Players tend to select an equilibrium that is "unique," merely because that uniqueness causes each player to expect every player to focus on it.[18]

---

18. For example, Schelling asked New Haven residents to name the place and time they would meet someone in New York City on a given day, if they had failed to communicate more specifically on the subject. Although there is an extremely large number of possibilities, over half the individuals named the same place—Grand Central Station—and almost everyone named the same time—noon.

Individuals may be aware of convergent expectations indicating where the focal points are settled.[19] Focal solutions might be guided by a "set of rules," like "tradition," which indicates "that everyone can expect everyone else to be conscious of as conspicuous a candidate for adoption," by "etiquette and social restraint," or by "clothing styles and motorcar fads" (Schelling, 1960, p. 91). Thus, the players' coordination is dependent on "their social perception and interaction," and shared "impressions, images, understandings" (p. 163). The skills that are required to find focal points can be taught and learned, but it may not be possible to reduce them to mechanistic rules.

By harmonizing expected responses, focal points reduce uncertainty, despite the presence of imperfect information, enabling individuals to coordinate their activities towards the achievement of their goals (Leeson et al., 2006). Lewis (1969) gives a more formal structure to characterize coordination problems. Players individually choose between alternative strategies, and the combinations of strategies chosen by the players determine a payoff for each of them. Payoffs can be interpreted as measures of the strength of self-interested desires. In pure coordination games, individuals' interests roughly coincide. In general, in order to have a sufficient reason for choosing a particular action, an agent needs to have a sufficient degree of belief that the other agent will choose a certain action. Agents prefer conformity over non-conformity, conditional on others' conforming. A certain action stands out as the one that most likely everyone will pick. Such an action is salient to the parties. In Lewis' (1969) argument, coordination may be achieved with the aid of mutual expectations about action. Past experience of a convention, or "precedent," is the source of such expectations. In the driving game, the drivers condition their choices on the history of play. That is, if everyone has been driving on the right side of the road up until now, everyone will continue to drive on the right; if everyone has been driving on the left, everyone will keep doing the same.

A specific convention is an actual regularity in the behavior of a given population. Conventions are empirically observable behavioral patterns in a recurrent situation of social interaction. A specific convention is assumed to be arbitrary, in the sense of being the realization of only one of the multiple potential regularities that could emerge in a recurrent situation of social interaction. There always exists an alternative behavioral regularity to the established convention. The arbitrariness is captured by the representation of convention as a stable solution to a recurrent coordination problem. Lewis (1969) justifies the stability of conventions through the reconstruction of the actual processes of reasoning of those involved in the interaction situation where a particular convention emerges. On the other hand, other studies,

---

19. In coordination games, the simplest way to create a focal point is by communication and agreement. Many experimental studies demonstrate that the players can increase their level of coordination in such games by engaging in "cheap talk" (Crawford, 1998).

such as Sugden (1986, 1989) and Skyrms (1996), base their explanation on the development of behavioral dispositions through an adaptive process. The process, sooner or later, leads individuals to coordinate on a particular convention.

Individuals have limited knowledge of their environment. By communicating rules of conduct with others, they learn to become more and more informed. As Hayek (1988) suggests:

> *Learning how to behave is more the source than the result of insight, reason and understanding. Man is not wise, rational and good, but has to be taught to become so.*

(Hayek, 1988, p. 21)

Individuals adjust their action plans according to the results of the combinations of their own actions with those of others. A mutual adjustment of individual action plans is necessary for achieving a social order. Individuals who belong to the same environment are characterized by shared beliefs. They share the knowledge of what to do and how to behave in certain situations; these rules of behavior, tacit or not, facilitate individual action planning by providing a stable and predictable framework within which individuals can act. Then the benefits to individuals who behave according to these rules depend on how many other individuals do so also. The more individuals adhere to the rules, the more predictable and reliable the rules become.

A regularity in behavior can be described as a convention within a population. When people interact with each other, their actions and the deliberate reasoning by which they choose their actions are interdependent. Lewis (1969) confines his analysis to a narrow class of coordination problems. Repeatedly, pairs of players drawn from the population engage in some game-like interaction. In these situations, the players' interests approximately coincide, and there are two or more equilibria. The formal structure of a pure coordination game for two players is represented by the following normal form. Player 1 chooses a strategy from the set $\{s_{11}, \ldots, s_{1n}\}$, where $n > 2$, and Player 2 chooses a strategy from the set $\{s_{21}, \ldots, s_{2n}\}$. For each strategy profile, there is a payoff to each player. For Lewis' purposes, it is sufficient to say that payoffs are real numbers. Let the chosen strategies be s_{1g} and s_{2h}. If $g = h$, each player receives one unit of utility, but otherwise each receives zero. There are n strict Nash equilibria, and both players are indifferent between all of them. With multiple equilibria, the outcome that emerges depends on features other than the payoffs. In this structure, there is complete symmetry between players, between strategies, and between equilibria. A pure coordination game is a game in which, at each pure strategy combination, the players all receive the same payoff. Player 1 and Player 2 each have n pure strategies, and receive exactly the same payoff at each of the n pure strategy combinations. Coordination games may exhibit multiple

Pareto-ranked equilibria (Cooper and John, 1988; Cooper, 1999). Cooper et al. (1990) provide experimental evidence that the Pareto superior equilibrium is not guaranteed in simple two-person coordination games with complete information.[20] As is often expressed in macroeconomics, an economy may be stuck at an inefficient equilibrium: a depression in aggregate economic activity arises when the economy falls into the trap of a low activity level Nash equilibrium.[21]

When people interact with each other, their actions, and the deliberate reasoning by which they choose their actions, are often interdependent. In Lewis (1969), the definition of convention requires not only that agents engaged in a game play a coordination equilibrium, but also that the agents have common knowledge that they all prefer to conform with the equilibrium, given that every agent conforms with this equilibrium. Lewis (1969) formulates the notion of common knowledge in order to define a convention.[22] A proposition p is common knowledge for a set of agents if, and only if:

1. each agent k knows p;
2. each agent i knows that each agent k knows that p, each agent j knows that each agent i knows that each agent k knows that p, and so on.

For Lewis (1969), a coordination equilibrium is a convention only if the players have common knowledge of a "mutual expectations criterion" (MEC):

MEC: Each agent has a decisive reason to conform to his or her part of the convention, given that he or she expects the other agents conform to their parts.

In general, in order to have a sufficient reason for choosing a particular action, an agent needs to have a sufficient degree of belief that the others will choose certain actions.[23] Such a sufficient degree of belief is justified by a system of mutual expectations. Lewis (1969) adds this common knowledge

---

20. In Van Huyck et al. (1990), when the number of players is large, coordination failure is almost certain to occur.

21. Global games have been used to analyze macroeconomic phenomena with strategic complementarities (Morris and Shin, 1998; Goldstein and Pauzner, 2005). In global games (Carlsson and van Damme, 1993), Nature first randomly chooses the coordination game to be played, and then each player observes a noisy, idiosyncratic signal of the selected game. As the noise in player's observations vanishes in such global games, the sequence of equilibria converges to the risk-dominant equilibrium of the game actually selected by Nature.

22. Since the publication of Lewis's *Convention*, many game theorists have given various common knowledge assumptions underpinning any solution concept for games. See Brandenburger and Dekel (1988) and Brandenburger (1992).

23. There are several arguments as to whether the formalization of a wider notion of rationality is necessary to endow the agents in Lewis' definition of convention with a sufficient reason to coordinate their behavior. For example, according to Bicchieri (2006), individuals only need empirical or plain expectations of the conformity of others to a particular convention to actually conform to that convention.

requirement to the definition of convention, to rule out cases in which agents coordinate as the result of false beliefs regarding their opponents.

If members of a group coordinate successfully, they will start to notice a regularity. Members develop a preference for conforming to this regularity if others do so as well. A general expectation that members will conform to the regularity is formed. Consequently, a stable pattern of behavior, or a convention, emerges that is based on the general expectation of each member about the way that others will behave. Suppose that members of a population P face some coordination problem in a recurrent situation where there is a history of conformity with a behavioral regularity R. Everyone prefers to conform with R if almost everybody else does the same. Then, R is a population-wide regularity in behavior by which members reliably pick a coordination equilibrium. Such a regularity R is a convention if it in fact occurs.

Lewis (1969, p. 78) devises a theory of convention holding that:

A regularity R in the behavior of members of a population P when they are agents in a recurrent situation S is a convention if and only if it is true that, and it is common knowledge in P that, in almost any instance of S among members of P:

1. *almost everyone conforms to R;*
2. *almost everyone expects almost everyone else to conform to R;*
3. *almost everyone has approximately the same preferences regarding all possible combinations of actions;*
4. *almost everyone prefers that any one more conforms to R; on condition that almost everyone conforms to R;*
5. *almost everyone would prefer that anyone more conforms to R' on condition that almost everyone conforms to R';*[24]

where R' is some possible regularity in the behavior of members of P in S, such that almost no one in almost any instance of S among members of P could conform to both R' and to R.

Conformity to the prevailing convention may not be explained solely from the premises of classical game theory, due to the Nash equilibrium selection problem inherent in any coordination problem. Lewis' (1969) theory of conventions provides a way to allow external information to play a role in reasoning about one's choice of strategy. To understand conventions, we must consider how they are meant to figure in the practical reasoning of people. For members, their appreciation of the practical force of conventions depends on their expectation, or implicit understanding, that the conventions are practiced in their community. Lewis (1969) adds an account of the psychological mechanism by which conformity is

---

24. Note that 5 requires that there be multiple equilibria.

maintained. Conventions are understood as salient regularities of behavior, common knowledge of which enables people to successfully negotiate the complex landscape of ordinary social interactions. Salience plays a crucial role in coordination problems. Facing a coordination problem, individuals recognize that an equilibrium has certain salient features, and each player expects them to be noticed by the other players too. Individuals try for a coordination equilibrium that is somehow salient: one that stands out from the rest by its uniqueness in some conspicuous respect. If everyone chooses the salient combination of actions in situations of the relevant type, a fortunate result is obtained.

Salience constitutes a focal point that facilitates coordination when purely rational considerations are insufficient to identify the best move. Focal points may be determined by cultural, cognitive, or even biological factors. Then, Lewis' account exposes the limitations of purely rational considerations for the analysis of collective behavior. Salience ultimately comes down to a "non-rational" (but not "irrational") propensity to choose in certain ways when reason gives out (Sugden, 1995). In order to understand how institutions emerge from individual interactions, we have to consider how a focal point facilitates coordination. Lewis' account highlights the importance of repeated play. In the case of conventions, salience is determined by past conformity (salience by precedent).

Lewis (1969) explains coordination by salience as follows:

> *How can we explain coordination by salience? The subjects might all tend to pick the salient as a last resort, when they have no stronger ground for choice. Or they might expect one another to have that tendency, and act accordingly; or they might expect each other to expect each other to have that tendency and act accordingly; and so on. Or—more likely—there might be a mixture of these. Their first- and higher-order expectations of a tendency to pick the salient as a last resort would be a system of concordant expectations capable of producing coordination at the salient equilibrium.*
>
> (Lewis, 1969, p. 35−36)

Lewis (1969) argues that precedent is a form of salience. In this case, a particular kind of salience has served as a coordination device of a similar coordination problem in the past. Each individual has some tendency to repeat actions that have proved to be successful in previous interactions. The precedent will make one of the coordination equilibria salient. It is because a regularity of behavior already exists that there is a precedent to follow. Each individual's expectation that others will do their parts, strengthened perhaps by replication using his or her higher-order expectations, gives him or her some reason to do his or her own part. When players have no reason to choose one strategy rather than another, their default choices tend to favor strategies with certain properties of salience.

Knowing this, rational players of coordination games choose salient strategies deliberately.[25] In Lewis' (1969) argument, the assumption of salience is used to explain how one particular convention is first selected from the set of possible conventions, and the assumption of common knowledge is used to explain why, once any convention has been selected, no one deviates from it (Skyrms, 1996).

Common knowledge is an infinite structure in logical space. For conventions, this infinity can be understood in terms of potential reason (Cubitt and Sugden, 2003). Although this may be fine for logical purposes, the question remains how do we, finite beings, connect this infinity? Convention can be viewed in a nearly Lewis' (1969) sense, but without the assumption of common knowledge in an evolutionary analysis. All that matters for players in a population is that they expect others to stick to their part in a specific equilibrium profile. Once the first order expectation is available, players will coordinate on the specific equilibrium profile. The special case of conventions is analyzed in signaling games (Skyrms, 1996, 2010). In Skyrms (1996, 2010), players are "programmed" to play specific strategies, without assuming rationality and common knowledge. The overall evolution of the population is directed by the replicator dynamics, and players never fail to coordinate on some signaling equilibrium. Which signaling equilibrium is selected depends on the initial conditions of the system (or on the presence of noise, or other kinds of random fluctuations), not on the players' cognitive or epistemic states. Skyrms' (1996, 2010) analysis provides us with a naturalization of the concept of salience. For example, many simple signals can be understood in cultural terms. Generally, culture-based signals can solve coordination problems without invoking common knowledge.

In his analysis of the origination of institutions, Hayek (1973) emphasizes the difference between the rules of purposively acting individuals and the cultural rules of a society. The former rules are innate behavioral regularities that all humans possess (genetic rules), and the latter are rules of conduct generally recognized by specific groups (cultural rules). That is:

> The chief points on which the comparative study of behavior has thrown such important light on the evolution of law are, first, that it has made clear that individuals had learned to observe (and enforce) rules of conduct long before such rules could be expressed in words; and second, that these rules had evolved because they led to the formation of an order of the activities of the group as a whole which, although they are the results of the regularities of the

---

25. This idea can be formalized in a "cognitive hierarchy" model, where level-k types respond to a distribution of lower types (Stahl and Wilson, 1995; Camerer et al., 2004). The anchoring Level 0 type exists mainly in the minds of higher types; Level 1 players assume their opponents are at Level 0, and try to match those choices; Level 2 players assume their opponents are at Level 0 or Level 1, and try to match their choices, and so on.

actions of the individuals, must be clearly distinguished from them, since it is the efficiency of the resulting order of actions which will determine whether groups whose members observe certain rules of conduct will prevail.

(Hayek, 1973, p. 74)

Klein (1997) distinguishes between "coordination" and "metacoordination." The distinction between coordination and metacoordination lines up with the distinction between conventions and social orders. According to Klein (1997):

There is nothing in Lewis's definition of convention to preclude the inferior coordination equilibrium as the social regularity. The ranking of coordination equilibria as "inferior" or "superior," whether in a Pareto sense or some other collective sense, is not a component of coordination and convention. It is, however, the fundamental component of metacoordination.

(Klein, 1997, p. 329−330, original emphasis)

How is the superior outcome achieved? It is achieved not by conventions within specific reoccurring situations, but by the "concatenation of affairs functioning as a complex social order" (Klein, 1997, p. 330). Social orders subsume conventions, and metacoordination is not achieved apart from matters of coordination. Coordination is an inherently interpersonal phenomenon. Even in the state of affairs where individual decisions are coordinated and go through as planned, it is still "discoordinated" if a better decision could have been selected in economic systems. Hence, metacoordination is required for participants to discover and exploit all available opportunities in transactions.

Respecting a set of rules of conduct conditions, to some extent, the coordination of individuals' actions. Individuals act according to rules of conduct, and they understand the actions of others. However, how is coordination "coordinated"?[26] Hayek took a balanced view of human nature: man is both conforming (rule-following) and "purpose-seeking." An agent "is successful not because he knows why he ought to observe the rules which he does observe, or is even capable of stating all these rules in words, but because his thinking and acting are governed by rules which have by a process of selection been evolved in the society in which he lives, and which are thus the product of the experience of generations" (Hayek, 1973, p. 11). The ways individuals perceive their interests that drive them to action are dependent on the institutional context within which they act. Pecuniary incentives are crucial to explain the expected patterns of individual's behavior in markets. A liberal political order is then characterized by the rule of law, and by the protection of property and of market exchange. Individuals

---

26. This is about what kind of game the players wish to play with each other (Vanberg, 2007). It is not the question of whether they can play a given game, but the question of how they, together with other players, may come to play a better game.

are expected to increase their efforts in a particular direction if the prospects of return improve. Hayek (1976, p. 109) defines "catallaxy" as "the special kind of spontaneous order produced by the market through people acting within the rules of the law of property, tort and contract." Wealth grows because people can innovate with their own property, and the returns to their efforts accrue to them.

7.5 CULTURE AS INFORMATION

7.5.1 Culture and Social Learning

What is culture? Culture is a broad concept; it ranges from underlying beliefs and assumptions, to visible structures and practices. The study of culture has been overwhelmingly the province of anthropology. An intuitive and representative definition of the term "culture" is provided by Harris (1971):

> *Cultures are patterns of behavior, thought and feeling that are acquired or influenced through learning and that are characteristics of groups of people rather than individuals.*

> (Harris, 1971, p. 136)

Numerous cross-cultural studies have demonstrated how peoples' lifestyles are influenced by cultural factors. As an example, individualism/collectivism describes the relationship between the individual and the prevailing collectivity in society. Individualism implies a loose social framework in which people focus on their goals, needs, and rights more than community concerns. In individualistic societies, individuals believe in their free will and dominance over their environment. On the other hand, collectivism is characterized by a tight social framework in which people value in-group goals and concerns. Collective societies foster a collective mind-set. Theories of culture differ in their focus on the various layers of culture. Schein (1992) provides three layers of cultural phenomena. The first (or the most external) level is the visible and audible behavioral patterns, and the constructed physical and social environment. The second (or the middle) layer is "espoused values," which includes strategies, goals, and philosophies. The third (or the deepest and invisible) level is that of basic assumptions and beliefs about human nature and relationship to the environment.

Culture refers to the set of learned traditions and living styles shared by the members of a group.[27] Culture includes the specific ways of thinking, feeling, and behaving of a human group. Culture thus means shared values and common meanings. Hayek (1988, p. 23) suggests as follows: "[s]haped by the environment in which individuals grow up, mind in turn conditions

---

27. According to Gintis (2007), anthropology treats culture as an expressive totality defining the life space of individuals, including symbols, language, beliefs, rituals, and values.

the preservation, development, richness, and variety of traditions on which individuals draw. By being transmitted largely through families, mind preserves a multiplicity of concurrent streams into which each newcomer to the community can delve." Each culture has its own set of core values and basic assumptions. Culture teaches preferences in life, and manifests itself in how people behave, think, and believe.

A practice will spread within society due to individual imitation. According to Richerson and Boyd (2005), culture is "information" capable of affecting individuals' behavior, which they acquire from others through teaching, imitation, and other forms of social transmission. The term "information," means any kind of mental state, conscious or not, that is acquired or modified by social learning and affects behavior.

Social learning is the mechanism whereby information is transmitted. Through learning from others, humans finesse the bounds of rationality (Boyd and Richerson, 1993). What we observe as cultural representations and practices are variants (of cultural traits), found in roughly similar forms in a particular place or group because they have resisted change and distortion through innumerable processes of acquisition, storage, inference, and communication. Much variation in human behavior cannot be understood without accounting for beliefs, values, and other socially-acquired determinants of behavior. In these models, the spread of specific variants of cultural representations (such as particular beliefs or concepts represented by the mind) is seen as partly analogous to the spread of alleles in a gene pool. The tools of population genetics can be applied to the spread of cultural traits, and allow us to predict their spread, given such parameters as the initial prevalence of a trait, the likelihood of transmission, and various biases. Each individual of the new generation chooses one cultural model from the previous generation, either from their parents or from an individual taken at random, and adopts his or her behavior (Boyd and Richerson, 1985). Human beings have developed means to preserve cultural forms across generations, by embodying them in tools, physical constructs, and language. Cultural transmission works most of the time as an inheritance system that allows the accumulation of what has been learned from generation to generation. Culture replicates the phenotypic structure of the parental generation, and behaves as a true system of inheritance. For Richerson and Boyd (2005), the most basic type of micro-event in cultural evolution is the adoption by an individual of some cultural variant. The collective phenomenon of culture is taken to be the evolving outcome of the aggregation of these micro-events. These evolutionary models do not assume that acquiring more information from the environment implies less information specified in the genome. On the contrary, acquiring more information from the environment requires richer cognitive dispositions to render such information sensible, and it, therefore, requires more genetic specification. Cumulative cultural transmission leads to a cultural evolutionary process with a great adaptive value.

Humans possess culture-specific adaptations that are the product of cultural evolution.[28]

Behavior is at least partly acquired via social learning. Social learning occurs when people are more likely to learn a behavior because of the behavior of other individuals. Social learning plays a key role in cultural transmission between generations. Hayek's theory of mind describes the human mind as an adaptive classification system by which individual behavior is shaped. Hayek (1988, p. 22) argues: "[w]hat we call mind is not something that the individual is born with, as he is born with his brain, or something that the brain produces, but something that his genetic equipment ... helps him to acquire, as he grows up, from his family and adult fellows by absorbing the results of a tradition that is not genetically transmitted."

The existence of culture presupposes a population capable of mental representations. People are likely to perceive bits of information that is germane to existing schemata—knowledge structures that represent objects or events and provide default assumptions about their characteristics, relationships, and entailments under conditions of incomplete information (DiMaggio, 1997). People experience culture as schematic structures that organize that information. As suggested by Denzau and North (1994, p. 15), culture can be regarded as "encapsulating the experiences of past generations of any particular cultural group." Culture collectively accumulates partial solutions to frequently encountered problems of the past, and works as a filter for processing and interpreting new sensory data.

The group is made up of individuals who have their own cognitive mechanisms. Culture enables individual agents to acquire adaptive behavior in different environments. Individuals gradually realize that it is an advantage for them to observe the common social rules shared by the group. This gives rise to interpersonal relations. Interpersonal relations emerge if individuals are interested in sharing common beliefs that make communication easier. Individual perception is then turned into shared mental models. Cognitive activities change with the passing of time, and they are influenced by personal experience. These changes allow individuals to adopt spontaneously the existing cultural and social representations.

The group members are already immersed within institutions representing a heritage of cultural knowledge. Vanberg (1994) claims that our cultural

---

28. Cosmides and Tooby (1992) claim that the blank-slate view of the mind is wrong, and that one cannot understand human cultural evolution without taking into account the effect of domain-specific psychological mechanisms shared by human beings. The human brain is densely packed with programs which cause intricate relationships between information and behavior. Many of these programs are functionally specialized to produce behaviors that solve particular adaptive problems, such as mate selection, language acquisition, family relations, and cooperation. Much of human behavior is thus generated by psychological adaptations that evolved to solve persistent problems in human ancestral environments.

inheritance can be seen as incorporating the accumulated experiences of a community. Vanberg (1994) writes:

> *Our genetic heritage can be seen as incorporating the accumulated experience of the species, as implicit 'hard-wired knowledge' that we share as members of the human species. Our socio-cultural heritage can be viewed as incorporating the accumulated experience of a cultural community, conjectural, largely tacit, 'knowledge' that is embodied in language, traditions and other cultural artifacts, and that we share with other members of our culture. Finally, the product of individual learning—which takes place on the basis of our genetic heritage and within the context of our socio-cultural heritage—is the accumulated experience of our own trials and errors, of success and failure. The whole process of learning, of the growth of knowledge, is seen as a process of correcting, adjusting and refining such 'conjectural knowledge' in the light of experience.*
>
> (Vanberg, 1994, p. 35, original emphasis)

The constitution of a particular set of action plans depends on a complex structure of beliefs that have evolved over time. From an individual point of view, individuals evaluate their action plans in terms of achievements. Individuals may revise the whole or parts of action plans, if they judge their plans not to be sufficiently effective. The acquisition of new knowledge is determined by the individual's intention to imitate the behavior of others.

The transmission of information goes from generation (parent), to generation (offspring), to generation (grand offspring). In biology, transmission genetics is the study of inheritance systems, and genetic transmission occurs vertically. Philosophers of biology tend to believe that biological information is not adequately captured by Claude Shannon's mathematical theory of communication.[29] In the original formulation of Shannon's theory of communication, according to Bergstron and Rosvall (2011, p. 164), "information is what an agent packages for transmission through a channel to reduce uncertainty on the part of a receiver." Bergstron and Rosvall (2011, p. 165) offer the following formulation of the transmission view of information.

Transmission view of information:

An object X conveys information if the function of X is to reduce, by virtue of its sequence properties, uncertainty on the part of an agent who observes X.

It is natural to interpret uncertainty reduction as being somehow connected to Shannon's theory of communication.

Cavalli-Sforza and Feldman (1973) initiate the subdiscipline of cultural anthropology known as gene—culture coevolution. They suggest that human behavior is a product of two different and interacting processes: genetic evolution and cultural evolution. In their book *Culture and the Evolutionary Process*,

---

29. Shannon and Weaver (1949) define the technical level of communications as the accuracy and efficiency of the communication system that produces information.

Boyd and Richerson (1985) label their perspective "dual-inheritance" theory, emphasizing that human beings have both a genetic and a cultural heritage.[30] According to their research, biological evolution has produced a unique capacity for culture in the human descent line, and culture itself began to evolve by natural selection. The process of natural selection acting on cultural variants transformed the human species. Human capacities are the product of an evolutionary dynamic involving the interaction between genes and culture. Human biology and culture are, therefore, intertwined inextricably.

Some mechanisms modify the forces of cultural evolution shaped by the transmission of human behavior. These mechanisms are classified as follows (Cavalli-Sforza and Feldman, 1981; and Boyd and Richerson, 1985):

1. **Random cultural variation:**
   Cultural transmission involves various kinds of "errors." The rate of culturally transmitted errors, in fact, seems much higher than that of genetic mutation.
2. **Cultural drift:**
   In small groups, cultural mutations may have a large impact, causing the respective culture to be less stable.
3. **Biased cultural transmission:**
   Humans are predisposed to adopt certain preexisting cultural variants, and these will increase in frequency. Boyd and Richerson (1985) distinguish between three types: (1) direct bias (the adoption of cultural variants depends on the properties, or attractiveness, of the variants); (2) indirect bias (the imitation of certain characteristics that are perceived to be associated with others that are regarded as attractive); and (3) frequency-dependent bias (the imitation of the majority is dominant).
4. **Guided (Lamarckian) variation:**
   Humans can, consciously and purposefully, change their behavior, rules, and norms through learning-by-doing and communication. Guided variation occurs when individuals transform cultural variants in a nonrandom (perhaps genetically adaptive) direction. Successful practices are passed through tradition, learning, and imitation. It is often referred to as the Lamarckian aspect of cultural evolution, because it is a source of purposeful creation of variety, and acquired characteristics may be passed on.[31] Guided variation can be contrasted with direct bias, where

---

30. Edward O. Wilson, in his book *Sociobiology: The New Synthesis* (1975), suggested that human beings could be understood in terms of the same principle that governed animal behavior. Some researchers, like Wilson, took the position that human behavior could be understood only if biological evolution were taken into account. Other researchers, however, argued that human beings were unique, and that it was useless to study human behavior in the same way that biologists studied animal behavior.
31. Lamarckian evolutionary processes differ from Darwinian ones, because the former admits the possibility of the inheritance of acquired traits.

individuals evaluate traits in the population and preferentially adopt certain traits over others.

5. Gene—culture coevolution:
    This refers to the interaction between the cultural evolution and biological evolution of the human species. Cultural traits have an impact on the survival and reproduction, or the fitness of individual, and are, in turn, influenced by these.

Boyd and Richerson's (1985) idea is that behaviors are culturally transmitted in a way that tends to homogenize the phenotype of individuals in each group. In the study of the evolution of culture, Boyd and Richerson (1985) have developed many theoretical tools, often in the form of mathematical models suitable for cultural change. Most models of cultural evolution use mean field ones whose dynamics derive either from epidemiology or from conventional population genetics. Their models explore the long-term, population-level consequences over many generations and in large, often structured populations. Each individual of the new generation chooses one cultural model from the previous generation, and adopts their behavior. Culture, therefore, replicates the phenotypic structure of the parental generation.[32] Boyd and Richerson (1985) consider, in particular, conformist-biased imitation, whereby individual agents imitate whatever behavior is the most frequent in their local group. Conformist-biased imitation generates a strong phenotypic homogeneity in groups. Boyd and Richerson's theory of the evolution of culture deals with the following three processes:

1. The evolution of cognitive abilities required for cultural learning and transmission.
2. Cultural change, that is, diachronic change in the kinds and frequencies of culturally transmissible variants.
3. The coevolution of culture and cognition, that is, how cultural change affects the evolution of cognitive abilities, and vice versa.

In order to understand cultural evolution, we need to know how the structures of cultural transmission give rise to cultural changes, and to recognize why these structures have evolved.

7.5.2 Culture and Path Dependence

Cultural differences persist because of differences in the physical and social environments. Traditions and behavioral differences, in general, persist only because differences in relevant features of the environment

---

32. Dawkins (1976) uses the term "meme" to describe a unit of human cultural transmission analogous to the gene, arguing that culture is a system of replication of memes.

persist. If individuals preferentially imitate the most common beliefs and behavior, then cultural variation can be stable, despite substantial mixing of people and ideas. However, culture can be changed by coming into contact with another culture. Whether acculturation occurs or not depends on the extent to which people are attracted to the other culture, and on how deeply they strive to maintain their own cultural identity (Berry, 1980). Learning how to behave according to rules is more the source than the result of reason. Thus, "[j]ust as instinct is older than custom and tradition, so then are the latter older than reason: custom and tradition stand between instinct and reason—logically, psychologically, temporally" (Hayek, 1988, p. 23).

Individuals' expectations and actions somehow seem to achieve the quite remarkable level of coordination regularity witnessed in many aspects of social life. Beliefs of different people are not completely diverse, and they possess a common structure which makes communication possible. Therefore, the rules of conduct which guide the individuals' behavior depend on the mental "common" structure. Culture coordinates the expectations of many agents about the actions, and it shapes and structures our daily patterns of behavior, guiding much of what we should do by prescribing what behavior is acceptable. Hayek (1973) emphasizes the difference between the rules of purposively acting individual agents and the cultural rules of a society. The former rules are the innate behavioral regularities all humans possess, while the latter are rules of conduct generally recognized by specific groups. As Hayek (1973) notes:

> The chief points on which the comparative study of behaviour has thrown such important light on the evolution of law are, first, that it has made clear that individuals had learned to observe (and enforce) rules of conduct long before such rules could be expressed in words; and second, that these rules had evolved because they led to the formation of an order of the activities of the group as a whole which, although they are the results of the regularities of the actions of the individuals, must be clearly distinguished from them, since it is the efficiency of the resulting order of actions which determine whether groups whose members observe certain rules of conduct will prevail.
>
> (Hayek, 1973, p. 74)

Individuals who belong to the same cultural environment tend to share common beliefs. As Denzau and North (1994) point out, individuals with common cultural backgrounds and experiences will share reasonably common mental structures. Agents who belong to the same cultural group are exposed to the same external representation of knowledge, which produces shared mental models. Culturally shared mental models expedite the process by which people learn directly from experiences, and facilitate communication between people. Culture thus provides shared collective understandings in shaping individuals' actions.

North (2005) argues that beliefs are mental constructs derived from learning through time:

> *Culture not only determines societal performance at a moment in time but, through the way in which its scaffolding constraints the players, contributes to the process of change through time. The focus of our attention, therefore, must be on human learning—on what is learned and how it is shared among members of a society and on the incremental process by which the beliefs and performances change, and on the way in which they shape the performance of economies through time.*

(North, 2005, viii)

Human actions are imprinted by their history. The resulting belief system constrains the repertoire of possible reactions to changes in the environment. As North (2005, p. 11) put it, "we choose among alternatives that are themselves constructions of the human mind." Then, "the belief system underlying the institutional matrix will deter radical change" (North, 2005, p. 77). Here, an institutional matrix is considered as a framework of interconnected institutions that together make up the rules of the economy. The structure formed by several institutional arrangements will define a set of interrelated incentives for individual agents. The presence of one particular institution may or may not be compatible with the presence of another. The conditions for the existence of an institution must be determined by taking into account a large set of institutional arrangements. The aggregate coherence given by a set of institutional arrangements is defined by their complementary character. The influence of one institution is reinforced when the other complementary institution is present. Complementary institutions reinforce each other. The complementary character is fundamental for defining the pattern of evolution of an economic system. Understanding institutional change necessitates an understanding of path dependence, in order to appreciate the nature of the limits it imposes on change. Path dependence includes features such as persistency and lock-in. We can say that "at any moment of time the players are constrained by path dependence—the limits to choices arising from the combination of beliefs, institutions, and artifactual structure that have been inherited from the past" (North, 2005, p. 80).

Arthur (1989, 1994) explains increasing returns in technology by using a model of path dependence.[33] Arthur (1994) has summarized the characteristics of increasing returns processes:

1. Unpredictability:

Because early events have a large effect and are partly random, many outcomes may be possible. We cannot predict ahead of time which of these possible end states will be reached.

---

33. In a Pólya urn experiment, after the first colored ball is randomly selected, the probability of all subsequent colors being selected depends on the proportion of colors in the urn. The final composition of the urn is entirely indeterminate before the first color has been selected. Once early random processes lead to the selection of certain colors, the system begins to stabilize around an equilibrium.

2. Inflexibility:

The farther into the process we are, the harder it becomes to shift from one path to another. In applications to technology, a given subsidy to a particular technique will be more likely to shift the ultimate outcome if it occurs early rather than late. Sufficient movement down a particular path may eventually lock in one solution.

3. Nonergodicity:

Accidental events early in a sequence do not cancel out. They cannot be treated (which is to say, ignored) as "noise," because they feed back into future choices. Small events are remembered.

4. Potential path inefficiency:

In the long-run, the outcome that becomes locked-in may generate lower payoffs than a forgone alternative would have.

The path-dependence idea stresses the importance of past events for future actions, or of foregoing decisions for current and future decision-making. Hence, decisions are conceived of as historically conditioned. The concept of path dependence has raised considerable interest in its various aspects, technological, economic, and institutional. The original idea of technological path dependence suggests that past decisions continue to affect any future development against other potential technological choices that are eventually more efficient (David, 1985).

A key finding of path dependence is a property of lock-in by historical events. Once an outcome begins to stabilize, it becomes progressively locked to attractors that are not always optimal. With increasing, an institutional pattern, once adopted, delivers increasing benefits with its continued adoption. Small chance events then have durable consequences in the long-run. Arthur (1989, 1994) and David (1985) expose increasing returns in technology by using a model of path dependence. They explain how technologies are locked-in through usage, to the point that they cannot be replaced by new and superior technologies. Initially, a technology was adopted as a result of a historical accident (or it was judged to be the best technology available). The technology has become more attractive through increasing returns. A new technology is invented (or discovered to have been available in the past). People realize that the new technology would have been superior to the currently dominant technology if adopted in the past. However, they lack the motive to replace the current technology with the new superior one. The QWERTY arrangement prevailed despite the fact, according to David (1985), that a superior keyboard arrangement, such as the DVORAK arrangement, existed. Through increasing returns, it became difficult to replace the QWERTY keyboard with superior technologies. As a consequence, the superior platform, DVORAK, was never adopted, and QWERTY persisted.

In a world of increasing returns to scale, initial and trivial circumstances can have important and irreversible influences on the ultimate market allocation of resources. An institutional pattern, once adopted, delivers increasing benefits with its continued adoption, and thus, over time, it becomes more and more difficult to transform the pattern or select previously available options, even if these alternative options would have been more efficient. Thus, an inefficient outcome can persist. The form of path dependence conflicts with conventional economics, where efficient outcomes are attained. Liebowitz and Margolis (1994) claim that, where there is a feasible improvement to be gained from moving onto a better path, economic agents are willing to pay to bring the improvement about.[34] However, such a switching cost must include the uncertainty concerning the cost of adopting the superior technology. It would be more efficient to adhere to the current technology than switch to the superior one.

The concept of "procedural rationality" in Herbert Simon (1976, 1978) offers a more plausible answer as to why agents cling to failing habits (Khalil, 2013). The notion of bounded rationality is traced back to the pioneering work of Herbert Simon, beginning in the 1950s. Simon (1957) embarks on an attempt to drastically revise the concept of economic agent by paying attention to the limits in the information and the computational capacities possessed by an economic agent. In Simon's view, simplified models capture the main feature of the problem, without capturing all its complexities. The decision-maker's simplified model of the world appears to be the key device of adaptation to bounded rationality. The term "procedural rationality" denotes a concept of rationality which focuses on the effectiveness of the procedures used to choose actions. This concept is different from that of "substantive rationality," such as "rationality as maximization," where choice is entirely determined by the agent's goals, subject to constraints. In discussing the concept of procedural rationality, Simon highlights the theory of heuristic search. People develop patterns of behavior based on their exposure to certain types of situations. These patterns of behavior can be referred to as habits. Once habits are developed, they tend to persist over time. Thus, habits and technologies are not chosen as the best available, but rather as whatever is encountered initially.

Societies do not evolve in an optimizing process looking for the best or the more efficient solution. According to Leibenstein (1979), the main argument against the maximization postulate is an empirical one—people frequently do not maximize:

> *In considering alternatives to conventional micro theory, one of the questions that arises is how to handle the usual maximization assumption. At present the maximization postulate has an unusually strong hold on the mind set of*

---

34. Liebowitz and Margolis (1990) suggest that there is little evidence to support the notion that QWERTY was, in fact, inferior to DVORAK. According to them, examples of alleged path-dependency externalities (where luck caused an inferior product to defeat a superior one), though charming, are empirically false myths.

economists. To examine why this is the case would take us too far afield into the sociology of the profession. Suffice it to say that in my view the belief in favor of maximization does not depend on strong evidence that people are in fact maximizers.

(Leibenstein, 1979, p. 493)

According to Leibenstein (1966), individuals supply different amounts of effort, where effort is a multidimensional variable, under different circumstances. The difference between maximal effectiveness of utilization and actual utilization is considered as the degree of X-inefficiency. According to Leibenstein (1979), the micro—micro problem is the study of what goes on inside a "black box." Leibenstein's approach aims to explain organizations' external activities through the study of their internal processes. In conventional economic theory, effort is assumed to be maximized in its quantity and quality dimensions, irrespective of institutional conditions. However, economic agents are not all the same and, therefore, are characterized by different objective functions. Whether X-efficiency is achieved or not depends on the economic agents' preferences. Moreover, the introduction of effort discretion into the modeling of the path-dependency theory allows for the persistence of multiple (both optimal and sub-optimal) equilibria. It helps to explain the development of inefficient economic regimes. X-inefficiency—the deviation of labor productivity from its maximum—can exist under competitive conditions. Altman (2000) introduces effort discretion into path-dependency modeling. Given effort discretion, incentives need not exist for agents to adopt superior economic regimes. The introduction of effort variability allows for the existence of sub-optimal equilibrium in the long-run. The existence of high- and low- productivity regimes might be a product of history.

From a cognitive economic perspective, creativity and path dependence are strictly linked, if we take into account the role of knowledge and learning in decision-making processes (Rizzello, 2004). Path dependence refers to a series of events that constitute a self-reinforcement process evolving into one of many potential states. Creativity, on the other hand, derives from the original processes of synaptic reorganization continuously activated by the brain. These processes allow each individual to perceive the world in a differentiated manner. The evolution of the brain, which certainly depends on its preexisting neural structure, is path-dependent.

A rule-following behavioral pattern produces a path-dependent process. Social interactions often lead to patterns of what are called "positive feedbacks." Generalized increasing returns due to institutional complementarities appear to be a source of multiple equilibria (Bowles, 2004). Later on, North (1990) applies increasing returns arguments to institutions more broadly. With self-reinforcing sequences, a particular institutional pattern is reproduced over time; initial steps in a particular direction induce further movement in the same direction, such that over time it becomes difficult or impossible to reverse direction. The established institution generates powerful inducements that reinforce its own stability. Path dependence in the evolution of belief systems results from a "common cultural heritage" which "provides a means of reducing the divergent mental models that people in a society possess and constitutes the means for the intergenerational transfer of unifying perceptions" (North, 2005, p. 27). Cognition may have more subjective aspects, while culture enables individuals to develop intersubjectively shared mental models. Understanding the culturally-derived norms of a specific society is crucial to the perception of institutional change. Societies that are more complex than tribal communities are characterized by different and competing interpretations of the social environment. Because culture reflects habituation which one acquires in one's group, it is slow to change. Some individuals are resistant to alter their belief systems, and hence their resulting behavior. Belief systems are the ideas and thoughts common to individuals that govern social interaction. Thus, different cultures imply different belief systems.

7.6 SUMMARY

Friedrich Hayek's thought largely rests on the concept of spontaneous order. Spontaneous orders in human affairs are patterns that arise as the unintended consequences of individual actions. Spontaneous orders are characterized by specific unpredictability (i.e., their detailed features are unpredictable), but also by general predictability (i.e., their general properties are predictable). People live in a world of expectations about interactions with others' actions. It is necessary to provide a causal explanation of how people are able to form expectations that are sufficiently reliable to facilitate the degree of plan coordination, if the possibility of social order is to be satisfactorily explained. The dissemination of knowledge is crucial in society. Knowledge is not given to anyone in its totality. Individual knowledge is order-dependent: people process signals and transform them into knowledge. Price movements signal changes to which individuals are motivated to adapt, even in the absence of any knowledge as to their causes. When individuals implement their plans formulated on the basis of their local knowledge, their actions generate changes in relative prices which summarize the knowledge for the scarcity of various kinds of resources. Then, the information provided by market prices enables individuals to form reasonably accurate expectations of others' plans.

The model of cognitive limits in human beings is grounded on the theory of the abstractness of the human mind. Social rules are behavioral patterns that individuals expect each other to follow. We are able to use social rules, but we are not necessarily able to spell them out. Relying on rules is a device we have learned to use, because our reason is insufficient to master the details of complex reality. There are always unexplained elements in the behavior of human beings. Individuals often fail to understand exactly how they ought to act. If rules are recognized as recurrent patterns of behavior, individuals act according to rules of conduct. These rules summarize and aggregate society's beliefs and attitudes; the rules are followed because individuals with limited cognition have to rely on them in exercising their actions. The notion of the abstractness of the human mind is combined with that of rule-governed behavior. By harmonizing expected responses, rules of conduct reduce uncertainty. The diffusion of shared behavioral patterns is necessary to obtain social order. Shared rules facilitate decision-making in complex situations by limiting the range of circumstances to which individuals have to pay attention.

REFERENCES

Altman, M., 2000. A behavioral model of path dependency: the economics of profitable inefficiency and market failure. J. Behav. Exp. Econ. 29, 127–145.

Arthur, W.B., 1989. Competing technologies, increasing returns, and lock-in by historical events. Econ. J. 99, 116–131.

Arthur, W.B., 1994. Increasing Returns and Path Dependency in the Economy. University of Michigan Press, Ann Arbor.

Aumann, R.J., 1976. Agreeing to disagree. Annals of Statistics 4, 1236–1239.

Bergstron, C.T., Rosvall, M., 2011. The transmission sense of information. Biol. Philos. 26, 159–176.

Berry, J.W., 1980. Social and cultural change. In: Triandis, H.C., Beislin, R.W. (Eds.), Handbook of Cross Cultural Psychology, Vol. 5. Allyn & Bacon, Boston, pp. 211–280.

Bicchieri, C., 2006. The Grammar of Society: The Nature and Dynamics of Social Norms. Cambridge University Press, Cambridge, MA.

Boettke, P., 2002. Information and knowledge: Austrian economics in search of its uniqueness. Rev. Austrian Econ. 15, 263–274.

Bowles, S., 2004. Microeconomics: Behavior, Institutions, and Evolution. Princeton University Press, Princeton, NJ.

Boyd, R., Richerson, P.J., 1985. Culture and the Evolutionary Process. University of Chicago Press, Chicago.

Boyd, R., Richerson, P.J., 1993. Rationality, imitation, and tradition. In: Day, R.H., Chen, P. (Eds.), Nonlinear Dynamics and Evolutionary Economics. Oxford University Press, Oxford, pp. 131–149.

Brandenburger, A., 1992. Knowledge and equilibrium in games. J. Econ. Perspect. 6, 83–101.

Brandenburger, A., Dekel, E., 1988. The role of common knowledge assumptions in game theory. In: Hahn, F. (Ed.), The Economics of Missing Markets. Information and Games. Clarendon Press, Oxford, pp. 46–61.

Caldwell, B., 2014. Introduction. In: Caldwell, B. (Ed.), The Collected Work of F. A. Hayek: The Market and Other Orders, Vol. 15. University of Chicago Press, Chicago.

Camerer, C.F., Ho, T., Chong, K., 2004. A cognitive hierarchy model of games. Q. J. Econ. 119, 861–898.

Carlsson, H., van Damme, E., 1993. Global games and equilibrium selection. Econometrica 61, 989–1018.

Cavalli-Sforza, L.L., Feldman, M.W., 1973. Cultural versus biological inheritance: phenotypic transmission from parents to children (a theory of the effect of parental phenotypes on children's phenotypes). Am. J. Hum. Genet. 25, 618–637.

Cavalli-Sforza, L.L., Feldman, M.W., 1981. Cultural Transmission and Evolution: A Quantitative Approach. Princeton University Press, Princeton, NJ.

Cooper, R.W., 1999. Coordination Games: Complementarities and Macroeconomics. Cambridge University Press, Cambridge, MA.

Cooper, R.W., John, A., 1988. Coordinating coordination failures in Keynesian models. Q. J. Econ. 103, 441–463.

Cooper, R.W., DeJong, D.V., Forsythe, R., Ross, T.W., 1990. Selection criteria in coordination games: some experimental results. Am. Econ. Rev. 80, 218–233.

Cosmides, L., Tooby, J., 1992. Cognitive adaptations for social exchange. In: Barkow, J., Cosmides, L., Tooby, J. (Eds.), The Adapted Mind. Oxford University Press, Oxford, pp. 163–228.

Crawford, V., 1998. A survey of experiments on communication via cheap talk. J. Econ. Theory 78, 286–298.

Cubitt, R., Sugden, R., 2003. Common knowledge, salience and convention: a reconstruction of David Lewis' game theory. Econ. Philos. 19, 175–210.

David, P.A., 1985. Clio and the economics of QWERTY. Am. Econ. Rev. 75, 332–337.

Dawkins, R., 1976. The Selfish Gene. Oxford University Press, Oxford.

Denzau, A.T., North, D.C., 1994. Shared mental models: ideologies and institutions. Kyklos 47, 3–31.

DiMaggio, P., 1997. Culture and cognition. Ann. Rev. Sociol. 23, 263–287.

Fleetwood, S., 1995. Hayek's Political Economy: The Socio-Economics of Order. Routledge, London.

Foss, K., Foss, N.J., 2006. The limits to designed orders: authority under "distributed knowledge" conditions. Rev. Austrian Econ. 19, 261–274.

Foster, J., 2006. Why is economics not a complex system science? J. Econ. Issues 40, 1069–1091.

Gintis, H., 2007. A framework for the unification of the behavioral sciences. Behav. Brain Sci. 30, 1–16.

Goldstein, I., Pauzner, A., 2005. Demand-deposit contracts and the probability of bank runs. J. Finance 60, 1293–1327.

Granovetter, M., 1985. Economic action and social structure: the problem of embeddedness. Am. J. Sociol. 91, 481–510.

Harris, M., 1971. Culture, Man and Nature: An Introduction to General Anthropology. Thomas Y. Crowell, New York.

Hayek, F.A., 1948. Individualism and Economic Order. University of Chicago Press, Chicago.

Hayek, F.A., 1960. The Constitution of Liberty. University of Chicago Press, Chicago.

Hayek, F.A., 1967. Studies in Philosophy, Politics, and Economics. Routledge & Kagan Paul, London & Henley.

Hayek, F.A., 1973. Law, Legislation, and Liberty, Vol. 1, Rules and Order. University of Chicago Press, Chicago.

Hayek, F.A., 1976. Law, Legislation, and Liberty, Vol. 2, The Mirage of Social Justice. University of Chicago Press, Chicago.

Hayek, F.A., 1978. New Studies in Philosophy, Politics, Economics and the History of Ideas. Routledge, London.

Hayek, F.A., 1988. The Fatal Conceit: The Errors of Socialism. University of Chicago Press, Chicago.

Hayek, F.A., 1994. In: Kresge, S., Wenar, L. (Eds.), Hayek on Hayek: An Autobiographical Dialogue. University of Chicago Press, Chicago.

Hayek, F.A., 2014. In: Caldwell, B. (Ed.), The Collected Works of F. A. Hayek, Vol. 15, The Markets and Other Orders. University of Chicago Press, Chicago.

Hodgson, G.M., Knudsen, T., 2010. Darwin's Conjecture: The Search for General Principles of Social and Economic Evolution. University of Chicago Press, Chicago.

Khalil, E.L., 2013. Lock-in institutions and efficiency. J. Econ. Behav. Organ. 88, 27–36.

Kirzner, I.M., 1973. Competition and Entrepreneurship. University of Chicago Press, Chicago.

Kirzner, I.M., 1979. Perception, Opportunity and Profit. University of Chicago Press, Chicago.

Kirzner, I.M., 1992. The Meaning of Market Process, Essays in the Development of Modern Austrian Economics. Routledge, London.

Kirzner, I.M., 1997. Entrepreneurial discovery and the competitive market process: an Austrian approach. J. Econ. Lit. 35, 60–85.

Klein, D.B., 1997. Conventions, social order, and the two coordinations. Const. Polit. Econ. 8, 319–335.

Lachmann, L.M., 1970. The Legacy of Max Weber. Heinnemann, London.

Lachmann, L.M., 1976. From Mises to Shackle: an essay on Austrian economics and the Kaleidic society. J. Econ. Lit. 14, 54–62.

Lachmann, L.M., 1986. The Markets as an Economic Process. Basil Blackwell, Oxford.

Leeson, P.T., Coyne, C.J., Boettke, P.J., 2006. Converting social conflict: focal points and the evolution of cooperation. Rev. Austrian Econ. 19, 137–147.

Leibenstein, H., 1966. Allocative efficiency vs. "X-efficiency". Am. Econ. Rev. 56, 392–415.

Leibenstein, H., 1979. A branch of economics is missing: micro-micro theory. J. Econ. Lit. 17, 477–502.

Lewis, D., 1969. Convention: A Philosophical Study. Harvard University Press, Cambridge, MA.

Lewis, P., 2012. Emergent properties in the work of Friedrich Hayek. J. Econ. Behav. Organ. 82, 368–378.

Lewis, P., 2015. Notions of order and process in Hayek: the significance of emergence. Camb. J. Econ. 39, 1167–1190.

Liebowitz, S.J., Margolis, S.E., 1990. The fable of the keys. J. Law Econ. 33, 1–25.

Liebowitz, S.J., Margolis, S.E., 1994. Network externalities: an uncommon tragedy. J. Econ. Perspect. 33, 133–150.

Littlechild, S.C., 1986. Three types of market process. In: Langlois, R.N. (Ed.), Economics as a Process. Cambridge University Press, Cambridge, MA.

Markose, S.M., 2005. Computability and evolutionary complexity: markets as complex adaptive systems (CAS). Econ. J. 115, F156–F192.

Maynard Smith, J., 1982. Evolution and the Theory of Games. Cambridge University Press, Cambridge, MA.

McKelvey, R.D., Palfrey, T.R., 1995. Quantal response equilibrium for normal form games. Games Econ. Behav. 10, 6–38.

Mises, L., 1935. Economic calculation in the socialist commonwealth. In: Hayek, F.A. (Ed.), Collectivist Economic Planning. Routledge & Sons, London.

Mises, L. ([1949] 1966). Human Actions, A Treatise on Economics. Chicago: Henry Regnery.

Morris, S., Shin, H.S., 1998. Unique equilibrium in a model of self-fulfilling currency attacks. Am. Econ. Rev. 88, 587–597.

Nelson, R.R., Winter, S., 1982. An Evolutionary Theory of Economic Change. Harvard University Press, Cambridge, MA.

North, D.C., 1990. Institutions, Institutional Change and Economic Performance. Cambridge University Press, Cambridge, MA.

North, D.C., 2005. Understanding the Process of Economic Change. Princeton University Press, Princeton, NJ.

O'Driscoll Jr., G.P., Rizzo, M.J., 1996. The Economics of Time and Ignorance, *2nd ed.* Routledge, London.

Osborne, M.J., Rubinstein, A., 1994. A Course in Game Theory. The MIT Press, Cambridge, MA.

Polanyi, M., 1951. The Logic of Liberty. University of Chicago Press, Chicago.

Polanyi, M., 1969. In: Grene, M. (Ed.), Knowing and Being: Essays by Michael Polanyi. University of Chicago Press, Chicago.

Richerson, P.J., Boyd, R., 2005. Not by Genes: How Culture Transformed Human Evolution. University of Chicago Press, Chicago.

Rizzello, S., 2004. Knowledge as a path-dependent process. J. Bioecon. 6, 255–274.

Schein, E.H., 1992. Organizational Culture and Leadership. Jossey-Bass, San Francisco.

Schelling, T.C., 1960. The Strategy of Conflict. Harvard University Press, Cambridge, MA.

Shannon, C.E., Weaver, W., 1949. The Mathematical Theory of Communication. University of Illinois Press, Urbana, IL.

Simon, H.A., 1957. Models of Man: Social and Rational. John Wiley, New York.

Simon, H.A., 1976. From substantive to procedural rationality. In: Latsis, S.J. (Ed.), Method and Appraisal in Economics. Cambridge University Press, Cambridge, MA, pp. 129–148.

Simon, H.A., 1978. Rationality as process and as product of thought. Am. Econ. Rev. 68, 1–16.

Skyrms, B., 1996. The Evolution of the Social Contract. Cambridge University Press, Cambridge, MA.

Skyrms, B., 2010. Signals: Evolution, Learning, and Information. Oxford University Press, Oxford.

Stahl, D.O., Wilson, P., 1995. On players' models of other players. Games and Economic Behavior 10, 218–254.

Sugden, R., 1986. The Economics of Welfare, Rights and Co-operation. Basil Blackwell, Oxford.

Sugden, R., 1989. Spontaneous order. J. Econ. Perspect. 3, 85–97.

Sugden, R., 1995. A theory of focal points. Econ. J. 105, 533–550.

Vanberg, V.J., 1994. Rules and Choice in Economics. Routledge, London.

Vanberg, V.J., 2007. Corporate social responsibility and the 'game of catallaxy': the perspective of constitutional economics. Const. Polit. Econ. 18, 199–222.

Van Huyck, J.B., Battalio, R.C., Beil, R.O., 1990. Tacit coordination games, strategic uncertainty, and coordination failure. Am. Econ. Rev. 80, 234–249.

Wilson, E.O., 1975. Sociobiology: The New Synthesis. Harvard University Press, Cambridge, MA.

Chapter 8

Understanding Institutional Evolution

8.1 INTRODUCTION

This chapter discusses the dynamic interaction between individuals and institutions. Theories of institutions can be classified into two broad approaches: institutions-as-rules and institutions-as-equilibria. According to the institutions-as-rules approach, institutions are conceived as rules or constraints that guide the actions of individuals engaged in social interactions. On the other hand, the institutions-as-equilibria approach views institutions as behavioral patterns or regularities.

Institutional change covers both the process of changing an existing institution and the establishment of a new institution. In any case, institutional structures and individual actions coevolve. In order to have a complete picture of institutions, we need to take both approaches, institutions-as-rules and institutions-as-equilibria, into consideration.

Coevolution between two different systems is a key concept in this chapter. There are some coevolutionary processes in biological, ecological, and economic studies. For example, in gene–culture coevolutionary theory, genes adapt to a fitness landscape of which cultural forms are critical, and the resulting genetic changes lay the basis for further cultural evolution (Gintis, 2007).

In ecological economics, Norgaard (1994) argues that social and ecological systems should be studied as coevolving systems. For Norgaard (1994), coevolution involves relationships between entities which affect the evolution of the entities. Norgaard (1994) describes the coevolutionary process as follows:

> *Thinking of the changes in social and environmental systems over time as a process of coevolution acknowledges that cultures affect which environmental features prove fit and that environments affect which cultural feats prove it. In this sense, coevolution accepts both environmental determinisms and cultural determinism while recasting them as a selection process.*

> (Norgaard, 1994, p. 71)

The Cognitive Basis of Institutions. DOI: https://doi.org/10.1016/B978-0-12-812023-1.00008-9

Norgaard (1994) criticizes economics (both neoclassical and Marxist) for treating nature as a passive pool exploited by advancing technologies. Human beings adapt to their external environments, actively transform them, and adapt to their transformations. Coevolution denotes the fact that evolutionary changes in one subsystem are a response to changes in the other subsystem with which it interacts.[1]

Nelson and Winter's (1982) evolutionary economic theory is also grounded on an explicit dynamic account of the interaction between mechanisms of variation (which constantly introduce variety, novelty, and heterogeneity among routines), and mechanisms of selection (which tend to reduce heterogeneity among routines). Furthermore, Nelson (1995) argues technological systems change arising through the coevolution of technologies, industrial organization, and institutions. For Nelson (1995), this is the fundamental process underlying economic progress:

> *It is clear that, somehow, in the now advanced industrial nations, there have been mechanisms that have made the coevolution of technology, industrial organization, and institutions more broadly, move in directions that have led to sustained economic progress. Private actions leading to "self organization" have been part of the story, but collective action has been as well. It is absurd to argue that processes of institutional evolution "optimize"; the very notion of optimization may be incoherent in a setting where the range of possibilities is not well defined, even if the issue of different interests could be resolved in this terminology . . .*

(Nelson, 1995, p. 83, original emphasis)

Changes in subsystems, such as knowledge, values, technologies, organizations, and environments, are analyzed as evolutionary processes. These subsystems interact and coevolve. Coevolutionary change is path-dependent, which highlights the lock-in of technologies, institutions, and environments, and explains why it is difficult to escape from prevailing practices. Coevolution, however, includes the generation of new variation, and hence long-term opportunities to break through path dependencies.

As a consequence, a large variety of institutions and industrial organizations can influence the growth trajectories across advanced economies. The concept of institutional complementarities is based on multilateral reinforcement mechanisms between institutional arrangements (Aoki, 2001). Each institution defines a set of constraints that determine an individual's strategies. The influence of one institution is reinforced when the other complementary institution is present.

---

1. Recently, coevolution denotes very different types of interactions: biological—cultural, ecological—economic, production—consumption, technology—preferences, behavior—institution, and human genetic—cultural (Durham, 1991; Gowdy, 1994).

The contemporary world in which we find ourselves is characterized by numerous interpretations of cultures. Culture shapes the core values and norms of its group members. A cultural group consists of individuals who share a set of beliefs that is recognizably different from those of other groups. There is a tendency for people to fail to recognize that their life is only one of many possible lives. People are subject to cultural bias, failing to adjust properly for possible information with different cultural backgrounds. Every social group has its own culture, and one individual belongs to one culture originally. Different groups are characterized by different systems of rules (traditions, customs, and practices). Individuals follow certain rules of conduct because others do the same in the same group.

Changes occur when groups with different cultural backgrounds begin to interact. Culture can change by coming into contact with another culture. Cultural differences between groups can generate adaptations at the individual level. Individuals choose either to imitate novel practices or to condemn them. Whether acculturation occurs or not depends on the extent to which people are attracted to the other culture, and on how deeply they strive to maintain their own culture. Social learning is the process by which experiences modify preexisting behavior and understanding. Each new conception links to a process of classification into categories. If no experience corresponds to any new information, a process of recombination is necessary to permit its classification (Hayek, 1952). The human nervous system may be modified to display novel actions. As individuals interact with members of the other group(s), they may experience the world through those other cultural members. The mechanism of cultural influence can be understood in the context of the human capacity to adapt to novel cultural systems. Cultural influences are not homogeneous within any culture, and one can expect significant differences in the degree of acquisition of cultural schemata.

Culture can be defined as socially transmitted information that has a lasting effect on an individual's behavioral phenotype (Boyd and Richerson, 1985). The essential feature of culture is social learning that results in the transfer of skills, thoughts, and feelings from person-to-person. There are several channels through which these transfers might occur: by simple imitation, by the transmission of beliefs, or by altering preferences directly. Cultural transmission allows societies to solve problems that no individual, or group of individuals, can solve at a point in time. However, human populations living in similar environments can exhibit different cultural patterns of behavior, and hold different beliefs and values that can be adaptive, sometimes neutral, or even maladaptive (Boyd and Richerson, 1985). The theory of dual inheritance in Boyd and Richerson (1985) takes for granted that culture is a part of human biology, but it assumes that culture acts as a second system of inheritance that uses transmission rules that are different from genetic inheritance. Then, much variation in human behavior cannot be understood without accounting for beliefs, values, and other socially acquired determinants of behavior. According to Boyd and Richerson (1985), each agent of the new generation chooses a cultural model from the previous generation, either from their parents or an individual taken at random, and adopts his or her behavior. As a consequence, the initial frequencies of the alternative cultural behaviors are maintained, and culture replicates the phenotypic structure of the parental generation.

When two groups of individuals having different cultures come into intercultural contact, several possibilities arise. First, two cultures may endure as two distinct cultures (Axelrod, 1997). Second, people in the minority group may adopt the convention of the majority group in the process of assimilation (Lazear, 1999). Third, two cultures may be modified to induce an eclectic convention. All agents, in the third case, make cognitive adjustments after interactions with members of the other group. Because the schemata people employ are socially learned, they can be altered through socialization. Agents in both groups have learned different practices from others, whereby a new custom may be established and shared. We need to revisit our mental models to fit what will be appropriate in a variety of challenging settings. A process of acculturation may begin and lead to cultural adaptations between both groups (Berry, 2005). Acculturation is the dual processes of cultural and individual changes that take place as a result of contact between two cultural groups. Such an internal revision of the rules of conduct provides each agent with the opportunities for understanding and communication. At the cultural level, changes in both groups emerge during the process of acculturation. At the individual level, on the other hand, individuals adapt to their new situations and change their action plans. Different cultures develop common traits, through mutual understanding and communication.

The cascade of cultural changes associated with modernization is the result of the momentous change that occurs early in the economic environment. Premodern cultures share certain beliefs and values that are different from those of economically developed cultures. With economic development, people begin to abandon such traditional beliefs and values.[2] For example, people adopt the idea that smaller families are better, even though their increasing wealth makes it easier to raise a large family. Individuals within modernizing populations begin to abandon traditional attitudes, such as the desire for a large family and passivity in the face of obligations to family,

---

2. Karl Polanyi (1957) highlights the uniqueness of the unbound market in its capacity to unhinge the social elements of society and to replace them with economic ones: "All types of societies are limited by economic factors. Nineteenth century civilization alone was economic in a different and distinctive sense, for it chose to base itself on a motive only rarely acknowledged as valid in the history of human society, and certainly never before raised to the level of a justification of action and behavior in everyday life, namely, gain. The self-regulating market was uniquely derived from this principle" (Polanyi, 1957, p. 30).

and replace them with modern attitudes, such as a desire to better oneself, and a recognition of the need to control family size. When a society begins to undergo economic development, new kinds of organizations for production and trade provide people with new means of making their living. Through social interaction, members of a group generate and share norms that coordinate group behavior. When societies modernize, different beliefs and values are likely to emerge.

The differentiation of knowledge is a condition of progress in human society. However, differences in the structure of understanding, through providing necessary frameworks for the construction of knowledge of distinct kinds, may create substantial obstacles to the integration of these distinct kinds of knowledge. Such differences in perception are identified in Hayek's cognitive theory. That is, "events which to our senses may appear to be of the same kind may have to be treated as different in the physical order, while events which physically may be of the same or at least a similar kind may appear as altogether different to our senses" (Hayek, 1952, p. 4). Human knowledge is conditioned by the structure of the human mind.[3] People learn not only explicitly, but also implicitly, from language, customs, traditions, and so on. Therefore, knowledge has a social dimension; people cannot articulate much of what they know. Hayek's cognitive theory helps explain cultural similarities and differences; the sensory order can be similar, but never identical, among individuals.

8.2 INSTITUTIONS AS RULES

The institutions-as-rules approach has its roots in the seminal contribution of Douglass North's 1990 book, *Institutions, Institutional Change and Economic Performance*. North's institutional approach is stated in the opening paragraphs:

> *Institutions are the rules of the game in a society or, more formally, are the humanly devised constraints that shape human interaction. In consequence they structure incentives in human exchange, whether political, social, or economic . . .. Institutions reduce uncertainty by providing a structure to everyday life. They are a guide to human interaction, so that when we wish to greet friends on the street, drive an automobile, buy oranges, borrow money, form a business, bury our dead, or whatever, we know (or can learn easily) how to perform these tasks.*

(North, 1990, pp. 3–4)

---

3. Relatively complex nervous systems will operate by multiple classifications. Specifically, the classes at one level may be grouped to form classes at a higher level. In the system of connections, as Hayek (1952, p. 51) describes, "the distinct responses which affect the grouping at a first level become in turn subject to a further classification (which also may be multiple in both the former senses)."

Hayek ([1967] 2014, p. 279) also argues as follows: "[n]ot every system of rules of individual conduct will produce an overall order of the actions of a group of individuals, and whether a given system of rules of individual conduct will produce an order of actions, and what kind of order, will depend on the circumstances in which the individuals act." Thus, according to North and Hayek, individuals are institutionally embedded or constrained. Institutions structure human agency. Rules indicate what to do or what not to do in a given circumstance.[4] Rules guide actual behavior, and can be considered as prescribed behavioral patterns.

Ideas and ideologies affect individual's behavioral patterns. That is, "ideas and ideologies shape the subjective mental constructs that individuals use to interpret the world around them and make choices" (North, 1990, p.111). Incentives, perceptions, and ways of thinking are "socially transmitted ... and are a part of the heritage we call culture" (North, 1990, p. 37). Individual action is, in that sense, socially and historically determined. An individual is constrained by the existing institutional structure. Institutions influence individual behavior in crucial ways.

Institutions are means by which individuals are able to gather sufficient information in order to coordinate. Institutions save knowledge and information, and they provide "points of orientation" likely to make actions and expectations relatively compatible (Lachmann, 1970). Regularity in social behavior specifies behavior in specific recurrent situations. People believe that certain actions will lead to certain outcomes. For Lachmann (1976), human action can be understood only in terms of the plan that gave rise to it. Individuals carry an image of what they want to achieve in their minds. Human action is the implementation of plans created to bring about their imagined ends. According to Lachmann (1976), plans are divergent because subjective expectations are based on the image that individuals form about an unknown, though not unimaginable, future:

> *Future knowledge cannot be had now, but it can cast its shadow ahead. In each mind, however, the shadow assumes a different shape, hence the divergence of expectations. The formation of expectations is an act of our mind by means of which we try to catch a glimpse of the unknown. Each of us catches a different glimpse ... Divergent expectations are nothing but the individual images, rather blurred, in which new knowledge is reflected, before its actual arrival, in a thousand different mirrors of various shapes.*
>
> (Lachmann, 1976, p. 59)

Divergence of plans is the consequence of the extension of subjectivism to expectations. The principle of subjectivism implies that different people

---

4. Searle (2005) claims that an institution is any system of constitutive rules of the form "X counts as Y in C," where X is a preinstitutional entity, Y is a status function, and C refers to the domain of application of the rule.

can interpret the same situation in different ways, and can also form different expectations on the basis of those expectations. Because of diverse experiences, individual agents will respond differently to the same objectively-defined stimulus. That is, "different men in identical situations may act differently because of their different expectations of the future" (Lachmann, 1970, p. 36). The problem of divergent expectations is more fundamental: even though the market involves an adequate means for the dispersion of knowledge regarding the appropriateness of "past" actions, the learning involved is in fact useless without knowledge of the "future." Some peoples' plans will invariably fail and need to be revised. There is no guarantee that the process of plan revision will always result in overall plan coordination. The economic configuration emerging from the interaction of individual plans is definitely one of disequilibrium. How do the plans of different people relate to one another? According to Lachmann (1970, p.49), "[t]he answer has to be sought in the existence, nature, and functions of institution." For Lachmann, a social structure is just a "recurrent pattern of events" (Lachmann, 1970, p. 23). Institutions serve to structure specific recurring interaction situations. Lachmann regards people as social beings whose beliefs, expectations, and plans are profoundly shaped by the institutional environment in which they live. Individual actions are guided by intersubjectively shared social structures. This implies that people are able to understand and anticipate how others will act in particular situations, thereby facilitating the formation of reliable expectations and mutually compatible plans. If people have no grounds for believing that a particular set of institutions will endure into the future, they will have no reason to orient their plans towards those institutions. Institutions play a crucial role in enabling people to formulate expectations that are sufficiently accurate for the successful coordination of plans. Lachmann (1970) describes these in the following ways:

> An institution provides means of orientation to a large number of actors. It enables them to co-ordinate their actions by means of orientation to a common signpost. If the plan is a mental scheme in which the conditions of action are co-ordinated, we may regard institutions, as it were, as orientation schemes of the second order, to which planners orient their plans as actors orient their actions to a plan ... The existence of such institutions is fundamental to a civilized society. They enable each of us to rely on the actions of thousands of anonymous others about whose individual purposes and plan we can know nothing. They are nodal points of society, co-ordinating the actions of millions whom they relieve of the need to acquire and digest detailed knowledge about others and form detailed expectations about their future action
>
> (Lachmann, 1970, p. 49–50)

People face great uncertainty when they assess the acceptability of their actions, make predictions about how others will act, and project likely results.[5] They have to form expectations in the context of ignorance about reality. However, expectations can be formed as a result of rule-governed creativity. Following rules helps people make decisions with some degree of certainty about which behavior is acceptable and which is not. The fact that expectations are informed by a set of common rules implies that people are able to anticipate how others will act in particular situations, and to expect that others will be able to do likewise in response to their own actions. This reduces the amount of information individuals must collect, and enhances their ability to make plans and to coordinate with each other. Institutions coordinate individual actions at a lower cost, because they reduce the volatility in the plans of others. In order for knowledge to disseminate among individuals (and therefore become a possible source of institutional change), there must be some underlying society that is knowledge-based.[6] It can be considered as a social network that shares beliefs, providing the information channels that circulate ideas. People learn not only explicitly, but also implicitly, from language, customs, traditions, and so on. Therefore, knowledge has a social dimension. Institutions are tightly interwoven with the reduction of uncertainty via knowledge. In his *The Counter-Revolution of Science*, Hayek, ([1952] 1979, p. 217, original emphasis) suggests as follows: "[t]here is a great deal of knowledge which we never consciously know implicit in the knowledge of which we are aware, knowledge which yet constantly serves us in our actions, though we can hardly be said to 'possess' it."

We cannot articulate much of what we know. It is important to realize that this tacit knowledge serves in the internal structure of action. "Knowing how" corresponds to tacit knowledge, while "knowing that" relates to conscious knowledge. Knowing how consists in using habits and following rules whose nature and definitions do not need to be explained in the individual's mind. Individual agents are creating institutions which they do not know in advance. In fact, Hayek ([1952] 1979, p. 150) contends that "many of the greatest things man has achieved are the result not of consciously directed thought, and still less the product of a deliberately coordinated effort of many individuals, but of a process in which the individual plays a part which he can never fully understand."

---

5. Humans have acquired the capacity to interpret others' actions in terms of mental states of others. This capacity is generally referred to as "theory of mind" (Gallese and Goldman, 1998). Humans develop theory of mind at an early age. A person with theory of mind has the ability to conceive of themselves, and of others, as intentional beings. To understand others' actions in terms of their mental states is a major ingredient in successful social interactions (Frith and Frith, 2001).

6. We can consider the relationship between individual and collective minds from the perspective of individual and collective rationality. According to Arrow (1974), organizations can acquire more information than any individual, and this type of collective rationality can extend the domain of individual rationality.

Institutions not only provide people with information, but also influence the very perception that people have of reality. Concerning the coordination of knowledge, Hayek (1948) argues as follows: "[h]ow can the combination of fragments of knowledge existing in different minds bring about results which, if they were to be brought about deliberately, would require a knowledge on the part of the directing mind which no single person can possess?" (Hayek, 1948, p. 79). The fragments of knowledge existing in different minds are not justified true beliefs. If all the dispersed bits of knowledge were true, there could be no contradictions between them and, therefore, no need for their resolution through coordination. The crucial problem is how individuals with different knowledge are able to coordinate their behavior, despite such differences. The rules, guiding individual actions as a whole, are abstract and unconscious. Individuals are therefore unable to explain the actions of others. However, they can understand the action of others, because others act according to a similar mode of categorization of the real world.

8.3 INSTITUTIONS AS EQUILIBRIA

In contrast to the approach just described in the previous section, the second characterization views institutions as equilibria of strategic games. Within the institutions-as-equilibria approach, institutions are roughly regularities of behavior. According to Schotter (1981), a social institution is defined as "a regularity in social behavior that is agreed to by all members of society, specifies behavior in specific recurrent situations and is either self-policed or policed by some external authority" (Schotter, 1981, p. 11). Such a regularity is described as an equilibrium in strategic, noncooperative games. It is also the product of a dynamic process in which interactions become patterned. Then, expectations and interpretations of behavior become generalized among individuals. In the institutions-as-equilibria approach, individuals, interacting with others, are assumed to continue to change their behavioral responses to the actions of others, until no improvement can be obtained in their expected outcomes from independent actions. For a rule to be an institution, it must be widely distributed in the minds of the group members. Repeated patterns of behavior create expectations about future behavior, and ensure a degree of predictability in social interaction. When others' behavior is relatively easy to predict, individual agents spend less time deciphering actions, and are better able to coordinate with others. Institutions depend on the individuals who reproduce, transform, or create them. To understand why regularized patterns of interaction exist, we need to ask why all agents would be motivated to produce a particular equilibrium. According to Aoki (2001), an institution is defined as a system of self-sustained, shared beliefs that motivate these behaviors. The institutions that create a social environment must be ultimately traceable to individual mind-sets, because all social phenomena are the consequence of individual actions and plans.

Here, equilibrium refers to a situation in which individual knowledge and expectations, and the actions based on these, are compatible with the actions of other individuals. The concept of equilibrium is explored from a subjective perspective. In an equilibrium state, individual plans are fully coordinated, and each plan is successfully executed. The problem is to explain how individuals' subjective beliefs come to conform to the objective data of the world around them. Hayek (1948) considers equilibrium as a relationship between individual actions:

> *I have long felt that the concept of equilibrium itself and the methods which we employ in pure analysis have a clear meaning only when confined to the analysis of the action of a single person and that we are really passing into a different sphere and silently introducing a new element of altogether different character when we apply it to the explanation of the interactions of a number of different individuals.*
>
> (Hayek, 1948, p. 35)

Unlike its counterpart in the physical world, Hayek (1948) realizes that the notion of equilibrium in economics must refer to the views of individuals. Hayek's concept of equilibrium has little to do with Walrasian general equilibrium theory. As Hayek (1948) points out, economic models usually explain equilibrium by assuming an equilibrium starting point. Hayek's concept of equilibrium allows the introduction of a fundamental analysis of the inter-temporal compatibility between individual plans. Hayek (1948) goes on to argue that "if we want to make the assertion that, under certain conditions, people will approach the [equilibrium] state, we must explain by what process they will acquire the necessary knowledge" (Hayek, 1948, p. 46). Then, Hayek (1948) suggests inter-temporal equilibrium as follows:

> *[S]ince equilibrium relations exist between the successive actions of a person only in so far as they are part of the execution of the same plan, any change in the relevant knowledge of the person, that is, any change which leads him to alter his plan, disrupts the equilibrium relations between his actions taken before and those taken after the change in his knowledge. In other words, the equilibrium relationship comprises only his actions during the period in which his anticipations prove correct. ... [S]ince equilibrium is a relationship between actions, and since the actions of one person must necessarily take place successively in time, it is obvious that the passage of time is essential to give the concept of equilibrium any meaning.*
>
> (Hayek, 1948, p. 36–37)

If equilibrium (as a relation between actions) exists at a point in time, individual plans are consistent with each other, and are completely coordinated. In the absence of any change, equilibrium will exist at every point in time. Such a state requires that the mutual beliefs and expectations of individuals are congruent with the objective facts of the world. An individual's plans and actions become the objective facts that others must take into account in their plans. We must discover how people come to know enough to carry out mutually consistent plans. As Hayek (1948) argues, equilibrium exists "if the actions of all members of the society over a period are all executions of their respective plans" (Hayek, 1948, p. 37). "Correct foresight" is not "a precondition which must exist in order that equilibrium may be arrived at. It is rather the defining characteristic of a state of equilibrium" (Hayek, 1948, p. 42). According to Hayek, there must be *ex ante* mutual compatibility among the different plans in the system. First, each agent must correctly anticipate the actions of others with whom he or she intends to transact. Second, each agent's plan must be based on the same expectations of external events. This condition requires convergence of expectations of external events; it rules out inconsistent plans based on divergent expectations. The recurrent patterns of conduct help realize plans by reducing the volatility in the plans of other agents. If a tendency exists toward equilibrium, individual expectations will be more and more correct. An equilibrating tendency can be a tendency of institutions to coordinate individual actions. Some institutional environments allow individuals to learn and act on the information that is dispersed throughout the society. "The only trouble," as Hayek (1948, p. 45, original emphasis) points out, "is that we are still pretty much in the dark about (a) the *conditions* under which this tendency is supposed to exist and (b) the nature of the *process* by which individual knowledge is changed."

In his pioneering study *Convention*, Lewis (1969) characterizes the notion of convention in terms of coordination equilibria, that is, equilibria of noncooperative games, such that every player strictly prefers that all conform to the equilibrium.[7] For Lewis (1969), a convention is a state in which players engaged in a game play a coordination equilibrium, and their preferences to conform to this equilibrium are common knowledge. A coordination equilibrium of a noncooperative game is a strategy combination such that all players would be strictly worse off if any player were to deviate from this equilibrium. If no one will wish to deviate from a coordination equilibrium, the convention corresponding to the equilibrium will be stable over time.

In order to explain how a particular convention might emerge, Lewis (1969) introduces the notion of "salience." A coordination equilibrium is salient to the players if it somehow stands out so that all expect each other to coordinate on this equilibrium. According to Lewis (1969), players do not necessarily need pre-play communication. Salience can result from the players correlating their strategies and their expectations with certain events

---

7. According to Crawford and Ostrom (1995), Lewis (1969), Ullmann-Margalit (1977), and Coleman (1987) are characterized as an institutions-as-norms approach. This approach assumes that many observed patterns of interaction are based upon the shared perceptions among a group of individuals of proper and improper behavior in particular situations.

external to the game. This particular convention is defined as a correlated equilibrium (Aumann, 1974, 1987). In a correlated equilibrium, players tie their strategies to various "state of the world," or other pieces of information which are not formally part of the game.[8]

Institutions are conceived as correlated equilibria of coordination games with multiple equilibria. In a correlated equilibrium, individual actions are statistically dependent on an event or signal sent by a correlation device. One interpretation of correlated equilibrium is that a correlation device (here, an institution) instructs people to take actions, sending signals to them regarding what they have to do in a given situation. An institution is a signaling mechanism that the individuals can use to coordinate their actions in social interaction. This suggests that correlation of actions in the population is not unrealistic. While the correlation required by predictable behavior may be asking too much, it is possible to obtain correlated equilibrium in the population.

In Aumann and Brandenburger (1995), their concern is the existence of a stable state of behavioral beliefs across players consistent with their possible behavioral choices. Aoki (2011) interprets the equilibrium theorem due to Aumann and Brandenburger (1995) in the context of societal games recursively played:

> *If the agents share a cultural belief and their payoff functions are mutually known up to the commonly cognizable states, then they share a stable identical behavioral belief up to the commonly cognizable states via a convergent quasi-public representation of their beliefs. Thus, a substantive form of institution entails as a convergent public representation. The converse is also almost true. That is, in order for the agents to share a stable identical behavioral belief on the commonly cognizably states via a substantive form of institution, they must be most likely to share a cultural belief.*

(Aoki, 2011, p. 27)

The societal games are being recursively played and expected to be played. In order for players' action choices to be mutually consistent and sustainable in equilibrium, each player need not know the details of the other players' intentions and choices. It is sufficient for the players to know only the salient features of the ways the game is repeatedly played. How is it possible that a cultural belief becomes commonly known among agents? In the context of societal games, a cultural belief will be historically shaped and commonly shared.[9] It is somewhat like a correlation device in correlated equilibrium.

---

8. In Vanderschraaf (1995), Lewis' (1969) conventions are shown to be correlated equilibria.
9. According to Aoki (2007), the following conceptualization of an institution is proposed; "An institution is self-sustaining, salient patterns of social interactions, as represented by meaningful rules that every agent knows and are incorporated an agent's shared belief about how the game is played and to be played" (Aoki, 2007, p. 6).

Correlated equilibrium allows players' actions to be statistically dependent on some random signals external to the model (Aumann, 1974, 1987). Correlated equilibrium only requires rationality and common priors, while Nash equilibrium requires stronger premises. Different players can take different actions potentially. However, their actions are conditional on some external signal. "Nature" first gives a publicly observable signal. Players' strategies assign an action to every possible observation. If no player has an incentive to deviate from the recommended strategy, the distribution is a correlated equilibrium. In general, social norms regulate the strategic interaction of social actors. If individuals play a game with several Nash equilibria, social norms can serve to choose the most socially desirable among these equilibria. Social norms play the role of a "choreographer" who leads people to take actions according to some commonly known probability distribution (Gintis, 2009). A social norm prescribes which strategy every player should choose depending on the observed signal, in such a way that it is rational for him or her to follow the suggestion, knowing that others are following it too.

A game is a description of a strategic interaction involving two or more players. A game is any situation with the following three aspects. First, there is a set of participants, whom we call the players. Second, each player has a set of options for how to behave. These options are referred to as the players' possible strategies. Third, for each choice of strategies, each player receives a payoff that can depend on the strategies chosen by every player. Everything that a player cares about is summarized in the player's payoffs.

A game G is defined as a triple (N, S, u) where:

$N = (1, 2, \ldots, n)$ is the set of the players of the game;
$S = S_1 \times \ldots \times S_n$ is the Cartesian product of the set S_i of available pure strategies for each player i; and
$u: S \rightarrow R$ is the payoff function mapping S into a n-tuple of real numbers for each strategy combination in S.

The triple (N, S, u) allows us to specify any strategic interaction. It is assumed that each player knows everything about the structure of the game.

Each player chooses a strategy to maximize their payoff, given their beliefs about the strategies used by the other players. Let $u_i (s_j; s^*)$ be the utility of player i playing strategy j, with other players playing the vector of strategies s^*. Player i playing s_j is rational if, for any other strategy k:

$$u_i(s_j; s^*) \geq u_i(s_k; s^*).$$

Nash (1950, 1951) proposed a principle for reasoning about behavior in games. The idea of Nash equilibrium is that if the players choose strategies that are best responses to each other, then no player has an incentive to deviate to an alternative strategy. The system is thus in a kind of equilibrium, with no force pushing it toward a different outcome.

A correlated equilibrium can be considered as a Nash equilibrium in an "extended game" where a choreographer or social norm is a player. The choices can be connected in some way; if players relate their choice of strategies with some jointly observable feature of their interaction, they can achieve a correlated equilibrium. One interpretation of correlated equilibrium is that a choreographer instructs people to take actions according to some commonly known probability distribution. The choreographer is considered as a social norm that assigns roles to individuals in society.

For any game (N, S, u), define Ω as the set of all possible worlds, and $\omega \in \Omega$ as the state of the world that actually obtains. Each player i possesses an information partition H_i on the states of the world. Then, H_i can be thought as player i's private information. Define $P = \cap_i H_i$ as the "common information partition," that is, the intersection of the information partitions of all players. $P \omega$ means that, at ω, each player knows which cell in P obtains. An event E is "public" when $P \omega \subseteq E$ since, for any $\omega \in P \omega$, everyone knows E, everyone knows that, and so on. Define a function x: $\Omega \to S$ as a system of strategies correlated with the state of the world, with $x(\omega) = (x_1(\omega), \ldots, x_n(\omega)) \in S$. Finally, define ($\Omega$, q) as the probability space over all possible states of the world, where q is a probability distribution over Ω.

A strategy profile $x^*(\omega) = (s_1^*, \ldots, s_n^*)$ forms a correlated equilibrium if, for each player i, for each state of the world ω and given the probability space (Ω, q):

$$E_i[u_i(x^*(\omega))|H_i](\omega) > E_i[u_i(s_i, x^*(\omega))|H_i](\omega).$$

Given each player's conjectures conditional on their information partition and the state of the world ω, no player should have interest to deviate from the strategy prescribed by $x^*(\omega)$. If the choices of actions by players are connected in some way (e.g., players relate their choices of strategies with some jointly observable feature of their interaction situation), they can achieve a correlated equilibrium. Correlated strategies are like different roles assigned to the parties, tying specific choices to observable events in the world. Any institutional elements (rules, conventions, norms, etc.) can be interpreted as a correlated equilibrium by linking an institutional element with the occurrence of a public event E. As indicated above, a public event is an event such that it refers to a world ω that belongs to the common information partition of the players. Therefore, E is common knowledge. A public event E is thus governed by an institutional element if, whenever E occurs, it is common knowledge that everyone follows the correlated equilibrium. Correlated equilibria define parts for group members to play in schemes that focus the expectations, and correlate the choices of members who find themselves in norm-governed situations of social interaction. There is a combination or scheme of actions of players such that it is common knowledge in the community that most members expect most other members to do their parts in the scheme, that most have a strong reason not to act unilaterally against this expectation, and that most regard general conformity to some such scheme as ultimately better than members going their own ways. A correlated equilibrium can capture the idea that players use the same strategy, while potentially performing differing actions.

An obvious solution in such circumstances is to adopt a correlation device. Whatever the nature of the correlated device (a convention or any other institutional element), the effect captured by the correlated equilibrium operates through the formation of the players' beliefs. A correlation device is a signaling mechanism that the players can use to coordinate their actions. A correlated equilibrium can be interpreted as consistent with a social norm. On the assumption that other players follow the signal, no one is served better by acting differently. This implies that the set of actions is a correlated equilibrium.

As Lewis (1969) argues, a form of salience is necessary such that every agent can derive common behavioral inference therefrom. Salience does not in general allow players to generate mutually supporting expectations which converge on the salient equilibrium. Nonetheless, salience does enable players to coordinate their actions. People often rely on "precedent." By projecting forward their experience of similar situations, people form their expectations about what others will do. Precedent is one important kind of salience. Individuals have some tendency to repeat actions that have proved successful in previous interactions. Precedent works through an individual's use of shared standards of inductive reasoning. Individual's choices generate equilibrium states of the game. The equilibrium states of the game are reproduced by the force of precedent. Each player need not know the choices of all other players in their entirety. It may be sufficient for individuals to share some rough ideas regarding how the game is played.

8.4 UNDERSTANDING COEVOLUTION

In order to clarify the notions of "micro" (or individual) level, and "macro" (or collective) level, and examine the relationships between them, consider Fig. 8.1 which builds on the framework proposed by the sociologist James Coleman (1990).

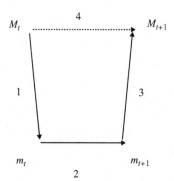

FIGURE 8.1 Macro–micro relations.

Fig. 8.1 makes a distinction between the macro-level (institutions) and the micro-level (individuals). The macro-level is patterns or regularities, and the micro-level is elements or properties responsible for them. While institutions represent phenomena that coordinate human activities at the macro-level, they are reproduced by individuals in the activities on the micro-level. As shown in Fig. 8.1, there are links between macro—macro (Relation 4) and macro—micro (Relation 1), micro—micro (Relation 2) and micro—macro (Relation 3). Fig. 8.1 describes the process through which a system is characterized by a state M_t at time t reaches a state M_{t+1} at time $t+1$. Social sciences are essentially committed to providing explanations of the relationship between these macro—macro states (Relation 4). Macro phenomena in social sciences must ultimately be explained by mechanisms that involve individuals' actions and their interactions. Hence, the micro elements, m_t and m_{t+1}, are responsible for the transition from a state M_t to a state M_{t+1}. These micro elements are constructed by individual's actions and their interactions. The macro—micro relation (Relation 1) indicates how individual's mental models are influenced by the macro state of the system. The micro—micro relation (Relation 2) indicates that individual's actions are explained in terms of the mental models. Furthermore, the micro—macro relation (Relation 3) indicates that some pattern or regularity at the macro-level is explained in terms of the actions of individuals. These explanations repeat the arrow of time: later macro phenomena are always being explained in terms of earlier macro phenomena. Institutions generate their own reproduction through Relations 1, 2, and 3. An individual's behavior reinforces the set of rules that produced it. Coleman (1994) argues:

> *Both the institutions through which the micro-to-macro link takes place, and those through which the macro-to-micro link takes place, may be taken as exogenous in rational choice theory, in studying the effects of particular institutional structures on individual actions or on systemic outcomes.*

> (Coleman, 1994, p. 171)

According to Coleman (1994), the influence of social structure is not limited to individual actions, but also affects the way that the actions of individuals combine to produce systemic outcomes. Instead of analyzing the macro—macro relation (Relation 4), we try to establish how macro-level events or conditions affect the individual (Relation 1), how the individual assimilates the impact of the macro-level events or conditions (Relation 2), and how a number of individuals, through their actions and interactions, generate macro-level outcomes (Relation 3). Thus, we explain changes and variations at the macro-level, by showing how macro states at one point in time influence the actions of individual agents, and how these actions generate macro states at a later time. Institutions generate a certain structure of rules. The structure of rules shapes human action, and generates a particular emergent outcome out of the interactions of the agents involved. The cycle then repeats. Thus, we have the relationship between individuals and the collectivity, an emergent property, which remains at the level of social structures.

Hayek's cognitive theory provides an account of a particular adaptive classifier system that produces a classification over a field of sensory inputs. Individual knowledge is the adaptive response, based on the classification that the brain has generated. According to Caldwell (2000, p.13), in Hayek's papers ("Rules, perception and intelligibility," "The theory of complex phenomena," and "Degrees of explanation"), the following three salient points emerge:

1. Complex orders occur within a variety of phenomena, from the individual brain all the way on up to a society;
2. Orders typically arise when elements contained in them follow abstract rules; and
3. Rules are often followed unconsciously; that is, the "agents" following them (even in cases where the "agent" is a human, and so is capable of speech) cannot explain what the rule is or why he or she is following it.

Hayek explicitly points to the mind-institution link. Perception involves the capacity to identify regularities or patterns. Cognitive activity functions as a mechanism of adaptation. Expectations are formed endogenously by virtue of an individual adapting behavior to achieve a closer fit with external reality. Therefore, expectations will ordinarily correspond to rules of conduct geared toward an individual's successful adaptation to the external environment. Rizzello and Turvani (2000) contend that institutions serve the individual's cognitive capacity, which develops in interaction with the institutions and determines their path-dependent character:

> *The process of knowledge acquisition as conceived by Hayek is path-dependent. The external data and information are recognized, interpreted, and learned on the basis of cognitive paths that "depend" on genetic features and the previous paths of neuronal classification. ... Each external datum is "interpreted." If it is already possessed ... the interpretation is routine and tacit. It will be included in those classification patterns which in earlier interrelations were effective and provided satisfactory outcomes. When we are faced with new information or a new external stimulus, which we are unable to interpret immediately, we tend to refer it to similar previously experienced situations. All of these paths are both tacit and idiosyncratic, but also heterogeneous and personal, since they depend on the previous experience of the individual. If, after an initial interpretation, the subsequent action is satisfying, the individual will work out a routine to be used whenever that information and situation ... recur. In this way he will have set in motion a self-reinforcing mechanism determining and absorbing state, which may be locally stable but not necessarily optimal.*
>
> (Rizzello and Turvani, 2000, p. 176, original emphasis)

Thus, human cognitive activity can be understood as an adaptive system which is input-transforming and knowledge-generating. The mind constructs a representation of the external world by organizing and interpreting sensory data. As Hayek (1952) suggests, the connections between the physiological elements are the primary phenomena which create the mental phenomena. Evolution establishes a set of possible patterns of connections on the basis of the organism's history.

The evolution of the order of actions, which can yield new rules, can lead to corresponding changes in the mind. Then, the mind will rearrange sensory experiences into new configurations that allow better predictions to be made about social reality. Institutions emerge as a result of individual agents who attempt to reduce uncertainty. Institutions determine the social structure of payoffs, which govern the conduct of economic activity. Due to changing circumstances, individuals adopt plans that can better coordinate economic activities. The criterion of fitness is confirmation of expectations as indicated by the success of individual actions. Expectations more consistent with social reality give an advantage to the individuals holding them. Players' new action choices based on such expectations generate satisfactory payoffs, and a new pattern of plays of the game becomes collectively recognized as the way the game is now being played. Thus, the key story is as follows:

perceived social reality → mental models → actions → altered perceived social reality.

It is a "disequilibrium" discovery procedure where there are unexploited opportunities in the form of new rules. Since knowledge is dispersed, different people will have access to different information and know different things. That is, two individuals, when placed in identical contexts, will make different choices, because their respective abilities to pick up the information that reaches them, and moreover to process it, are different. This implies that "much of the particular information which an individual possesses can be used only to the extent to which he himself can use it in his own decisions. Nobody can communicate to another all that he knows, because much of the information he makes use of himself will elicit only in the process of making plans of action" (Hayek, 1988, p. 77). Action plans are intrinsically well-constructed, but there are mistakes in the interpretation of information provided by the environment. Due to the limited nature of the human mind, agents cannot grasp all of the relevant information provided by processes of social interaction. Any agent will discover only opportunities related to his or her own prior knowledge. To some extent, the concept of ignorance accounts for the gap between the opportunities available in society and the opportunities perceived by individual agents. The gap exists because of the dispersion of knowledge among agents who are considered to be subjective. Interrelated plans are therefore discoordinated. The discovery of opportunities can be considered as the perception of a new framework about how to adapt in particular contexts. The agent's ability to identify opportunities depends on his or her subjective understanding of external environmental conditions. As individuals gain new experiences and update their stock of knowledge, those ways of doing things will crystalize over time into institutions, where new specific meanings are derived from the interactions between the parties involved.

Individuals may coordinate to new behavioral patterns, realizing an inconsistency between following original rules and the expected results from their action plans. Individual behavior in the same group shows a high degree of similarity, because of common perceptions of the external environment. Human action obeys action plans made by individuals in accordance with their mental models. To trade with agents in the other group needs the establishment of new sets of rules. While people want to follow their original conventions, new sets of rules may enhance social interactions and facilitate economic exchanges. Once different courses of action have become desirable to their goals, people select some of them. However, some people may be frustrated, because they have to start behaving in a different manner. Individuals care about the inconsistency with their original modes of behavior. According to cognitive dissonance theory (Festinger, 1957), people usually experience an uncomfortable feeling of cognitive dissonance when they act in a counter-attitudinal manner.[10] The theory assumes that dissonance is a psychological state of tension that people are motivated to reduce. In order to reduce the uncomfortable feeling, they often change their original attitudes. Changes are caused by the violations of original practices, and the establishment of new ones or additional enforcement mechanisms to achieve favorable outcomes. In the beginning, individuals face the choice of whether or not to support a change in existing rules. The expected costs and benefits associated with institutional change (change in rules) play an important role. The original conventions constrain the repertoire of possible reactions to changes in the external environment. Information on expected benefits from alternative sets of rules and the costs of changing existing rules would affect the likelihood of institutional change. Individuals may modify their mental models and alter their perceptions of the effects of alternative rules. Shifts in mental models change individual's action plans, which in turn leads to cultural evolution. Cultural evolution is an endogenous phenomenon with a cognitive dimension.

In the first instance, the sensory representation of the environment, and of the possible goal to be achieved in his environment, will evoke a movement pattern generally aimed at the achievement of the goal. But at first the pattern of

---

10. Akerlof and Dickens (1982) constitute the model of cognitive dissonance in economics. In their model, agents select their beliefs to minimize the dissonance experienced.

movement initiated will not be fully successful. The current sensory reports about what is happening will be checked against expectations, and the difference between the two will act as a further stimulus indicating the required corrections. The result of every step in the course of action will, as it were, be evaluated against the expected results, and any difference will serve as an indicator of the corrections required.

(Hayek, 1952, p. 95)

Micro-level actions and their interactions generate emergent patterns at the macro-level.[11] The connections between the components are variable, and the components themselves are changed by their interactions. For Hayek, the mind is a spontaneous order, or the emergent outcome of the rule-governed interactions of a myriad of neurons. "What we call 'mind'," as Hayek (1952, p. 16) writes, "is thus a particular order of a set of events taking place in some organism and in some manner related to but not identical with, the physical order of events in the environment." In Hayek's ([1967] 2014) essay, "Notes on the evolution of systems of rules of conduct," the concept of "the overall order of actions" is suggested; "it [the overall order of actions] is so not only in the trivial sense in which a whole is more than the mere *sum* of its parts but presupposes also that these elements are related to each other in a particular manner. It is more also because the existence of those relations which are essential for the existence of the whole cannot be accounted for wholly by the interaction of the parts but only by their interaction with an outside world both of the individual parts and the whole" (Hayek, [1967] 2014, p. 282, original emphasis). Furthermore, Hayek ([1967] 2014) argues:

Societies differ from simpler complex structures by the fact that their elements are themselves complex structures whose chance to persist depends on ... their being part of the more comprehensive structure. We have to deal here with integration on at least two different levels, with, on the one hand, the more comprehensive order assisting the preservation of ordered structures on the lower level, and, on the other, the kind of order which on the lower level determines the regularities of individual conduct assisting the prospect of the survival of the individual only through its effect on the overall order of the society. This means that the individual with a particular structure and behaviour owes its existence in this form to a society of a particular structure, because only within such a society has it been advantageous to develop some of its peculiar characteristics, while the order of society in turn is a result of these regularities of conduct which the individuals have developed in society.

(Hayek, [1967] 2014, p. 288)

---

11. Within Austrian economics, emergence is typically conflated with the notion of spontaneous order (Harper and Lewis, 2012; Rosser, 2012). Austrian economists use the concept of emergence to explain how the interplay between the actions of numerous, independent individuals can generate an order which is not part of anyone's intentions.

At the higher levels, there are emergent properties that arise only as a result of the relations between lower-level entities.[12] Emergent properties are characteristic of spontaneous order. A spontaneous order is created when a population of individuals interact together. Their interactions create coherent patterns at the population level that are not simple aggregates of properties that the individuals possess in isolation.

It is important to explain individual actions and institutions in a unified framework. As Rizzello and Turvani (2002) suggest:

> *Relying on pre-existing institutions and norms to explain behavior and, there-fore, referring to the concepts of instinct and habit is not sufficient because what institutional theory tries to understand is their existence, their content, and their functions. Habits and customs no longer exist as specific forms of knowing and of social action and they need to be explained in terms of some functional property for society and economy particularly.*
>
> (Rizzello and Turvani, 2002, p. 201)

Preexisting social structures constitute the macro-structural context in which current action takes place. Social structures provide conceptual frameworks in terms of which people are able to interpret and understand the social relations in which they participate. Hodgson (2004) singles out the structure and agency aspects of institutions as follows:[13]

> *Emergent institutions help to pattern individual behavior. Consequently, individuals develop and reinforce habits consistent with that behavior, upon which revised beliefs and preferences transpire. These revised beliefs or preferences lead to further actions, which may affect institutions, and so on. This gives two-way mechanisms of reconstitute interaction, from individuals to institutions and back to individuals.*
>
> (Hodgson, 2004, p. 14)

Thus, agency and structure are mutually constitutive. For Hodgson (2006, p. 6, original emphasis), institutions are "both objective structures 'out there' and subjective springs of human agency 'in the human head'." Due to changing circumstances, individuals adopt new plans to improve coordination. By trial and error, successful new plans crystallize into new institutions.

---

12. According to Lewis (2015, p. 125), a complex system has the following four attributes. First, it is composed of a set of elements which are related to, and interact with, one another in a particular way. Second, it exhibits properties which are different from those of its parts taken in isolation. Third, these properties are often novel in the sense that it is hard to predict them from our prior knowledge of the elements. Fourth, it is adaptive in the sense that it adjusts to the broader environment in which it is situated via an evolutionary process involving variation, selection, and reproduction.
13. For Hodgson (2004), the phenomena that link agency and institutional structure are habits and processes of habituation. Habits themselves are formed through repetition of action and thought.

According to Greif and Kingston (2011), these two approaches, the institutions-as-rules and the institutions-as-equilibria, can be seen as complementary:

> *The institutions-as-rules approach seems appropriate for studying the development of institutions within an established structure that can enforce the rules, for example, in a stable democracy within which basic market-supporting institutions are already well-established. The institutions-as-equilibria approach might be better suited to studying the institutional foundations of markets and democratic political structures, and other situations in which enforcement of the "rules" must be considered as an endogenous outcome rather than taken as given.*

(Greif and Kingston, 2011, p. 41, original emphasis)

In the institutions-as-rules approach, institutions are considered as a set of rules. On the other hand, in the institutions-as-equilibria approach, the role of rules is to coordinate behavior. Institutions then guide individual behavior and thought, and lead to behavioral regularities.

An individual constructs a context of action through a cognitive interpretation process. Individual behavior is governed by cognitive (intrapersonal) rules. Cognitive rules belong subjectively to one person. People assimilate the external stimuli according to their own classifying principles. An external stimulus cannot be perceived unless it is connected to something whose meaning they already know. Once a new stimulus is classified and perceived, it will create a "routine" between the synapses. Thus, the more frequent the stimulus, the more consolidated the synapses. Personal knowledge is thus the outcome of the "neuro-cognitive" mechanisms by affecting individual mental models.

Personal knowledge is also deeply influenced by the social and cultural context. The context is first of all characterized by the presence of other individuals. People gradually realize that it is advantageous for them to obey social norms, which gives rise to interpersonal relationships. It is easy to imagine that the interpersonal relations emerge since individuals obey social norms and share common social rules, and this turns individual mental models into shared mental models. Social norms are established by affecting individual mental models. The process of feedback between mental processes and institutions is explicitly characterized by mutual influence.

As North (2005) emphasizes, the degree to which beliefs are changing via deliberate modification is quite restricted:

> *The institutional structure inherited from the past may reflect a set of beliefs that are impervious to change either because the proposed changes run counter to that belief system or because the proposed alternation in institutions threatens the leaders and entrepreneurs of existing organizations.*

(North, 2005, p. 157)

"[T]he key to human evolutionary change is," as North (2005, viii) says, "the intentionality of the player." An intention is a representation of a future course of action to be performed. It is not simply an expectation of future actions, but a proactive commitment to bringing them about. According to North (2005), consciousness can lead the construction of a set of rules. Institutional change requires an understanding of beliefs that are at the heart of human consciousness.

Institutions affect individual choices (Relation 1), individuals choose their actions (Relation 2), and institutions emerge through individual actions (Relation 3). Institutional evolution can be conceptualized as a chain of successive macro—micro—macro transformations. Institutions are both the rules that underlie individual behavior and patterns of behavior. The concept of institutions also has both the mental dimension and the emergence dimension. In the mental dimension, institutions guide individual behavior and thought. Individual agents have not only expectations, but also shared mental models. Institutions can be viewed as structures exogenous to agents, or as cognitive media embedded solely in individual agents. In the emergence dimension, on the other hand, the equilibrium state is generated as the result of actions chosen by individual agents. The process of institutional formation occurs through repetition of behavior. Institutions are strictly connected with the characteristics of the human mind as well as with the interrelation of each human mind with others.

Individual activities tend to follow generally repetitive patterns of behavior. Institutional evolution is the result of an interaction between an individual's perceptions of alternative rules and the action plans according to these rules. The coordination acts as an intersubjective learning procedure, in which various ideas dispersed among individuals are constantly tested against one another. This response is adaptive; due to changing circumstances, the agents adopt their plans so that they can better coordinate economic activities. The way an individual responds to external conditions is fully dependent on the particular classification the brain has generated. Institutions emerge as a result of adaptive responses by agents in the face of the perceived external environment. Old institutions could modify into new ones gradually. New rules might be discovered and disseminated through decentralized adjustment processes. Mental models can be modified by feedback from altered perceived reality, as a consequence of peoples' altered actions.

This chapter draws on ideas of coevolution of individual's mental models and institutions. Knowledge is something produced by the mind's classificatory operation. As Hayek (1952) argues, "[a]ll we can perceive of external events are therefore only such properties of these events as they possess as members of classes which have been formed by past 'linkages'. That qualities which we attribute to the experienced objects are strictly speaking not properties of that objects at all, but a set of relations by which our nervous system classifies them" (Hayek, 1952, p. 142, original emphasis). At the level of the individual, the cognitive processes enable the individual to adjust his or her actions to external reality. Individuals adjust their actions to achieve a better fit with reality. In this sense, the mind is endogenous to the individual's environment, which implies that expectations are also endogenous to the individual's environment.

Institutions do not evolve just by themselves; the direction in which they are changed depends on the beliefs of those who change them. Shifts in mental models change individual's plans and actions, which, in turn, leads to institutional evolution. Thus, a key to understanding institutional evolution is an understanding of how individuals modify their mental models. Institutional evolution is an endogenous phenomenon with a mental or cognitive dimension.

8.5 CULTURAL EVOLUTION

The development of human society can take place over various layers. For example, Hayek (1979) distinguishes the following three layers in social evolution. The first layer is the instinct one of biological evolution during human phylogeny. At this level, primitive forms of social behavior, values, and attitudes are genetically fixed, as a result of natural selection processes. The second layer of evolution is the reason one of human intelligence and its products, that is, knowledge and the numerous ways of recording, transmitting, and processing it. The propagation, elaboration, and storing of knowledge have accelerated scientific and technological progress enormously. At the third layer between instinct and reason, cultural evolution takes place. From the cultural evolutionary process, the rules of conduct, morals, and traditions emerge. Through the process, human interactions are shaped into orderly forms of civilization.

Evolutionary models traditionally deal with the change in the nature of ensembles of elements. Evolution through time would be a consequence of variation among members of the ensemble at any instant of time. Different individuals have different properties, and the ensemble is characterized by the collection of these properties. Thus, the distribution of properties changes as some types become more common. Evolution is a two-step process which consists of the origin of new variants and the replacement of older by newer variants. Evolution fundamentally depends on variation. Endler and McLellan (1988) classify the major evolutionary processes into the following six groups:

1. Mutational processes: The processes that generate variation are collectively called mutation in the broad sense.
2. Constraining processes: These are the processes that restrict the kinds of variation which are possible or likely.
3. Frequency-changing processes: These are the processes that change the frequency of variants.

4. Adaptive processes: These processes improve an organism's fit with its environment, or they improve the efficiency or responsiveness of its internal functions.

5. Rate-determining processes: These processes are caused by factors that affect the rate of evolution, without necessarily varying within populations.

6. Direction-determining processes: These processes are caused by factors that affect the direction of evolution, without necessarily varying within populations.

First of all, the fourth category suggests that some behavioral patterns are better adapted to prevailing evolutionary pressures, and consequently increase in relative significance compared to less adapted ones. Agents interact in a particular environment in a way that the differential advantage depends on the characteristics of others and the specification of the environment. According to Metcalfe (1998, p. 23), the remaining processes can be interpreted in economic terms as follows. The first category covers the entire field of innovation which is carried out by existing firms, or associated with the creation of new firms, together with the processes determining rates of entry and exit into and out of a population. The second category suggests that processes restrict and guide the possible patterns of variation in behavior. There is always a place of inertia and constraint in any evolutionary argument. The third category leads us towards the dynamics of resource allocation in the market context. The waves of evolutionary changes are transmitted through markets. The fifth and sixth processes in the list cover the overall framework of institutions and behavioral norms which shape innovation and the way in which markets transmit changes.

In abstract terms, Hayek (1967, p. 32) refers to evolution as the idea "that a mechanism of reduplication with transmittable variables and competitive selection of those which prove to have a better chance of survival will in the course of time produce a great variety of structures adapted to continuous adjustment to the environment and to each other." In a cultural context, evolution is "a process in which practices which had first been adopted for other reasons, or even purely accidentally, were preserved because they enabled the group in which they had arisen to prevail over others" (Hayek, 1973, p. 9). The notion of cultural evolution to which Hayek refers is pre-Darwinian, and was discovered by the Scottish moral philosophers of the 18th century whom Hayek has sometimes called "Darwinians before Darwin" (Hayek, 1973, p. 23). It is the individualistic perspective of the 18th-century Scottish thinkers David Hume, Adam Ferguson, and Adam Smith. The selection mechanism that is operating in cultural evolution works on the behavioral pattern, or the rules that govern behavior, directly. Vromen (1995) identifies the three mechanisms that are at work in cultural evolution. First, individual trial-and-error learning is the selection mechanism. As long as practices bring about satisfactory results, individuals maintain these practices; otherwise, they will try out something else. Second, we come to the mutation mechanism. New trials are experiments, performed either intentionally or unintentionally. Third, imitation provides the transmission (or replication) mechanism. Models of cultural evolution begin with the transmission of cultural traits from one set of agents to others, by analogy with the transmission of genes. Parents may transmit traits to their children, or teachers to their pupils. Cultural transmission leads to a cultural evolution process, in which various forces determine which cultural items are preferentially transmitted in a population.[14] The forces of cultural evolution can be defined as those that are capable of altering the frequencies of cultural traits over time. Culture can be considered as what a group learns over a period of time.

"With the diversity of human experiences in different environments," as Denzau and North (1994, p.15) state, "there exists a wide variety of patterns of behavior and thought." Cultural evolution accounts for the increasingly large numbers of individuals who can live together and create what we call civilization. To understand how civilization has developed, we must comprehend underlying processes arising from many individual actions which nobody directs or foresees. Some cultures have developed in completely different directions from that of western civilization:

> There are, undoubtedly, many forms of tribal or closed societies which rest on very different systems of rules. All that we are here maintaining is that we know only of one kind of such system of rules, undoubtedly still very imperfect and capable of much improvement, which makes the kind of open or 'humanistic' society possible where each individual counts as an individual and not only as a member of a particular group, and where therefore universal rules of conduct can exist which are equally applicable to all responsible human beings.
>
> (Hayek, 1976, p. 27, original emphasis)

Hayek's (1988) concept of the "extended order" refers to the process by which a human society develops the capacity to cope with increasing degrees of complexity.[15] The growing complexity is built on the foundations of

---

14. Restricting attention to discrete cultural traits, Bisin and Verdier (2000) deal with cultural trait formation.

15. North et al. (2009) divide all of human history into three social orders. First came the foraging order of small social groups. About 10 millennia ago the "limited access" orders arose. Limited access orders are characterized by (1) slow-growing economies vulnerable to negative shocks; (2) polities without generalized consent of the governed; (3) relatively small numbers of organizations; (4) small, centralized governments; and (5) a predominance of social relationships organized along personal lines. A couple centuries ago, "open access" orders emerged. Open access orders are characterized by: (1) political and economic development; (2) economies that rarely experience negative economic growth; (3) rich and vibrant civil societies with lots of organizations; (4) bigger, more decentralized governments; and (5) widespread impersonal social relationships.

social rules that coordinate the disparate interests, actions, and knowledge of individuals across society. Coordination of groups larger than hunting and gathering bands requires cultural evolution of learnt rules of conduct. In cultural evolution, the things getting varied or retained are not genes, but ways of acting. Acquired characteristics may be inherited in cultural evolution through social learning, but not in biological evolution. Hayek (1988) argues as follows that "the extended order resulted not from human design or intention but spontaneously: it arose from unintentionally conforming to certain traditional and largely *moral* practices, ... which have nonetheless fairly rapidly spread by means of an evolutionary selection—the comparative increase of population and wealth—of those groups that happened to follow them" (Hayek, 1988, p. 6, original emphasis). The individual actions are not motivated by the desire to produce social institutions. It then follows that the rules should be allowed to evolve organically.

Caldwell (2000) describes the following criticisms for Hayek's writings on cultural evolution:

- Hayek's analysis of the evolutionary process is too pessimistic, leaving little room for attempts to improve the institutional or constitutional setting.
- Hayek's endorsement of group selection as the mechanism by which cultural institutions are selected is inconsistent with his methodological individualism.
- Group selection itself has been discredited among biologists on grounds that are germane to its applications in the social sciences.

The secondary literature, especially, has become quite large. According to Hayek (1988), "cultural evolution operates largely through group selection" (Hayek, 1988, p. 25). The concept of group selection is considered to be a weak point in Hayek's system of ideas (Sugden, 1993). It is the idea that cultural traits and behavioral features are naturally selected on the basis of advantages and disadvantages for the groups of people who practice them. The theory of group selection implicitly assumes that specific conventions are tied to specific social groups.[16] Thus, a convention can spread only as a consequence of the expansion of the social group to which it belongs. The main criticism is that Hayek's analysis of cultural evolution that operates at the group level is inconsistent with his methodological individualism (Vanberg, 1994). Vanberg (1994) defines methodological individualism as the guiding principle that aggregate social phenomena can be and should be explained in terms of individual actions, their interrelations, and their (largely unintended) combined effects. Hodgson (1993) argues that it is

---

16. Group selection, in biology, refers to a process of natural selection that favors traits that increase the fitness of one group relative to other groups (Wilson, 1997). Such behavior is the result of Darwinian selection, but not selection rooted solely in the characteristics of individuals.

necessary to abandon methodological individualism in order to keep group selection.[17]

Vanberg (1994) detects an inconsistency in Hayek's writings about cultural evolution:

> [T]here is a tacit shift in Hayek's argument from the notion that behavioral regularities emerge and prevail because they benefit the individuals practicing them, to the quite different notion that rules come to be observed because they are advantageous to the group.
>
> (Vanberg, 1994, p. 83, original emphasis)

For Hayek, however, the individual and collective rules evolve according to different conditions. Hayek (1967) writes:

> [T]he systems of rules of individual conduct and the order of action which results from the individuals acting in accordance with them are not the same thing.
>
> (Hayek, 1967, p. 67)

Individuals interact with others and their environment.[18] They learn from experience: they acquire, communicate, and utilize knowledge in society. A mutual adjustment process of individual action plans makes it possible to realize a social order. The individual rules of conduct can, under certain circumstances, result in a certain order of actions. Under other circumstances, they produce a different order. The outcome of an evolutionary process can be said to reflect the product of a kind of social learning.

In Hayek's theory of cultural evolution, societies not only are subject to group selection but have developed through a process in which individuals choose the rules that form the social order (Gick and Gick, 2001). "Individual selection refers to the perception of rules that are slightly different from already existing ones and hence leads to the *creation* of new rules" (Gick and Gick, 2001, p. 156, original emphasis). It is not the regularity of the individual rules of conduct that determines the preservation of a group of

---

17. Methodological individualism is right to remind us that the social world is ultimately the result of many individuals interacting with one another, and that any theory that fails to accept this basic premise rests on mysterious metaphysical assumptions.

18. The mind is an active process by which we set out to make sense of our particular social environment, and is itself changed as a result of that interaction through feedback processes. Baron-Cohen (1995) proposes that the theory of mind inference system has evolved to promote strategic social interaction. Social exchange depends on the ability to infer the content of other peoples' desires, goals, and intentions. According to the "social brain" hypothesis, the extraordinary cognitive abilities of humans evolved as a result of the demands for social interactions (Dunbar, 1992, 1998). There is a relationship between mean social group size for a species, and its relative neocortex volume. The social brain hypothesis states that the computational capacity of a species' brain, principally reflected in the volume of its neocortex, sets a limit on the number of individuals who can be held together in a coherent social group. Language came about primarily as a tool of social intelligence.

individuals but rather the resulting order of actions. In the evolutionary process, there is no central mechanism which coordinates a shift in the rules perceived by individuals. New rules undergo some kind of decentralized selection process, as a consequence of which some spread through the population. The role of individuals is necessary to innovate practices. That is, individual minds conceive of problems and new ways to solve those problems, and individuals choose whether or not to follow a new rule. The individual actions are always the joint product of (internal) motivations and (external) events acting on individual's behavior. Cognition takes place by structuring individual minds in social contexts, and systems of rules literally "epitomize" the cognitive capabilities of humans that evolve socially. "Individual action and social emergence of rules appear to be two sides of the same coin: the mind" (Rizzello and Turvani, 2002, p. 199).

Individuals typically have the opportunity to change their rules. When the rules change, it is because of individual decisions. Changes in the rules of individuals may lead to a change in the resulting social order. Vromen (1995) presents a reinterpretation of Hayek's statements on cultural evolution as follows:

> The replacement of the pair of terms "group" and "individuals" by the pair of "order" and "rules" brings out clearly, I think, that what is accounted for in Hayek's notion of cultural evolution is the evolution of rules. What counts is the "differential reproductive success" of rules, not of individuals.
>
> (Vromen, 1995, p. 173, original emphasis)

Individuals, not groups of individuals, are responsible for the emergence, maintenance, and alteration of rules. Many traditional rules suppress our behavior, and may well seem irrational to us. A rule suppressing some behavior will be effective only if people generally expect others to act in a way which makes it effective. "Most of the steps in the evolution of cultures were made possible by some individuals breaking some traditional rules and practicing new forms of conduct—not because they understood them to be better, but because the groups which acted upon them prospered more than others and grew" (Hayek, 1979, p. 161).

Different forms of cultural tastes can lead to different types of network relations. As individuals interact with members of the other group, they may learn how to behave from those other cultural members. Then, individuals become connected and integrated into larger social networks. On the other hand, a particular form of culture may lead to social connections that are sustained by a restricted form of social ties. The interpersonal relations emerge because individuals are interested in sharing common elements that make communication easier, which turns individual perceptions into shared mental models. Institutional change is an endogenous phenomenon that starts in the minds of individuals. However, the evolution of beliefs may be consolidated, and establish the path of past achievements (path-dependence).

Path-dependent perceptions set up a bound for cultural evolution, anchoring it to what has already been ingrained into habits (Patalano, 2010). As Hayek (1979) argues:

> *The structures formed by traditional human practices are neither natural in the sense of being genetically determined, nor artificial in the sense of being the product of intelligent design, but the result of a process of winnowing or sifting, directed by the differential advantages gained by groups from practices adopted for some unknown and perhaps purely accidental reasons.*
>
> (Hayek, 1979, p. 155)

Tradition is a product of cultural evolution. Culture consists of sets of rules embodied in traditions. Cultural evolution reflects a process in which "what proved conducive to more effective human effort survived, and the less effective was superseded" (Hayek, 1967, p. 111). The Great Society is "one which we would choose if we knew that our initial position in it would be decided purely by chance" (Hayek, 1976, p. 132). However, the Great Society is no mere extension of the tribal society. Society in the large is fundamentally different from society in the small. People did not design culture, but culture developed as a result of repeated human interaction, as an unintended consequence thereof. Selection of rules can stabilize arbitrary behavior within a group, generating a population of multiple groups. Different groups might have different informal rules. Large societies are indeed pluralistic, comprised of smaller groupings, each with its own rules. While individuals may not themselves understand the rationale of the particular rules they follow, the net effect is that the capacities of the group to adapt to certain conditions are enhanced.

The accumulated stock of knowledge of past experiences is the underlying source of path-dependence. If individuals have cognitive tendencies of a particular kind, we would expect this to make an impact on how institutions perform and which institutions emerge over time. If human beings did not have the capacity to transmit norms and habits, institutions could not exist. Norms and habits are transmitted from one generation to another. The process of social evolution itself is incremental, because norms and habits embodied in culture are inherited to future generations. North (1990) suggests the following view of incremental institutional change:

> *[I]nstitutions typically change incrementally rather than in discontinuous fashion. How and why they change incrementally and why even discontinuous changes (such as revolutions or conquest) are never completely discontinuous are a result of the imbeddedness of informal constraints in societies ... These cultural constraints not only connect the past with the present and future, but provide us with a key to explaining the path of historical change.*
>
> (North, 1990, p. 6)

Axelrod (1997) argues that culture is taken to be what social influence influences. In fact, humans acquire large parts of their behavioral repertoire via forms of social learning (basically imitation). Understanding how people learn from others is important, not only for understanding individual decisions, but also for comprehending patterns of change and variation among human groups. Differences among individuals can exist because they acquire different behavior as a result of some form of social learning. In Axelrod's (1997) analysis, two groups that are already culturally similar are more likely to interact and, therefore, to be even more culturally similar. On the contrary, two neighboring groups with zero cultural similarity are unlikely to interact and, therefore, will have no tendency to be more culturally similar. The emergence of conventions can be accelerated if the population has a neighborhood structure and individuals adjust their behavior over time by imitation within their own neighborhood (Boyer and Orléan, 1992; Eshel et al., 1999). The structure is imposed by setting individuals at the nodes on a lattice, a torus in which all the edges join up with their opposite edges, and limiting their interactions to the individuals at adjacent nodes. It is interpreted as a social lattice in which interactions are restricted to certain classes of individuals who are represented as adjacent nodes on the lattice. Players may be more likely to interact with some players than others. A local interaction system describes a set of players and specifies which players interact with which other players. In a biological context, Hamilton (1964) demonstrates that cooperation can be sustained in viscous populations. Viscosity, in biological populations, is the tendency of individuals to have a higher rate of interaction with their close relatives than with more distantly related ones. There is a spatial or genetic clustering. The degree of clustering can be defined in terms of a viscosity parameter which refers to the probability of interacting with a neighbor playing a similar strategy. Myerson et al. (1991) formulate this idea in terms of a biological game. A strategy in a biological game is chosen through mutation by a kin group or genotype. Taking the limit as the parameter of viscosity tends to zero, Myerson et al. (1991) define a set of fluid population equilibria which is always nonempty and includes all evolutionary stable strategies. Local interactions may allow some forms of behavior to spread in certain dynamic systems.[19] What makes local interactions with imitation conducive to the emergence of conventions is that individuals expressing a particular behavior become increasingly likely to have neighbors who express the same behavior.

Interactions between agents form macroscopic patterns. An understanding of social orders can be achieved by viewing them in the light of the activities of myriad individual agents. A complex system is explained by both its constituent elements and the connections by which they are related. In a

---

19. Blume (1995) considers local interaction systems in which locations are on an m-dimensional lattice, and there is a translation invariant description of the set of neighbors.

dynamic analysis of emergence, the fundamental issue is that connections are changing, which makes connections the prime variables. It is impossible for agents to know all the feasible connections between the elements that constitute a present and future system. According to Mises ([1949] 1966, p. 31), "experience which the sciences of human action have to deal with is always an experience of complex phenomena." Agents may vary according to their cognitive characteristics, access to information, and goals. A striking feature of the Austrian theory of institutions is that human agents are creating institutions which they do not know of in advance. Carl Menger, founder of the Austrian school of thought, insisted that even unintended, organic social forms, including law, language, money, and markets, could be analyzed by reducing them to the analysis of individuals pursuing their interests. Following Vanberg (1994), the individualistic evolutionary theory developed by Menger is composed of the interaction of the following two processes. The first is "a process of variation," in which new ways of behavior are generated by means of individual choices. The second is a "process of selection," which explains how a practice will spread in society due to individual imitation. Aggregate variables can be seen as emergent outcomes of agent-level interactions. The concept of emergence is used to explain how the interplay between the actions of numerous independent agents can generate an order. In agent-based models, the agents interact with each other, and sometimes form networks through bottom-up processes. Agents' actions and their interactions create an irreversible process whose outcome is the occurrence of the emergent structure. The agents make decisions based on limited information. According to Marcy and Willer (2002, p. 146), agent-based models impose the following four assumptions:

1. Agents are autonomous:

    The system is not directly modeled as a globally integrated entity. Systemic patterns emerge from the bottom-up, coordinated not by centralized authorities, but by local interactions between autonomous decision-makers.

2. Agents are interdependent:

    Interdependence may involve processes like persuasion, sanctioning, and imitation, in which agents influence others in response to the influence that they receive. Interdependence may also be indirect, as when agent's actions change some aspect of the environment, which in turn affects the actions of other agents, such that the consequences of each agent's decisions depend in part on the choices of others.

3. Agents follow simple rules:

    Global complexity does not necessarily reflect the cognitive complexity of individuals. Agents follow rules, in the form of norms, conventions, protocols, moral and social habits, and heuristics. Although the rules may be quite simple, they can produce global patterns that may not be at all obvious, and are very difficult to understand.

4. Agents are adaptive and backward-looking:

When interdependent agents are also adaptive, their interaction can generate a "complex adaptive system" (Holland, 1995). Agents adapt by moving, imitating, replicating, or learning, but not by calculating the most efficient action.

In complex systems, scale matters, because individual elements are interdependent and self-organizing. The nature of relations among individuals changes as the number of elements in the system grows. Emergent order comes from the processes that require little or no foresight, but considerable experimentation and trial-and-error. Adaptation is the outcome of intentional experimentation, followed by selection and learning. Learning defines the process by which agents organize their rule-based structure through its interaction with the external environment. Evolution is a force that is complementary to, and may even substitute for, rationality (Alchian, 1950). Alchian (1950) rejects the assumption of profit maximization outright, arguing that there can be no definite criterion of rational behavior under conditions of uncertainty. Firms survive not because they achieve maximum profits according to the criterion of rational behavior, but because they are well-suited to the particular conditions of their environment and make positive profits. Structure is a heritable characteristic of the evolutionary process. People very often act without knowledge or confidence in the efficacy of their decisions. In order to act under uncertainty, people do not have to completely understand what they are doing. Individuals are fundamentally ignorant of the future, and it is essentially impossible to predict what is going to happen. No one knows or says how his or her actions lead to any goal or outcome.

8.6   SUMMARY

Theories of institutions can be classified into two broad approaches: institutions-as-rules and institutions-as-equilibria. Institutional structures and individual actions coevolve. In order to have a complete picture of institutions, we need to take both approaches, institutions-as-rules and institutions-as-equilibria, into consideration.

According to the institutions-as-rules approach, institutions are conceived as rules that guide the actions of individuals engaged in social interactions. Institutions structure human agency. Rules indicate what to do, or what not to do, in a given circumstance. On the other hand, the institutions-as-equilibria approach views institutions as behavioral patterns or regularities. Such a regularity is described as an equilibrium in strategic, noncooperative games. It is also the product of a dynamic process, in which interactions become patterned. In the institutions-as-equilibria approach, individuals, interacting with others, are assumed to continue to change their responses to the actions of others, until no improvement can be obtained in their expected outcomes from independent actions. For a rule to be an institution, it must be widely distributed in the minds of the group members. The institutional framework orders the interaction. Only the existence of institutions leads to individual representations of a given interaction that are intersubjectively compatible and, therefore, make the process of action and response predictable and expectable. Repeated patterns of behavior create expectations about future behavior, and ensure a degree of predictability in social interaction. When another's behavior is relatively easy to predict, individual agents spend less time deciphering actions and are better able to coordinate with others. Institutions depend on the individuals who reproduce, transform, or create them. To understand why regularized patterns of interaction exist, we need to ask why all agents would be motivated to produce a particular equilibrium. The institutions that create a social environment must be ultimately traceable to individual mind-sets, because all social phenomena are the consequence of individual actions. In Hayek's theory of cultural evolution, societies have developed through a process in which individuals choose the rules that form the social order. New rules undergo some kind of decentralized selection process, as a consequence of which some spread through the population.

REFERENCES

Akerlof, G.A., Dickens, W.T., 1982. The economic consequences of cognitive dissonance. Am. Econ. Rev. 72, 307–319.

Alchian, A.A., 1950. Uncertainty, evolution, and economic theory. J. Polit. Econ. 58, 211–221.

Aoki, M., 2001. Towards a Comparative Institutional Analysis. The MIT Press, Cambridge, MA.

Aoki, M., 2007. Endogenizing institutions and institutional changes. J. Inst. Econ. 3, 1–31.

Aoki, M., 2011. Institutions as cognitive media between strategic interactions and individual beliefs. J. Econ. Behav. Organ. 79, 20–34.

Arrow, K.J., 1974. Limits of Organization. W. W. Norton, New York.

Aumann, R.J., 1974. Subjectivity and correlation in randomized strategies. J. Math. Econ. 1, 67–96.

Aumann, R.J., 1987. Correlated equilibrium as an expression of Bayesian rationality. Econometrica 55, 1–18.

Aumann, R.J., Brandenburger, A., 1995. Epistemic conditions for Nash equilibrium. Econometrica 63, 1161–1180.

Axelrod, R., 1997. The dissemination of culture: a model with local convergence and global polarization. J. Confl. Res. 41, 203–226.

Baron-Cohen, S., 1995. Mindblindness: An Essay on Autism and Theory of Mind. The MIT Press, Cambridge, MA.

Berry, J.W., 2005. Acculturation: living successfully in two cultures. Int. J. Intercult. Relat. 29, 697–712.

Bisin, A., Verdier, T., 2000. Beyond the melting pot: cultural transmission, marriage, and the evolution of ethnic and religious traits. Q. J. Econ. 115, 955–988.

Blume, L., 1995. The statistical mechanics of best-response strategy revision. Games Econ. Behav. 11, 111–145.

Boyd, R., Richerson, P.J., 1985. Culture and the Evolutionary Process. University of Chicago Press, Chicago.

Boyer, R., Orléan, A., 1992. How do conventions evolve? J. Evol. Econ. 2, 165–177.

Caldwell, B., 2000. The emergence of Hayek's ideas on cultural evolution. Rev. Austrian Econ. 13, 5–22.

Coleman, J., 1987. Norms as social capital. In: Radnitzk, G., Bernholz, P. (Eds.), Economic Imperialism. Paragon, New York.

Coleman, J., 1990. Foundations of Social Theory. Harvard University Press, Cambridge, MA.

Coleman, J., 1994. A rational choice perspective on economic sociology. In: Smelser, N.J., Swedberg, R. (Eds.), The Handbook of Economic Sociology. Princeton University Press, Princeton NJ, pp. 165–180.

Crawford, S., Ostrom, E., 1995. A grammar of institutions. Am. Polit. Sci. Rev. 89, 582–600.

Denzau, A.T., North, D.C., 1994. Shared mental models: ideologies and institutions. Kyklos 47, 3–31.

Dunbar, R.I.M., 1992. Neocortex size as a constraint on group size in primates. J. Hum. Evol. 22, 469–493.

Dunbar, R.I.M., 1998. The social brain hypothesis. Evol. Anthropol. 6, 178–190.

Durham, W.H., 1991.). Coevolution: Genes, Culture, and Human Diversity. Stanford University Press, Stanford.

Endler, J.A., McLellan, T., 1988. The processes of evolution: toward a newer synthesis. Ann. Rev. Ecol. Syst. 19, 395–421.

Eshel, I., Sansone, E., Shaked, A., 1999. The emergence of kinship behavior in structured populations of unrelated individuals. Int. J. Game Theory 28, 447–463.

Festinger, L., 1957. A Theory of Cognitive Dissonance. Row, Peterson, Evanston.

Frith, U., Frith, C., 2001. The biological basis of social interaction. Curr. Dir. Psychol. Sci. 10, 151–155.

Gallese, V., Goldman, A., 1998. Mirror neurons and the simulation theory of mind-reading. Trends Cogn. Sci. 2, 493–501.

Gick, E., Gick, W., 2001. F. A. Hayek's theory of mind and theory of cultural evolution revisited: toward an integrated perspective. Mind Soc. 3, 149–162.

Gintis, H., 2007. A framework for the unification of the behavioral sciences. Behav. Brain Sci. 30, 1–61.

Gintis, H., 2009. The Bounds of Reason: Game Theory and the Unification of the Behavioral Sciences. Princeton University Press, Princeton, NJ.

Gowdy, J.M., 1994. Coevolutionary Economics: The Economy, Society and the Environment. Kluwer Academic Publishers, Boston.

Greif, A., Kingston, C., 2011. Institutions: rules or equilibria? In: Schofield, N., Caballero, G. (Eds.), Political Economy of Institutions, Democracy and Voting. Springer, Berlin, pp. 13–43.

Hamilton, W.D., 1964. The genetic evolution of social behavior. J. Theor. Biol. 7, 1–52.

Harper, D.A., Lewis, P., 2012. New perspectives on emergence in economics. J. Econ. Behav. Organ. 82, 329–337.

Hayek, F.A., 1948. Individualism and Economic Order. University of Chicago Press, Chicago.

Hayek, F.A., 1952. The Sensory Order: An Inquiry into the Foundations of Theoretical Psychology. University of Chicago Press, Chicago.

Hayek, F.A., 1952/1979. The Counter-Revolution of Science, 2nd ed Liberty Press, Indianapolis.

Hayek, F.A., 1967. Studies in Philosophy, Politics, and Economics. Routledge & Kagan Paul, London & Henley.

Hayek, F.A., 1973. Law, Legislation and Liberty, Vol. 1, Rules and Order. University of Chicago Press, Chicago.

Hayek, F.A., 1976. Law, Legislation and Liberty, Vol. 2, The Mirage of Social Justice. University of Chicago Press, Chicago.

Hayek, F.A., 1979. Law, Legislation and Liberty, Vol. 3, The Political Order of a Free People. University of Chicago Press, Chicago.

Hayek, F.A., 1988. The Fatal Conceit: The Errors of Socialism. University of Chicago Press, Chicago.

Hayek, F.A., 1967/2014. In: Caldwell, B. (Ed.), *The Collected Works of F. A. Hayek, Volume 15: The Markets and Other* Orders. University of Chicago Press, Chicago.

Hodgson, G.M., 1993. Economics and Evolution: Bringing Life Back into Economics. University of Michigan Press, Ann Arbor.

Hodgson, G.M., 2004. *The Evolution of Institutional Economics: Agency, Structure, and Darwinism in American Institutionalism.* Routledge, London.

Hodgson, G.M., 2006. What are institutions? J. Econ. Issues 40, 1−25.

Holland, J., 1995. Hidden Order: How Adaptation Builds Complexity. Perseus, Reading, MA.

Lachmann, L.M., 1970. The Legacy of Max Weber. Heinnemann, London.

Lachmann, L.M., 1976. From Mises to Shackle: an essay on Austrian economics and the Kaleidic society. J. Econ. Lit. 14, 54−62.

Lazear, E.P., 1999. Culture and language. J. Polit. Econ. 107, S95−S126.

Lewis, D., 1969. Convention: A Philosophical Study. Harvard University Press, Cambridge, MA.

Lewis, P., 2015. An analytical core for sociology: a complex, Hayekian analysis. Rev. Behav. Econ. 2, 123−146.

Marcy, M.W., Willer, R., 2002. From factors to actors: computational sociology and agent-based modeling. Ann. Rev. Sociol. 28, 143−166.

Metcalfe, J.S., 1998. Evolutionary Economics and Creative Destruction. Routledge, London.

Mises, L., 1966. Human Action: A Treatise on Economics. Henry Regnery, Chicago, Original Work Published 1949.

Myerson, R.B., Pollock, G.B., Swinkels, J.M., 1991. Viscous population equilibria. Games Econ. Behav. 3, 101−109.

Nash, J., 1950. Equilibrium points in n-person games. Proc. Natl. Acad. Sci USA 36, 48−49.

Nash, J., 1951. Non-cooperative games. Ann. Math. 54, 286−295.

Nelson, R.R., 1995. Recent evolutionary theorizing about economic change. J. Econ. Lit. 33, 48−90.

Nelson, R.R., Winter, S., 1982. An Evolutionary Theory of Economic Change. Harvard University Press, Cambridge, MA.

Norgaard, R.B., 1994. Development Betrayed: The End of Progress and a Coevolutionary Revisioning of the Future. Routledge, London.

North, D.C., 1990. Institutions, Institutional Change and Economic Performance. Cambridge University Press, Cambridge, MA.

North, D.C., 2005. Understanding the Process of Economic Change. Princeton University Press, Princeton, NJ.

North, D.C., Wallis, J.J., Weingast, B.R., 2009. Violence and Social Orders: A Conceptual Framework for Interpreting Recorded Human History. Cambridge University Press, Cambridge, MA.

Patalano, R., 2010. Understanding economic change: the impact of emotion. Const. Polit. Econ. 21, 270−287.

Polanyi, K., 1957. The Great Transformation. Beacon, Boston.

Rizzello, S., Turvani, M., 2000. Institutions meet mind: the way out of an impasse. Const. Polit. Econ. 11, 165–180.

Rizzello, S., Turvani, 2002. Subjective diversity and social learning: a cognitive perspective for understanding institutional behavior. Const. Polit. Econ. 13, 197–210.

Rosser Jr., J.B., 2012. Emergence and complexity in Austrian economics. J. Econ. Behav. Organ. 81, 122–128.

Schotter, A., 1981. The Economic Theory of Social Institutions. Cambridge University Press, Cambridge, MA.

Searle, J.R., 2005. What is an institution? J. Inst. Econ. 1, 1–22.

Sugden, R., 1993. Normative judgments and spontaneous order: the contractarian element in Hayek's thought. Const. Polit. Econ. 4, 393–424.

Ullmann-Margalit, E., 1977. The Emergence of Norms. Clarendon, Oxford.

Vanberg, V.J., 1994. Rules and Choice in Economics. Routledge, London.

Vanderschraaf, P., 1995. Convention as correlated equilibrium. Erkenntnis 42 (1), 65–87.

Vromen, J.J., 1995. Economic Evolution: An Inquiry into the Foundations of the New Institutional Economics. Routledge, London.

Wilson, D.S., 1997. Human groups as units of selection. Science 276, 1816–1817.

Index

Printed in the United States
By Bookmasters